# Principles of remote sensing

# Principles of remote sensing

Paul J. Curran

Copublished in the United States with
John Wiley & Sons, Inc., New York

**Longman Scientific & Technical**
Longman Group UK Limited,
Longman House, Burnt Mill, Harlow
Essex CM20 2JE, England
*and Associated Companies throughout the world*

*Copublished in the United States with*
*John Wiley & Sons, Inc., 605 Third Avenue, New York, NY 10158*



*First published 1985*
*Second impression 1985*
*Reprinted by* Longman Scientific & Technical *1986, 1988, 1989*

**British Library Cataloguing in Publication Data**

Curran, Paul J.
Principles of remote sensing.
1. Remote sensing
I. Title
621. 36'78 G70.4

ISBN 0-582-30097-5

**Library of Congress Cataloging-in-Publication Data**

Curran, Paul J., 1955 –
Principles of remote sensing.

Bibliography: p.
Includes indexes.
1. Remote sensing. I. Title.
G70.4.C87 1986 621.36'78 86 – 18946
ISBN 0-470-20392-5 (USA Only)

Set in 9½/11 pt. Linotron 202 Century Schoolbook
Produced by Longman Group (FE) Ltd
Printed in Hong Kong

To Helen

# Contents

*'and what is the use of a book' thought Alice, 'without pictures or conversation?'* (Carroll 1865)

# Preface

This book is produced for the increasing number of undergraduate and graduate courses in remote sensing. These courses are being taught to students with a wide range of backgrounds, all of whom require a text that can be used as a springboard into the remote sensing literature that now exists for their field of interest. This book aims to satisfy this requirement by providing a synthesis of the remote sensing principles that will be of value to environmental scientists with backgrounds in either pure or applied science. This is achieved by rejecting a 'cook-book' approach and concentrating upon three themes: first, that of the *relationship* between *properties* of our environment and remotely sensed *images* of our *environment*. Second, the techniques of *acquiring* remotely sensed *images* and third, the procedures of *processing* and *interpreting* remotely sensed images in order to locate, identify and study areas of our environment.

The book starts by providing a view of remote sensing in the context of the subjects past and future. Chapter 2 is the link upon which all other chapters are dependent, for it provides a discussion of the interactions between electromagnetic radiation and the Earth's surface. Chapters 3, 4 and 5 aim to acquaint the reader with a specific group of remote sensors and platforms. Photography from aircraft is covered in Chapter 3; Chapter 4 provides an introduction to the use of other sensors like multispectral scanners; thermal infrared linescanners and sideways-looking radars from aircraft and Chapter 5 provides an introduction to their use from satellites. In Chapter 6, various techniques for the processing of both photographic and digital image data are outlined. The text ends with a selection of reference material.

This book is a development of my research and teaching in remote sensing. I therefore wish to thank the many individuals and organisations who have made my research and teaching both possible and pleasurable. In particular, Dr Len Curtis, Dr John vanGenderen, Dr John Hardy, Dr Ted Milton, Dr Tim Munday, Dr John Townshend, Dr Neil Wardley, the Natural Environment Research Council, the National Remote Sensing Centre, the Nature Conservancy Council, the Remote Sensing Society, the Royal Society and the staff and students of the Universities of Bristol, Reading and Sheffield.

As it is difficult to create something by consensus I must lay claim to being entirely responsible for everything in this book including the errors and any annoying or interesting bits you may find. To help me to get the book to press, I am most grateful to the staff of Sheffield University, notably the formidable team of Anita Fletcher, Joan Dunn, Carole Elliss and Penny Shamma for typing the manuscribble, Paul Coles for skilfully preparing nearly all of the artwork, Dave Maddison and John Owen for producing a vast amount of photographic material, Peter Morley for obtaining many of the images and Julia Dagg for ordering several hundred references through the interlibrary loan service and the University of Sheffield Research Fund for a grant towards manuscript preparation.

Thanks also to Dr Iain Stevenson of Longman who asked me to start this book.

Final thanks go to Helen for living through it all, the dedication is a grossly inadequate measure of my gratitude.

**Paul Curran**
Sheffield, South Yorkshire
June 1983

# Acknowledgements

Permission to reproduce remotely sensed images and their interpretations is gratefully acknowledged. Every effort has been made to contact organisations and individuals to obtain permission and I apologise if any have been omitted from this list. Numbers in parentheses refer to Figure numbers in the text.

S. Birch, G. Foody, S. Laffoley and F. Wells, Sheffield University (1.2, 6.34); Cambridge Instruments (6.3); Cambridge University Collection (3.21, 3.24); Canada Centre for Remote Sensing (4.14); Carl Zeiss, Jena (5.32); Carl Zeiss, Oberkochen Ltd (3.3, 3.32); Cartographic Engineering Ltd (3.31, 3.32); CNMHS Spadem (3.1); Dr R. Cochrane, Auckland and the *International Journal of Remote Sensing* (5.13); Clyde Surveys Ltd (3.8, 4.11, 6.19); Deutsches Museum (3.2); Electricity Supply Board, (4.11); Environmental Research Institute of Michigan (6.15); European Space Agency (4.20, 4.23); Eurosense (4.3, 6.13); Goodyear Aerospace Corporation and Aero Service Corporation (4.25, 4.26, 4.28); Dr. C. Gurney, Washington D C (6.7); Hunting Geology and Geophysics Ltd (5.6); Hunting Technical Services Ltd (3.14); International Instrumentation Marketing Company Ltd (3.4); Intertech Remote Sensing Ltd (4.14); McDonnell Douglas Astronautics Company (5.4); Meridian Airmaps Ltd (3.9, 3.10, 3.24, 6.2); Merseyside County Council (Plate 1); Ministry of Agriculture, Fisheries and Food (3.11, 4.10); T. Munday, Durham (6.30); Ministry of Defence (3.24, 4.12, 6.10); Natural Environment Research Council (4.7, Plate 3); National Oceanic and Atmospheric Administration (5.29); Spectral Africa (Pty) Ltd (4.9); Spectral Data Corporation (6.8); Survey and General Instrument Co. Ltd (3.31); University of Dundee (5.27, 5.28); Victor Hasselblad, Aktiebolag (5.4); West Air Photography (3.12); Wild Heebrugg Ltd (3.26); Dr T. H. Lee Williams, Kansas (3.20).

I am also pleased to acknowledge the source of the following non-copyright material: European Space Agency (5.31), National Aeronautics and Space Administration (5.3, 5.5, 5.6, 5.9, 5.12, 5.14, 5.15, 5.19, 5.24, 5.26, 6.7, 6.25, Plate 4 and the cover); National Oceanic and Atmospheric Administration (5.16, 5.17, 5.25, 5.30, 6.20).

**Paul Curran**

# Chapter 1 Remote sensing today

'... *remote sensing is a reality ... whose time has come. It is too powerful a tool to be ignored in terms of both its information potential and the logic implicit in the reasoning processes employed to analyse the data. We predict it could change our perceptions, our methods of data analysis, our models and our paradigms.* (Estes *et al.* 1980)

## 1.1 Introduction

If you visit one of the early eighteenth century steelworks in Sheffield you will walk along cobbled paths from the cottages to the furnaces, from the furnaces to the forges and from the forges to the grinding sheds. For the men who toiled in these steelworks, Sheffield must have seemed very large, as to visit the city centre would have taken a day and a trip to London would have been unthinkable. Three generations and a transportation network later, the environment would still have appeared so vast that its abuse would not have been seen as a problem. In the twentieth century, as aircraft span the globe and the media inform, the world's finite size is all too apparent, and as a finite world has but limited amounts of minerals, energy, food and space, there is obvious cause for alarm. The first step to managing man's use of these resources is to map and monitor the Earth and to do this workers turned to aerial photographs (Ch. 3) and later to images from the novel sensors carried by aircraft (Ch. 4) and satellites (Ch. 5).

In 1960 when the name remote sensing was first coined (Fischer 1975) it simply referred to the observation and measurement of an object without touching it. Since that date remote sensing has taken discipline dependent meanings, in the environmental sciences of geography, geology, botany, zoology, civil engineering, forestry, meteorology, agriculture and oceanography it usually refers to *the use of electromagnetic radiation sensors to record images of the environment which can be interpreted to yield useful information.*

To help you pick your way through the remote sensing techniques that are available, the broad structure of both the subject and this book is illustrated in Fig. 1.1. For example, if you are interested in the use of digital data collected by a thermal infrared linescanning sensor onboard a satellite, then you will need to refer to portions of Chapters 2, 5 and 6. Given that the book cannot cover everything, three remote sensing techniques that are either little used, or straddle the borders of environmental science have been omitted. These are first, the use of sensors that do not employ electromagnetic radiation, for example, those which use force fields (e.g. magnetometers) or acoustic waves (e.g. side-scan sonars). Second, the use of sensors that do not produce images (e.g. altimeters) and third, the many meteorological techniques that are designed for the monitoring of short-term weather as opposed to long-term climate. For further details refer to Colwell (1983a).

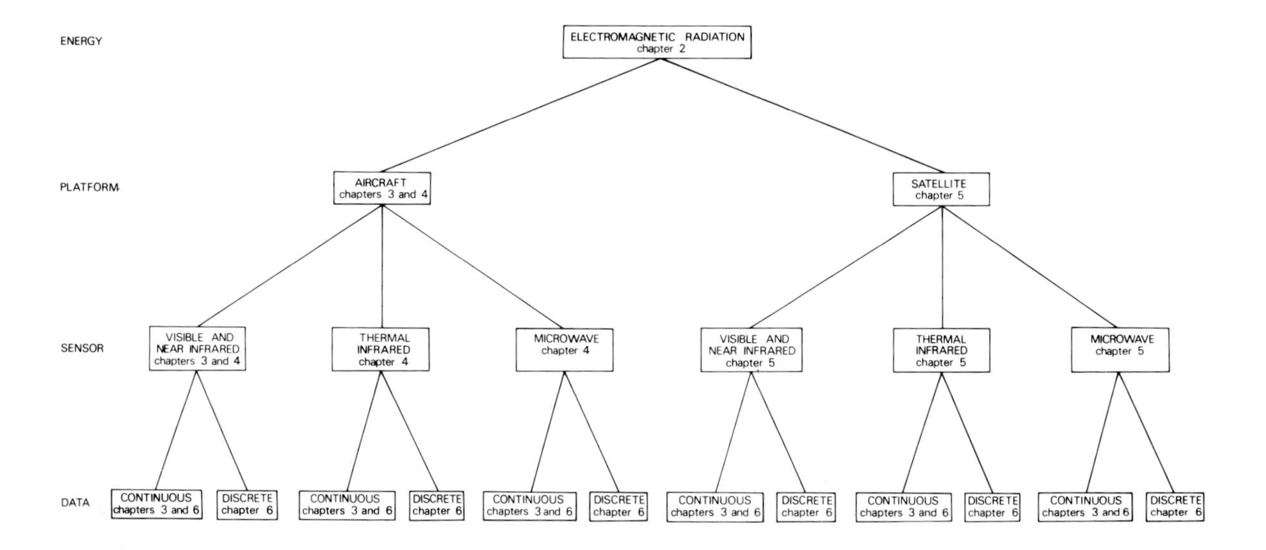


**Fig. 1.1**
The structure of remote sensing and this book.

## 1.2 Recent developments in remote sensing

During the 1960s and 1970s the new subject of remote sensing changed in both content and organisation. The 1960s were formative years in which visual interpretation of black and white aerial photographs paralleled research into the use of data from the new aircraft and satellite borne sensors. The developments of these years were reported in organisational reports, proceedings of symposia and photogrammetric journals. By the late 1960s a wide range of photographic emulsions were being used regularly and the results of experiments using data from both thermal infrared and microwave sensors onboard aircraft and cameras onboard satellites were starting to appear in the now expanding literature (Appendix B). Remote sensing, especially non-photographic remote sensing, grew rapidly after the successful launch of the Earth Resources Technology satellite (later renamed Landsat 1) in 1972 (Fischer *et al.* 1976). This satellite which carried sensors capable of providing synoptic views of the Earth's surface every 18 days proved to be the harbinger of many of the interpretation techniques that are in use today.

These general trends are illustrated in relation to the structure of this book in Fig. 1.2. In recent years there has been a general broadening of the subject coupled with three particularly rapid developments first, the operational use of remotely sensed data, second quantification and third, expansion of remote sensing education and training.

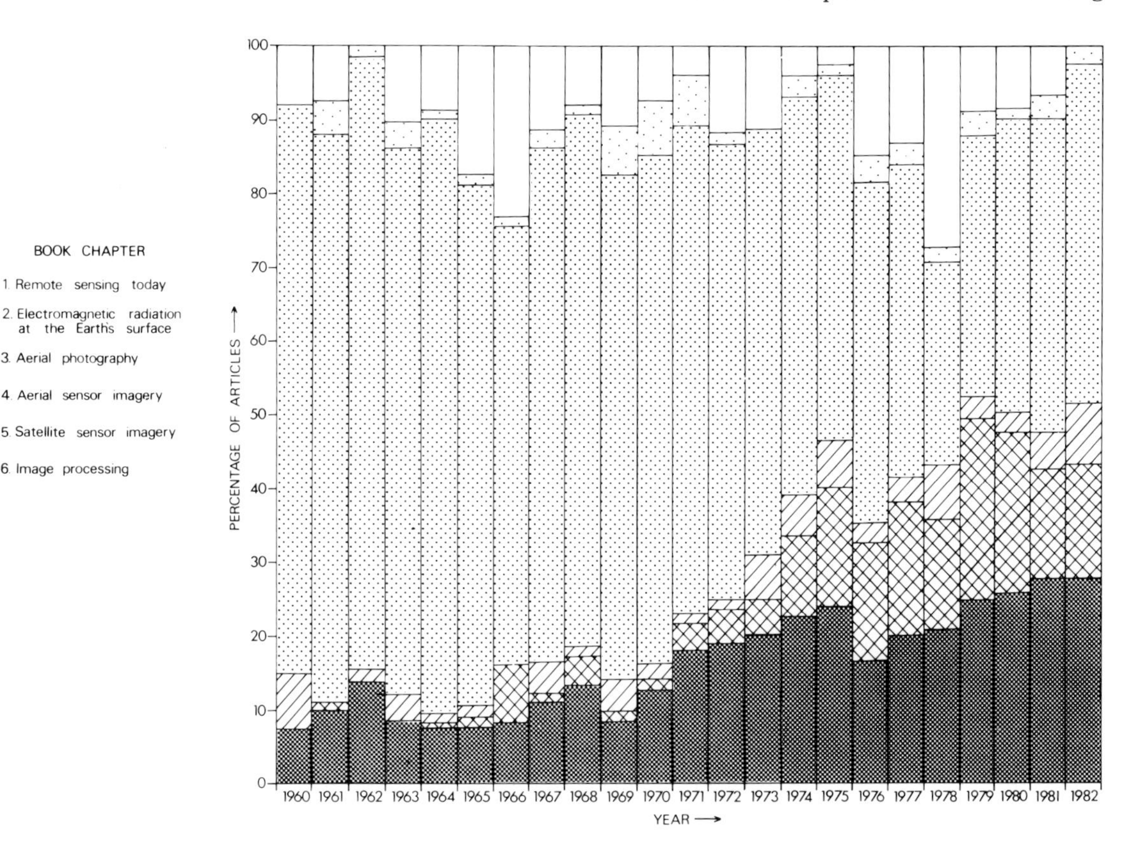


**Fig. 1.2**
Articles published in the journals *Photogrammetric Engineering* (1960–74) and *Photogrammetric Engineering and Remote Sensing* (1975–82). The articles have been classified according to the chapter contents of this book. (Data, courtesy S. Laffoley, Sheffield University)

## 1.2.1 Operational use of remote sensing

Each remote sensing technique starts life as an experiment and only becomes operational when it has been shown to work effectively. Therefore each remote sensor (Fig. 1.1) will have a range of applications from experimental to operational (Spann *et al.* 1981). In recent years the number of operational applications has been increasing, especially where the remotely sensed data are subsidised by governments. This increase is hardly surprising as the cost to the user of subsidised remotely sensed data is considerably less than the more traditional data sources. Nowhere is this more evident than in the use of meteorological satellite sensor data (sect. 5.5) which can undercut the cost of data derived entirely from meteorological stations to a third in Europe, a quarter in USA and a sixteenth in the Middle East (Barrett and Curtis 1982). Likewise the use of Landsat satellite sensor data (sect. 5.4.2) can reduce the production cost of a land-cover map by around a twentieth (Barrett and Curtis 1982) and the cost of a forest survey by around a half (Watkins 1978).

In many applications today there is either no subsidy or the benefits of remotely sensed data are so large that the cost of data purchase is insignificant. Within the first category is the use of panchromatic aerial photographs for cartography, false colour near infrared aerial photographs for forest surveys and thermal infrared linescanner imagery for the mapping of urban temperatures. Within the second category fall the increasing number of exciting global atmospheric, oceanic and agricultural studies, for example, the World Weather Watch (WWW) and the Global Atmospheric Research Project (GARP), that rely on remotely sensed images as their prime data source (Robinove 1975; Barrett and Curtis 1982). The benefits to be gained from global remote sensing are well illustrated by the techniques of agricultural inventory (Heiss *et al.* 1981). These were developed to satisfy USA's need for information on the world's cereal production prior to speculating on the world market (Slater 1980). The first comprehensive inventory was called the Large Area Crop Inventory Experiment (LACIE) and was undertaken from 1974 to 1977 (MacDonald 1979). Wheat production of the USSR, Latin America, China, Australia and India was estimated by multiplying the crop area derived from Landsat sensor data by the estimated crop yield derived from meteorological satellite sensor data (Curran 1980a). These estimates of wheat production were obtained region by region six to eight weeks before harvest with an accuracy of considerably better than 90% at the 90% significance level (Barrett and Curtis 1982). This information was worth around $200 million to the USA's agricultural industry each and every year (Slater 1980). A figure which should be compared with the $172 million it cost to launch the satellite Landsat 1 (Lovell 1977) and the $470 million it cost to launch the satellite Landsat 4 (NOAA 1982b). In 1980 LACIE was replaced by a larger monitoring program called AgRISTARS (Agriculture and Resources Inventory Surveys Through Aerospace Remote Sensing). This program uses airborne remote sensing data, satellite remote sensing data and a cunning blend of visual and machine interpretation techniques to achieve seven operational goals. These goals are (i) the early warning of changes affecting production and quality of renewable resources, (ii) forecast of commodity production, (iii) inventory of renewable resources, (iv) classification and inventory of land use, (v) estimation of land productivity, (vi) assessment of conservation practices and (vii) detection and evaluation of pollution. By the mid-1980s it is hoped that the majority of these will be operational (AgRISTARS 1981, 1982, 1983).

### 1.2.2 Quantification

The 'quantitative revolution' in the environmental sciences did not noticeably affect remote sensing until the 1970s. Quantification is now evident in the increasing use of computer based image processors (sect. 6.3), (Swain and Davis 1978; Remote Sensing Society 1983), for the analysis of satellite sensor images in digital form (Fig. 1.1). This is exemplified by sales of the Landsat satellite sensors images: in 1973 for every 8,000 Landsat sensor images sold in photographic format only one was sold in computer tape format. By 1978 this ratio had dropped from 8,000 : 1 to 39 : 1 (Table 1.1) and is expected to level out at around 10 : 1 by the late 1980s.

**Table 1.1** Number of Landsat (MSS and RBV) scenes sold by the EROS data centre, USA.

| *Fiscal year* | *Number of scenes* | | |
|---|---|---|---|
| | *Photographic format* | *Computer tape format* | *Photographic product to digital product ratio* |
| 1973 | 81,071 | 10 | 8,107 : 1 |
| 1974 | 157,178 | 228 | 689 : 1 |
| 1975 | 197,654 | 729 | 271 : 1 |
| 1976 | 297,253 | 3,299 | 90 : 1 |
| 1977 | 130,100 | 1,887 | 69 : 1 |
| 1978 | 110,723 | 2,853 | 39 : 1 |
| 1979 | 134,482 | 2,982 | 45 : 1 |
| 1980 | 128,433 | 4,139 | 31 : 1 |
| 1981 | 128,775 | 4,351 | 29 : 1 |
| 1982 | 118,858 | 5,224 | 23 : 1 |

*Sources*: NASA (1982a, 1982c); NOAA (1982a).

### 1.2.3 Education and training

Education and training in remote sensing is required at both an introductory and advanced level. At an introductory level, they help would-be users to both comprehend the significance of remotely sensed data and define their data requirements in terms of the platforms, sensors and interpretation techniques used in remote sensing (Fig. 1.1). At an advanced postgraduate level they provide the basis of a career in remote sensing.

Education and training at the introductory level is lacking the world over, for example, in the UK there are few short courses, remote sensing is absent from school syllabi and less than half of the Universities offer courses in remote sensing (Curran and Wardley 1983). By contrast education and training at the advanced postgraduate level is increasing rapidly (Estes *et al.* 1980). For example, in the UK the number of students working for a higher degree in remote sensing has increased fivefold in under a decade (Curran and Wardley, 1983). Within the next few years there is likely to be a large increase in education and training at the introductory level (Royal Society 1983) and the development of a professional structure at the advanced postgraduate level (Jensen and Dahlberg 1983).

## 1.3 Social and legal implications of remote sensing

Remote sensing has developed so quickly that it has outstripped our social and legal system (Estep 1968). The benefits of remote sensing are now so clear that the problems of remote sensing are infrequently articulated. When they are they tend to focus on the implications of moving funds from the traditional areas of public spending into remote sensing research, the degree of political power that remote sensing can provide to resource consumers (Stoebner 1976), the social efficacy of certain remote sensing techniques (Yanchinski 1980) and the right of one state to collect and disseminate remotely sensed data of another state (Stowe 1976; Lins 1979; Remote Sensing Society 1983).

The last is undoubtedly the most pressing problem for while remote sensing of one state by another is not illegal under international law, every state can invoke their legal right to territorial sovereignty and destroy the sensing vehicle. At present no state is likely to use this sanction as all states can obtain benefit from remotely sensed data. Unfortunately for the operators of satellite based sensing systems, the perceived benefits of remote sensing are closely allied to the exclusivity of the data (Kaltenecker and Lafferranderie 1977). The question of exclusivity is a vexed one and it has fallen to the United Nations Committee for the Peaceful Use of Outer Space to provide a forum for debate. At one extreme are the arguments of certain Latin American countries like Brazil and Argentina who would like the sensing state to ask permission to sense from the sensed state, to provide remotely sensed data to the sensed state and not distribute the remotely sensed data to third party states. The USSR and France do not argue for prior permission but do want the sensing state to provide the remotely sensed data to the sensed state and not distribute these remotely sensed data to third party states. At the other extreme are the USA, UK and Japan who argue for no legislation and an unrestricted dissemination of data.

To reconcile these views the United Nations Committee on the Peaceful Use of Outer Space is striving for compromise along the lines discussed by Jasentuliyana and Lee (1979). This would involve (i) openness of all remote sensing, (ii) priority data acquisition by sensed states, (iii) location of receiving stations in the sensed states which require them, (iv) prompt dissemination of environmentally transient data to states that require them, (v) increased education and training in remote sensing, (vi) monitoring of all remote sensing by the United Nations and (vii) agreements on data continuity (Matte and DeSaussure 1976).

## 1.4 Status of remote sensing

The growth of remote sensing like any other discipline tends to follow a sigmoid or logistic curve (Fig. 1.3), (Price 1963; Crane 1972). Stage one is a preliminary growth period with small absolute increments of literature and little or no social organisation. Stage two is a period of exponential growth when the number of publications double at regular intervals and specialist research units are established. Stage three is a period when the growth rate begins to decline and although annual increments remain constant, specialisation and controversy increase. Stage four is a final period when the rate of growth approaches zero, the specialist research units and social organisation break down and the subject reaches maturity (Jensen and Dahlberg 1983).

The location of remote sensing in this framework varies between countries. In most of the developing countries remote sensing is in stage one, in most of Europe remote sensing is in stage two (Curran

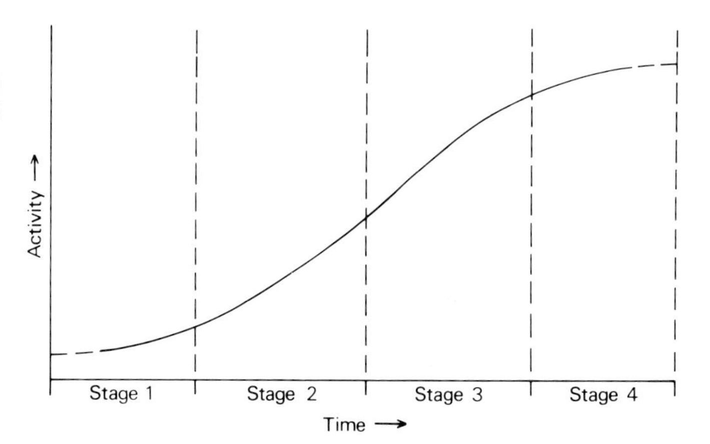


**Fig. 1.3**
The developmental stages of a scientific discipline. (Modified from Jensen and Dahlberg 1983)

and Wardley 1983) and in the USA it is just beginning to enter stage three (Jensen and Dahlberg 1983).

The ultimate goal of remote sensing is to reach the fourth stage of maturity when reliable information can, as a matter of routine, be generated for the management of our fragile planet. As the following chapters illustrate, that goal is now in sight.

## 1.5 Recommended reading

**Barrett, E.C.** and **Curtis, L.F.** (1982) *Introduction to Environment Remote Sensing* (2nd edn). Chapman and Hall, London; New York.

**Johannsen, C.J.** and **Sanders, J.L.** (1982) *Remote Sensing for Resource Management.* Soil Conservation Society of America, Iowa.

**Lillesand, T. M.** and **Kiefer, R. W.** (1979) *Remote Sensing and Image Interpretation.* Wiley, New York.

**Lintz, J.** and **Simonett, D.S.** (eds) (1976) *Remote Sensing of Environment.* Addison-Wesley, Reading, Massachusetts; London.

**Richason, B. F.** (ed.) (1978) *Introduction to Remote Sensing of the Environment.* Kendall/Hunt, Dubuque, Iowa.

**Sabins, F. F.** (1978) *Remote Sensing: Principles and Interpretation.* Freeman Hall, San Francisco.

**Siegal, B. S.** and **Gillespie, A. R.** (eds) (1980) *Remote Sensing in Geology.* Wiley, New York.

**Simonett, D.S.** (1983) Development and principles of remote sensing: *In* Colwell, R.N. (ed.) *Manual of Remote Sensing* (2nd edn). American Society of Photogrammetry, Falls Church, Virginia, pp. 1–35.

# Chapter 2 Electromagnetic radiation at the Earth's surface

*'Almost all of the turmoil of the world around us is beyond the reach of the sensors we are born with.'* (Parker and Wolff 1965)

## 2.1 Introduction

It is not possible to interpret fully a remotely sensed image or to discuss the design of a remote sensor until the way in which radiation interacts with the Earth's surface is understood. The aims of this chapter are first, to outline the remote sensing system and the link between electromagnetic radiation and each part of that system. Second, to present the terminology and units necessary to describe electromagnetic radiation and third, to discuss the way in which electromagnetic radiation interacts with vegetation, soil, water and urban areas.

## 2.2 A remote sensing system

A remote sensing system using electromagnetic radiation has four components: a source, interactions with the Earth's surface, interaction with the atmosphere and a sensor (Fig. 2.1).

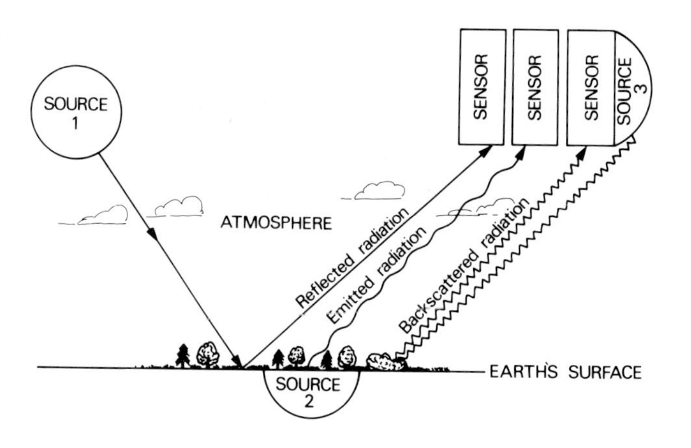


**Fig. 2.1**
A remote sensing system.

**Source:** the source of electromagnetic radiation may be natural like the Sun's reflected light or the Earth's emitted heat, or man-made, like microwave radar.

**Earth's surface interaction:** the amount and characteristics of radiation emitted or reflected from the Earth's surface is dependent upon the characteristics of the objects on the Earth's surface.

**Atmospheric interaction:** electromagnetic energy passing through the atmosphere is distorted and scattered.

**Sensor:** the electromagnetic radiation that has interacted with the surface of the Earth and the atmosphere is recorded by a sensor, for example a radiometer or camera. It helps to keep these four components of a remote sensing system in mind during later discussions of particular remote sensing systems.

### 2.2.1 Electromagnetic energy

The link between the components of the remote sensing system is electromagnetic energy.

Energy provides the ability to do work and by doing work energy is usually transferred from one point to another by conduction, convection or radiation. In remote sensing we are primarily concerned with energy transfer by means of radiation. Energy that is radiated behaves in accordance with basic wave theory which tells us that an electromagnetic wave is equally and repetitively spaced in time, moves with the velocity of light and has two force fields that are orthogonal to each other, one of which is electric and the other is magnetic. Three measurements are used to describe electromagnetic waves, these are wavelength ($\lambda$) in micrometres ($\mu$m), which is the distance between successive wave peaks, frequency ($v$) in hertz (Hz) which is the number of wave peaks passing a fixed point in space per unit time and velocity ($c$) in m $s^{-1}$ which within a given medium is constant at the speed of light. As wavelength has a direct and inverse relationship to frequency, an electromagnetic wave can be characterised by either its wavelength or its frequency because they are interchangeable. However, for reasons of custom wavelength rather than frequency is the measurement by which electromagnetic waves are usually identified in remote sensing. For further details refer to Feinberg (1968).

### 2.2.2 The electromagnetic spectrum

Electromagnetic radiation occurs as a continuum of wavelengths and frequencies from short wavelength, high frequency cosmic waves to long wavelength, low frequency radio waves (Fig. 2.2). The wavelengths that are of greatest interest in remote sensing are visible and near infrared radiation in the waveband 0.4–3 $\mu$m, infrared radiation in the waveband 3–14 $\mu$m and microwave radiation in the waveband 5–500 mm. How electromagnetic energy in these particular wavebands interacts with the surface of the Earth is discussed in section 2.5, with further details in Janza (1975) and Suits (1975, 1983).

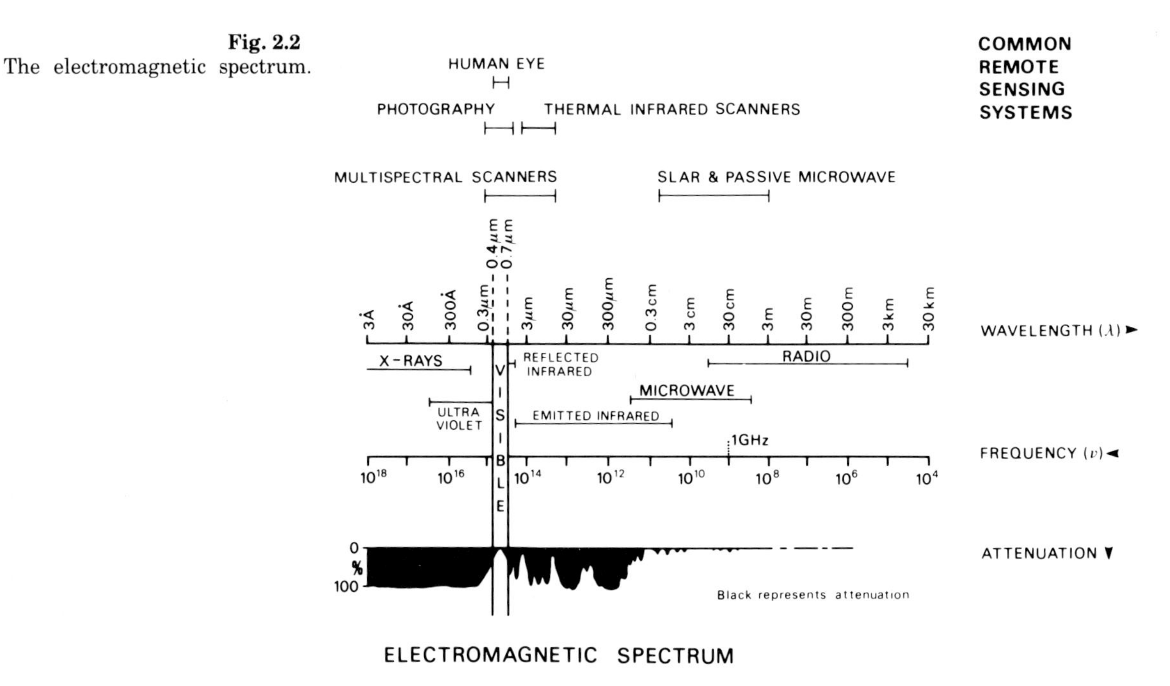


**Fig. 2.2**
The electromagnetic spectrum.

## 2.3 Remote sensing terminology and units

To describe a remote sensing system a collection of measurement units and radiometric terms are required. These units and terms are discussed here for reasons of continuity but you may wish to skip over this section and refer back to it, as and when required.

Unfortunately, for everyone involved in remote sensing there are many terms and units available to describe and measure each phenomenon. In recent years workers have achieved a degree of standardisation by adopting the measurement units of the Système International d'Unites (SI units), the physical and chemical terms and units of The Royal Society and the radiometric terms proposed by the American National Bureau of Standards (Royal Society 1975; Nicodemus *et al.* 1977; ASP 1978; Wolfe and Zissis 1978).

The standard terms and units you will require in order to understand the current literature are grouped for convenience in seven tables. Tables 2.1 to 2.3 contain basic physical terms and units, Tables 2.4 and 2.5 contain radiometric terms and units, Table 2.6 contains terms and units particular to the sensing of thermal infrared wavelengths, and Table 2.7 contains terms and units particular to the sensing of microwave wavelengths. In Table 2.1 it is assumed that you will be familiar with the first three terms but may not have used the last seven terms. The temperature scale has an SI unit with the name of kelvin (symbol K, not °K); however, the Celsius scale (°C) is often used in its place, as one degree Celsius is equal to one kelvin. To convert from degrees Celsius to kelvins you simply add 273.15 degrees: for example, 10 °C = 283.15 K. The two angular measurements given in Table 2.1 are used in remote sensing to define the angles at which radiation arrives at and departs from the Earth's surface; the plane

**Table 2.1**
Base and derived SI quantities and units.

| *Quantity* | *SI unit* |
|---|---|
| Length ($l$) | metre (m) or ångström (Å) |
| Time ($t$) | second (s) |
| Mass ($m$) | kilogram (kg) |
| Temperature ($T$) | kelvin (K) |
| Plane angle ($\propto$) | radian (rad) |
| Solid angle ($\Omega$) | steradian (sr) |
| Force ($m\ lt^{-2}$) | newton (N) |
| Energy ($m\ l^2\ t^{-2}$) | joule (J) |
| Power ($m\ l^2\ t^{-3}$) | watt (W) |

angle measurement records these angles in two dimensions while the solid angle measurement records these angles in three dimensions. The plane angle measurement has the unit of a radian ($\propto$), which is the angle at the centre of a circle formed by two radii ($r$) cutting off an arc equal in length to the radius (Fig. 2.3). The unit radian is equal to an angle of 57.3° and is defined in formula [2.1]

$$\alpha = \frac{\pi r}{r} \qquad [2.1]$$

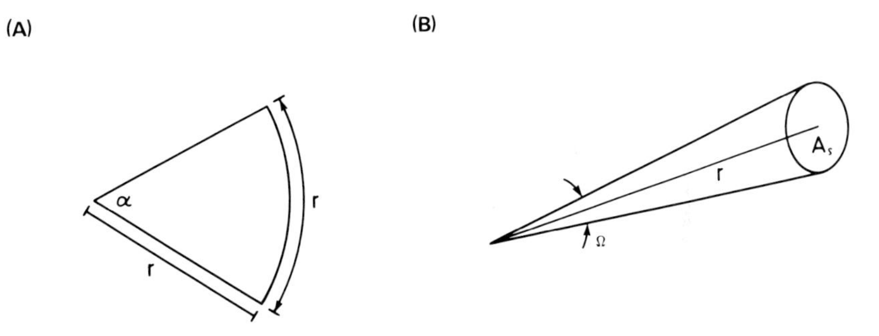


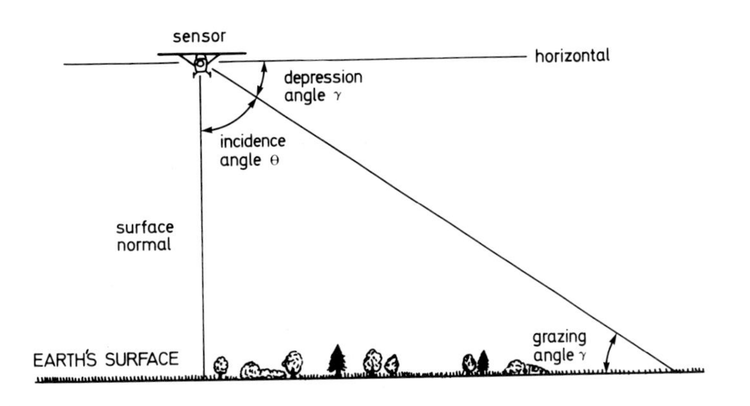


**Fig. 2.3**
Angle definitions: (A) the radian, (B) the solid angle and (C) sensor to Earth angles.

Solid angle measurement has the unit of a steradian, (Ω), (pronounced without the e) which is the solid angle subtended by an area ($A_s$) on the surface of a sphere (Fig. 2.3) divided by the square of the radius ($r$) of the sphere. This is defined in formula [2.2].

$$\Omega = \frac{AS}{r^2} \qquad [2.2]$$

**Table 2.2**
Other quantities and units.

| *Quantity* | *Units* |
|---|---|
| Area ($A$) | square metre ($m^2$) |
| Volume ($V$) | cubic metre ($m^3$) |
| Frequency (ν) | hertz or cycle $s^{-1}$ (Hz) |
| Wavelength (λ) | metre (m) |
| Incidence angle: angle between line of sight and vertical (θ) | degree (°) |
| Grazing or depression angle: angle between line of sight and horizontal (γ) | degree (°) |
| Density (D) | kilogram per cubic metre (kg $m^{-3}$) |

Table 2.2 contains seven commonly used quantities and units in remote sensing that have no SI derivation. Important terms in this list are the angular terms which are summarised in Fig. 2.3.

**Table 2.3**
Standard unit prefix notation for SI units.

| *Prefix* | *Notation* | *Multiplier* |
|---|---|---|
| tera | T | $10^{12}$ |
| giga | G | $10^{9}$ |
| mega | M | $10^{6}$ |
| kilo | k | $10^{3}$ |
| hecto | h | $10^{2}$ |
| deca | da | 10 |
| deci | d | $10^{-1}$ |
| centi | c | $10^{-2}$ |
| milli | m | $10^{-3}$ |
| micro | μ | $10^{-6}$ |
| nano | n | $10^{-9}$ |
| pico | p | $10^{-12}$ |

Table 2.3 contains standard unit prefix notation for SI units. For example, the average wavelength of green light is $0.55 \times 10^{-6}$ m; as this is cumbersome to write and comprehend a prefix is customarily employed. The prefix (μ) represents $10^{-6}$ and so we can describe the average wavelength of green light as being 0.55 micrometres or 0.55 μm.

In Table 2.4 the first five terms are general radiometric quantities used to describe the total radiation budget of a surface; *radiant energy*

**Table 2.4** Radiometric quantities and units.

| *Quantity* | *Defining expression* | *Units* |
|---|---|---|
| Radiant energy ($Q$) | – | joule (J) |
| Radiant density ($W$) | $\frac{dQ}{dv}$ | joule per cubic metre (J $m^{-3}$) |
| Radiant flux ($\Phi$) | $\frac{dQ}{dt}$ | watt (W) |
| Radiant exitance ($M$) | $\frac{d\Phi}{dA}$ (out) | watt per square metre (W $m^{-2}$) |
| Irradiance ($E$) | $\frac{d\Phi}{dA}$ (in) | watt per square metre (W $m^{-2}$) |
| Radiant intensity ($I$) | $\frac{d\Phi}{d\Omega}$ | watt per steradian (W $sr^{-1}$) |
| Radiance ($L$) | $\frac{d^2\Phi}{d\Omega(dA\cos\theta)}$ | watt per square metre per steradian (W $m^{-2}$ $sr^{-1}$) |
| Spectral radiant exitance ($M\lambda$) | $\frac{dM}{d\lambda}$ (out) | watt per square metre per micrometre (W $m^{-2}$ $\mu m^{-1}$) |
| Spectral irradiance ($E\lambda$) | $\frac{dE}{d\lambda}$ (in) | watt per square metre per micrometre (W $m^{-2}$ $\mu m^{-1}$) |
| Spectral radiance ($L\lambda$) | $\frac{dL}{d\lambda}$ | watt per square metre per steradian per micrometre (W $m^{-2}$ $sr^{-1}$ $\mu m^{-1}$) |

($Q$) is the total energy radiated in all directions; *radiant density* ($W$) is the total energy radiated by a unit area in all directions; *radiant flux* ($\Phi$) is the total energy radiated in all directions for a unit of time; *radiant exitance* ($M$) is the total energy radiated in all directions by a unit area in a unit time, and *irradiance* ($E$) is the total energy radiated onto a unit area in a unit time. The terms *radiant intensity* ($I$) and *radiance* ($L$) in Table 2.4 refer to radiation within a given angle of observation; where ($I$) is the total energy per solid angle of measurement and ($L$) is the total energy radiated by a unit area per solid angle of measurement. Radiance ($L$) is one of the most important radiometric terms in remote sensing (Fig. 2.4) because it describes what is actually measured by a sensor. The last three terms in Table 2.4 are examples of radiometric terms that have been restricted to narrow wavelength bands by the addition of the word spectral. For example, *spectral radiance* ($L\lambda$) refers to the energy within a wavelength band, radiated by a unit area per unit solid angle of measurement.

Knowledge of the radiant energy ($\Phi$), or radiant exitance ($M$) of the Earth's surface is insufficient to characterise an object in remote sensing, as both of these vary with incoming irradiance ($E$) or temperature ($T$). Therefore, unitless ratio terms are customarily used to describe the radiant energy ($\Phi$) or radiant exitance ($M$), as a proportion of irradiance ($E$) or the emissivity of a standard source. Four of the more common units are to be found in Table 2.5. Emissivity ($\varepsilon$) is the ratio of the radiant exitance of the surface ($M$) with a perfectly emitting surface called a blackbody ($M_{blackbody}$). Reflectance ($\rho$) is the ratio of incident ($\Phi_i$) to reflected ($\Phi_r$) radiant flux, absorbance ($\alpha$) is the ratio of incident ($\Phi_i$) to absorbed ($\Phi_a$) radiant flux and transmittance ($\tau$) is the ratio of incident ($\Phi_i$) to transmitted ($\Phi_t$) radiant flux, formula [2.4].

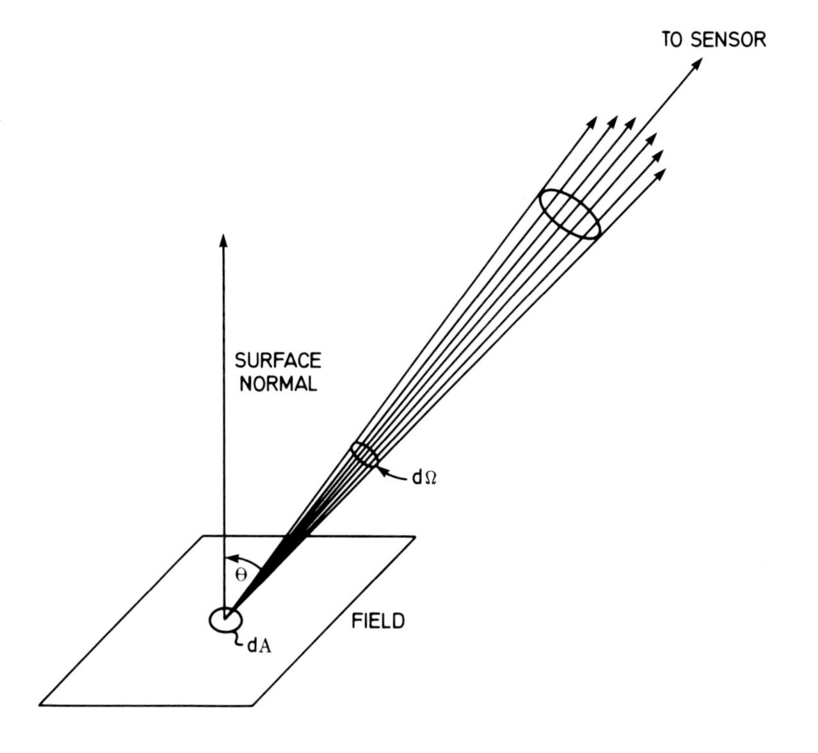


**Fig. 2.4**
Concept of radiance: the sensor will be receiving a radiant flux (Φ) with a solid angle (dΩ) from an area within the field (dA).

**Table 2.5**
Unitless radiometric quantities.

| *Quantity* | *Defining expression* |
|---|---|
| Emissivity (ε) | $\frac{M}{M_{\text{blackbody}}}$ |
| Reflectance (ρ) | $\frac{\Phi_r}{\Phi_i}$ |
| Absorbance (∝) | $\frac{\Phi_a}{\Phi_i}$ |
| Transmittance (ζ) | $\frac{\Phi_t}{\Phi_i}$ |

where: $M$ and $M_{\text{blackbody}}$ are respectively the radiant exitance of the measured object and that of a blackbody at the same temperature as the object

$\Phi_i$ = incident flux
$\Phi_r$ = reflected flux
$\Phi_a$ = absorbed flux
$\Phi_t$ = transmitted flux

Reflectance is an important but an often misused term in remote sensing. The two common mistakes are first, to use the word reflectance in place of the word radiance and second, to use reflectance without adequate definition. If reflectance is to be recorded within a limited wavelength range then it should be described as spectral reflectance (ρλ) and if it is to be recorded at a particular angle then this should also be stated.

To describe the angular nature of reflectance, two rather broad hemispherical and directional terms are used. Hemispherical refers to an angle of incidence or collection of radiant flux over a hemisphere and directional refers to the incidence or collection of radiant flux for

**Fig. 2.5** Angular nature of reflectance measurements. A graphical description of hemispherical and directional radiation and collection. (Modified from Judd 1967)

INCIDENCE

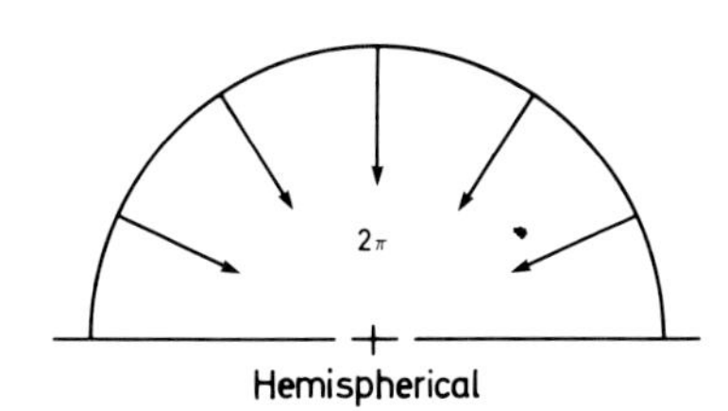


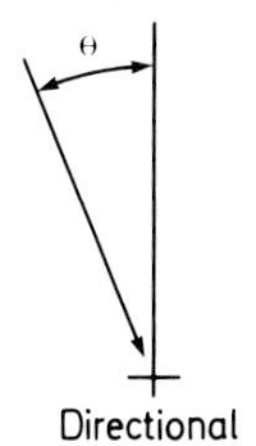


COLLECTION

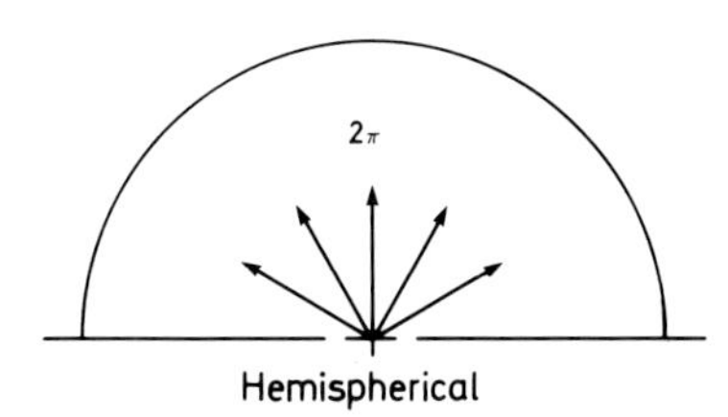


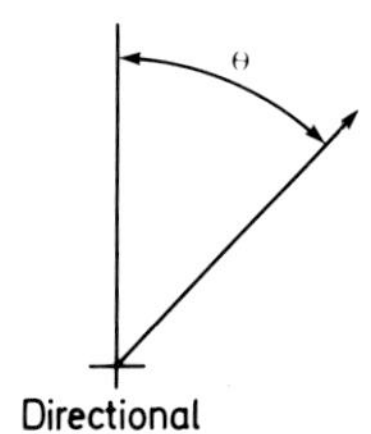


one direction only (Fig. 2.5). In remote sensing literature the majority of spectral reflectance measurements will be either bihemispherical (usually shortened to hemispherical) where the angles of incidence and collection are hemispherical, as would be the case in laboratory studies of reflectance, or bidirectional where the angles of incidence and collection are directional, as would be the case with satellite sensor measurements of radiance on a sunny day. The ideal measurement for use in remote sensing is the bidirectional reflectance distribution function (BRDF) which is the bidirectional reflectance at all possible angles of collection. As this is very difficult to measure, researchers make do with a simplified measurement – the bidirectional reflectance factor (BRF). The BRF is a ratio between the spectral radiance ($L\lambda$) at an angle ($\theta$) to the object of interest and a diffuse reflector at an angle ($\theta$) within the scene. As this measure approximates bidirectional reflectance it is usually called bidirectional reflectance instead of BRF (Silva 1978).

Table 2.6 contains four properties of materials that are used to describe the thermal qualities of objects upon the Earth's surface. Thermal capacity ($c$) is the amount of energy required to raise 1 gram of material by 1 °C, thermal conductivity ($k$), is the amount of energy that will pass through 1 $cm^3$ of material, in a unit time, when the two

**Table 2.6** Thermal properties.

| *Quantity* | *Defining expression* | *Units* |
|---|---|---|
| Thermal capacity ($c$) (specific heat) | — | $J\ kg^{-1}\ K^{-1}$ |
| Thermal conductivity ($k$) | — | $Wm^{-1}\ K^{-1}$ |
| Thermal diffusivity ($K$) | $K = \frac{k}{cD}$ | $m^2\ s^{-1}$ |
| Thermal inertia ($P$) | $P = \sqrt{Dck}$ | $Wm^{-2}\ K^{-1}\ s^{\frac{1}{2}}$ |

opposite ends of the material have a 1 °C difference in temperature, thermal diffusivity ($K$) is temperature change per unit volume of material per unit time and thermal inertia ($P$) is a measure of a material's resistance to a change in temperature. Thermal inertia is the quantity that is most amenable to determination by remote sensing as discussed in section 2.5.5.3.

Table 2.7 contains the two units used in connection with remote sensing at microwave wavelengths (Fig. 2.2); they are the scattering coefficient $\sigma^\circ$ and the return parameter ($\gamma$) and are both used to define the amount of electromagnetic radiation that is scattered back to the microwave sensor.

**Table 2.7**
Measurements of backscatter at microwave wavelengths.

| *Quantity* | *Defining expression* | *Units* |
|---|---|---|
| Scattering coefficient ($\sigma^\circ$) | $\frac{\sigma}{A}$ | decibel (dB) |
| Return parameter ($\gamma$) | $\frac{\sigma}{A_1}$ | decibel (dB) |

For definitions see page 19.

# 2.4 Sources and types of electromagnetic energy used in remote sensing

For the sake of convenience the electromagnetic spectrum will be divided into 'natural' radiation at visible and thermal infrared wavelengths and 'man-made' electromagnetic radiation at microwave wavelengths.

## 2.4.1 Natural electromagnetic energy

Of interest in remote sensing is visible light and near and middle infrared radiation (Fig. 2.2) which are both reflected by the Earth's surface and also middle and thermal infrared radiation which are both emitted by the Earth's surface. Wave theory (sect. 2.2.2) describes how such electromagnetic energy moves but not how the electromagnetic energy interacts with matter; this explanation is achieved by using the particle theory which suggests that electromagnetic radiation is composed of many discrete units called quanta or photons. The wave theory and particle theory when used together help us to understand the relationships between the wavelength ($\lambda$), frequency ($\nu$), radiant energy ($Q$) and radiant exitance ($M$) from sources of electromagnetic radiation that are at different temperatures ($T$). The relationships that are of value in remote sensing are summarised for a blackbody in Fig. 2.6. with the relevant formulae in Table 2.8. What these relationships tell us when combined is that all objects with a temperature above the zero point of the kelvin scale, absolute zero (−273.15 °C), radiate electromagnetic radiation. A hot object like a fire or the Sun will rapidly radiate high frequency short wavelengths of electromagnetic radiation and these short wavelengths will be high in energy and therefore easy to sense remotely. By contrast, a cool object like the

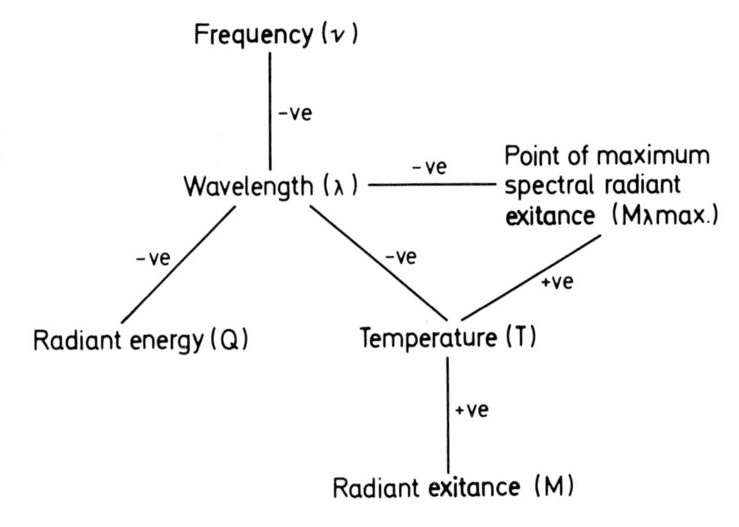


**Fig. 2.6**
The interrelationships between temperature, wavelength frequency, radiant energy, radiant exitance and the point of maximum spectral radiant exitance.

| *Relationship between* | *Form* | *Defining expression* | *Description* |
|---|---|---|---|
| λ and ν | −ve | $\lambda = \frac{c}{\nu}$ | Waves with a short wavelength have a high frequency |
| λ and $Q$ | −ve | $\lambda = \frac{hc}{Q}$ | Waves with a short wavelength have high energy |
| λ and $T$ | −ve | (a*) $M\lambda\text{max} = \frac{b}{T}$ | Waves with a short wavelength are exited from objects with a high temperature |
| $M$ and $T$ | +ve | (b*) $M = kT^4$<br>(d*) | The radiant exitance increases with temperature |
| $M_\lambda$ and λ and $T$ | see Fig. 2.7 | (c*) $M\lambda = \frac{c_1}{\lambda^5(e^{c2/\lambda}T_{-1})}$ | Hot objects radiate short wavelengths strongly |

*Notes*:

a* Wien's law
b* Stefan Boltzmann's law
c* Planck's law
d* To calculate $M$ for a real surface use Kirchhoff's law
$M_{real} = \varepsilon M_{blackbody}$
e Base of natural logarithms (2.718)

*Constants*:

c Velocity of light ($3 \times 10^8$ m $s^{-1}$)
$c_1$ First radiation constant ($3.74 \times 10^{-16}$ $Wm^2m$)
$c_2$ Second radiation constant (0.014 m K)
h Planck's constant ($6.6 \times 10^{-34}$ W $s^2$)
b Wien's constant (2,898 μm K)
k Stefan Boltzmann's constant ($5.67 \times 10^{-8} Wm^{-2}K^{-4}$)

**Table 2.8**
Electromagnetic radiation: the relationships for a blackbody between wavelength (λ), frequency (ν), radiant energy ($Q$), temperature ($T$), radiant exitance ($M$) and maximum spectral radiant exitance ($M_{\lambda max}$).

Earth will slowly radiate low frequency long wavelengths of electromagnetic radiation and these long wavelengths will be low in energy and therefore harder to sense remotely. This is illustrated in part by Fig. 2.7. In graph (a) a blackbody at the same temperature as the Sun (6,000 K or 5,727 °C) has a very high radiant exitance ($M$) at short wavelengths (λ) within the visible part of the spectrum and a high radiant exitance ($M$) at longer wavelengths (λ) within the thermal infrared part of the spectrum. In graph (b) a blackbody at the same temperature as the Earth (300 K or 27 °C) has a low radiant exitance ($M$) at long wavelengths (λ) within the thermal infrared part of the spectrum. This explains why the Sun radiates visible and thermal wavelengths strongly while the Earth radiates thermal infrared wavelengths weakly (Barrett and Curtis 1982; Suits 1983).

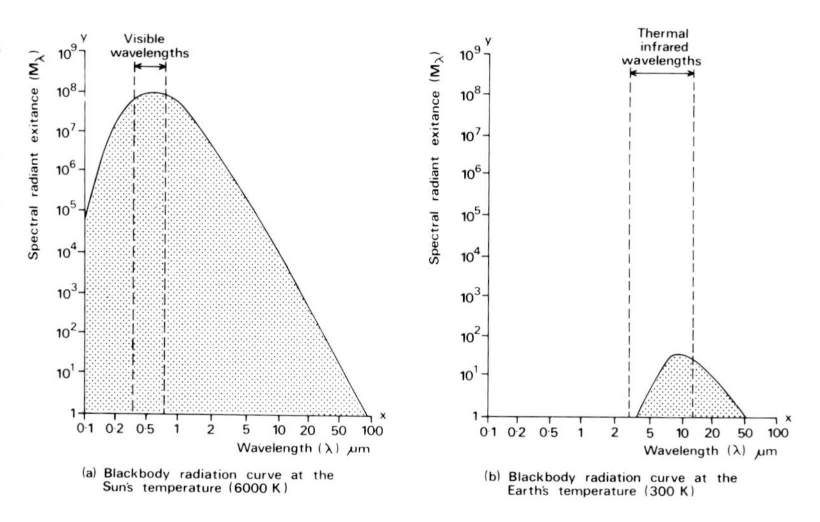


**Fig. 2.7**
The spectral distribution of energy radiated from blackbodies at the temperature of the Sun and the Earth. Note that spectral radiant exitance ($M\lambda$) is given on the y axis while the total radiant exitance ($M$) is given by the shaded area under the radiant exitance curves.

## 2.4.2 Man-made electromagnetic energy

Of interest in remote sensing is radiation of microwave wavelengths (Fig. 2.2) as this is used in imaging radar systems (sect. 4.4). A radar transmits short pulses of energy at microwave wavelengths towards the ground and then records, first, the time it has taken for the pulse to reach the object and return, and second, the strength and origin of the 'backscatter' or 'echo' received from objects within the sensor's field of view. This is illustrated in Fig. 2.8 where the single pulse is shown to move out from the aircraft at the speed of light; after $5 \times 10^{-6}$ seconds the pulse reaches the cereal crop in the field and is backscattered weakly to the aircraft and after $7 \times 10^{-6}$ seconds the pulse reaches the farmhouse and is backscattered strongly to the aircraft. The radar system records first, that there are two areas located $2 \times 10^{-6}$ seconds away from each other and second, that the backscattering properties of these two areas are different. It is the degree of backscattering that contains information on the objects within the scene and it is this information that is of most value in environmental remote sensing. The degree of backscattering is calculated using the radar equation, formula [2.3].

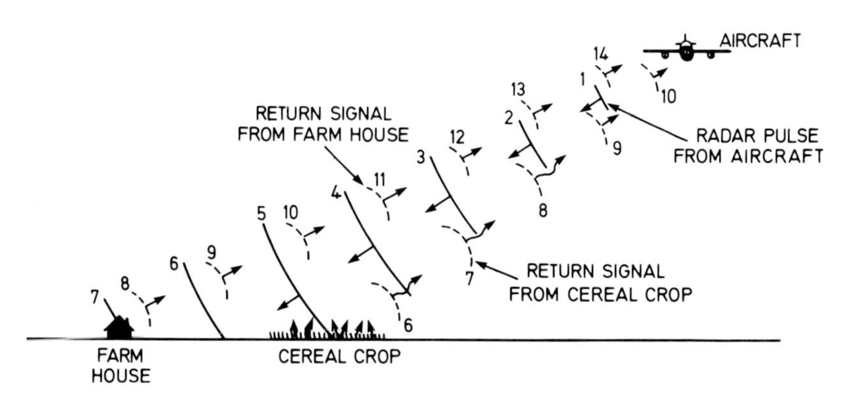


**Fig. 2.8**
An illustration of the propagation of one radar pulse towards the ground surface and its return to the sensor. The time periods are given in microseconds ($10^{-6}$ s).

$$\Phi_r = \frac{\Phi_t \, G_t^2 \, \lambda^2 \, \sigma}{(4\pi)^3 R^4} \qquad [2.3]$$

where
$\Phi_r$ = received power
$\Phi_t$ = transmitted power
$G_t$ = antenna gain (setting of antenna for radiation receipt)
$R$ = distance from transmitter to target
$\sigma$ = effective backscatter
$\lambda$ = wavelength

} parameters of the radar system

The radar equation is rather simplifed as it assumes that the sensor will be recording one type and area of surface whereas in practice it will be recording several surfaces of different area. In order to overcome this the radar backscatter is defined by one of two possible normalised units, either the scattering coefficient ($\sigma°$) or the return parameter ($\gamma$). The scattering coefficient ($\sigma°$) is the effective backscatter ($\sigma$) divided by the area of surface illuminated ($A$). The return parameter ($\gamma$) is the effective backscatter ($\sigma$) divided by the area of the radar beam ($A_1$), (Table 2.7). The return parameter is less dependent on incident angle ($\theta$) and is therefore the most suitable unit; however it is the more difficult to calculate, therefore the scattering coefficient ($\sigma°$) is usually used to define the radar backscatter (de Loor 1976; Ulaby *et al.* 1981). Because both $\sigma°$ and $\gamma$ are normalised quantities and vary over several orders of magnitude they are measured in decibels (dB), where a bel (B) is the $\log_{10}$ ratio of the power of one electric current to another.

The two other characteristics of man-made electromagnetic energy that are of importance are the wavelength and the polarisation.

### 2.4.2.1 Wavelength

A radar system although initially developed to operate at microwave wavelengths can in theory use any waveband from ultra-violet ($10^{-7}$ m) to radio (100 m); however, the majority of radar systems operate in the microwave region of the electromagnetic spectrum (Fig. 2.2) at a wavelength of between 5 mm and 500 mm. A code exists to identify the particular wavelength in which a radar is sensing. The basis of this code was developed by British scientists during the Second World War in order to confuse German intelligence services and it has been confusing students of radar ever since. To add to the problems researchers in the USA and NATO have developed their own codes (Easams 1972) and these overlap in part with the British code (Fig. 2.9).

At present X wavebands (British) are used for the majority of remote sensing, although future satellite remote sensing will probably operate in the longer L and C wavebands as the longer wavelengths require less power and ($\sigma°$) in the C waveband has been demonstrated to have a high correlation with vegetation cover.

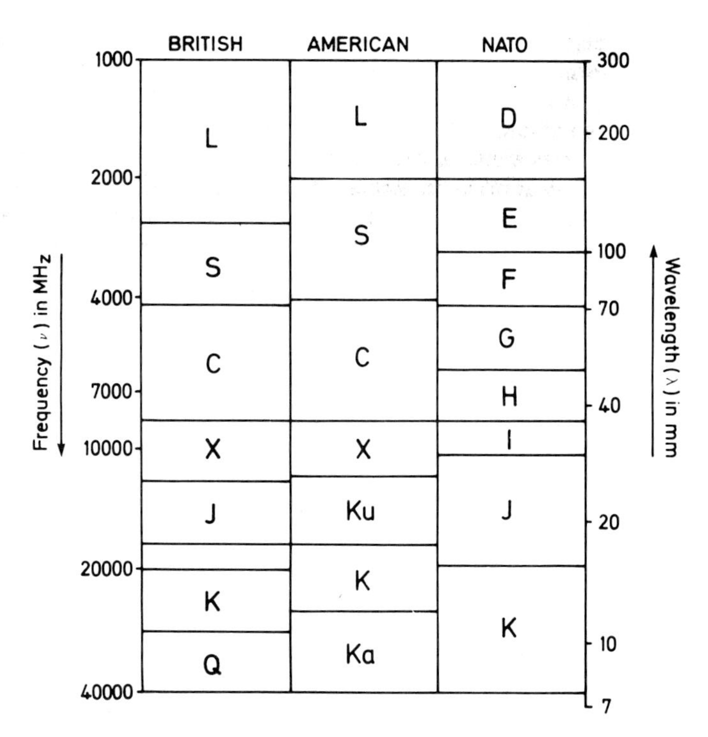


**Fig. 2.9** Radar waveband designations.

### 2.4.2.2 Polarisation

Radar signals are man-made, mass produced waves and like the majority of man-made, mass produced things they are all exactly the same and vibrate not only at the same wavelength but also in the same plane. The early radars transmitted microwaves with a horizontal polarisation and received microwaves with a horizontal polarisation and this is called HH radar. However, rough surfaces within the terrain actively depolarise electromagnetic energy and so many radar systems now record the vertically polarised return from a horizontally transmitted signal and this is called HV radar. The HH radar or like polarised radar records a maximum return from smooth surfaces at a low angle of incidence and the HV radar or cross polarised radar records a maximum return from rough surfaces. These systems are often used together as discussed in section 2.5.10.2 and illustrated in Fig. 4.20. For further details refer to MacDonald (1980) and Ulaby *et al.* (1981; 1982).

## 2.5 Earth surface interactions with electromagnetic radiation

Before a remotely sensed image can be fully interpreted a knowledge of how electromagnetic radiation interacts with the Earth's surface is needed. Therefore, the majority of this chapter will involve a discussion of the interaction of electromagnetic radiation with the four main components of a remotely sensed scene: vegetation, soil, water and urban areas. Emphasis will be placed on vegetation, soil and water, for urban areas contain all three with such spatial complexity that meaningful generalisations are difficult.

The wavelengths of electromagnetic radiation that have proved to be of particular value in environmental remote sensing will be stressed and these are: (i) reflected radiation in visible, near infrared and middle infrared wavebands, (ii) emitted radiation in middle and thermal infrared wavebands, and (iii) reflected radiation in microwave wavebands.

### 2.5.1 The interaction of visible, near infrared and middle infrared wavelengths with the Earth's surface

The spectral radiant flux ($\Phi\lambda$) incident on the Earth's surface is either reflected ($\rho\lambda$), absorbed ($\alpha\lambda$) or transmitted ($\tau\lambda$). As no energy is lost in this process, formula [2.4] applies.

$$\Phi\lambda = \rho\lambda + \alpha\lambda + \tau\lambda \qquad [2.4]$$

If the proportion of the spectral radiant flux ($\Phi\lambda$) that is reflected ($\rho\lambda$), absorbed ($\alpha\lambda$) or transmitted ($\tau\lambda$) is very dissimilar for different features on the Earth's surface, then we can identify features on the basis of their spectral properties. This is something we do every day of our lives for example, as shown in Fig. 2.10, white paper has high reflectance at all wavelengths, purple paper has high blue and red reflectance, red paper has high red reflectance, a cupro-nickel coin (2p) has increasing reflectance with wavelength and the author's fairly clean index finger has relatively high red reflectance. However, whether a remote sensor, (e.g. a camera in an aircraft or a radiometer in a satellite) can detect such variations in reflectance between objects is dependent upon four interrelated factors. First, the radiometric resolution of the sensor; second, the amount of atmospheric scatter; third, the surface roughness of the objects and fourth, the spatial variability of reflectance within the scene.

First, sensors vary in their ability to detect differences in radiance. For example, on board the satellite Landsat 4 the scanning radiometer called the Thematic Mapper can detect 256 levels of radiance and the scanning radiometer called the multispectral scanner can detect 64 levels of radiance. Therefore subtle differences in radiance recorded by the Thematic Mapper could defy detection by the multispectral scanner (sect. 5.4.2).

Second, atmospheric scatter increases the amount of radiance received by the sensor for each object and as a result the contrast between objects is reduced (sect. 2.6).

Third, surface roughness is of importance because the surface needs to be rough enough to allow radiation to interact with the surface of the objects. If the surface of the object is smooth and radiation is reflected without interaction, then little information will be transmitted to the sensor. Fortunately the majority of the Earth's surface appears rough at visible and near infrared wavelengths. To determine how rough a surface is at a particular wavelength Rayleigh's criterion of surface roughness is used (formula 2.5).

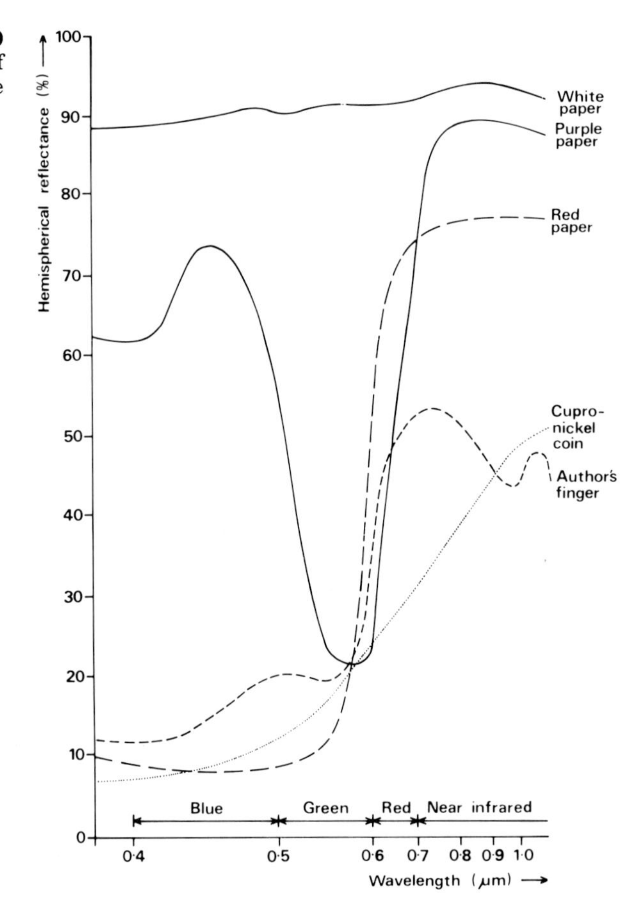


**Fig. 2.10**
The hemispherical reflectance of five objects recorded in the laboratory.

$$ht \leqslant \lambda/(8 \cos\theta) \qquad [2.5]$$

where $ht$ height variation above a plane in wavelengths
$\lambda$ = wavelength
$\theta$ = angle of incidence
(Hoffer 1978; Sabins 1978).

Objects which appear to be equally rough from all angles of observation are called Lambertian surfaces. As the majority of objects on the Earth's surface are non-Lambertian their angle of observation must be specified (sect. 2.3).

Fourth, the spatial variability of the scene is of importance because every sensor has what is termed 'a point spread function'. This means that the radiance recorded from an area of ground also contains radiance from the surrounding areas. For example, Forster (1980) reports that for a Landsat MSS image (e.g. Fig. 5.9) only 52% of the radiance recorded for each picture element or 'pixel' can be attributed to the area of that pixel. This observation, although alarming at first sight is of little importance because in the majority of remotely sensed scenes the probability of a pixel having the same radiance as its neighbours is very high (Steiner and Salerno 1975). This can cause problems in urban areas where radiance has high spatial variability due to the variety of land cover (Table 2.9) and the presence of shadow (Townshend 1981a).

**Table 2.9**
The radiance ($Wm^{-2}\ sr^{-1}$) and a radiance ratio for six components of an urban scene. Data are derived from a 1972 Landsat multispectral scanner image of Sydney, Australia.

| | *Waveband* | | | |
|---|---|---|---|---|
| | *Green* (0.5–0.6 μm) | *Red* (0.6–0.7 μm) | *Near infrared* (0.7–0.8 μm) | *Ratio Near infrared/Red* |
| House | 8.0 | 6.8 | 6.8 | 0 |
| Road | 8.4 | 6.0 | 5.3 | 0.9 |
| Concrete | 9.6 | 9.0 | 8.7 | 1.0 |
| Tree | 4.9 | 2.5 | 6.0 | 2.4 |
| Grass | 7.6 | 5.2 | 8.2 | 1.6 |
| Water | 6.4 | 3.0 | 1.4 | 0.5 |

*Source*: Modified from Forster 1980.

## 2.5.2 The interaction of visible, near infrared and middle infrared wavelengths of electromagnetic radiation with vegetation

The spectral reflectance of a vegetation canopy varies with wavelength. To understand why a canopy reflects more of certain wavelengths than of others it helps to first consider the hemispherical reflectance properties of an individual leaf (sect. 2.3). A leaf is built of layers of structural fibrous organic matter, within which are pigmented, water-filled cells and air spaces. Each of the three features – pigmentation, physiological structure and water content have an effect on the reflectance, absorbance and transmittance properties of a green leaf (Jensen 1983) as indicated in Figs 2.11 and 2.12. These will be discussed in turn.

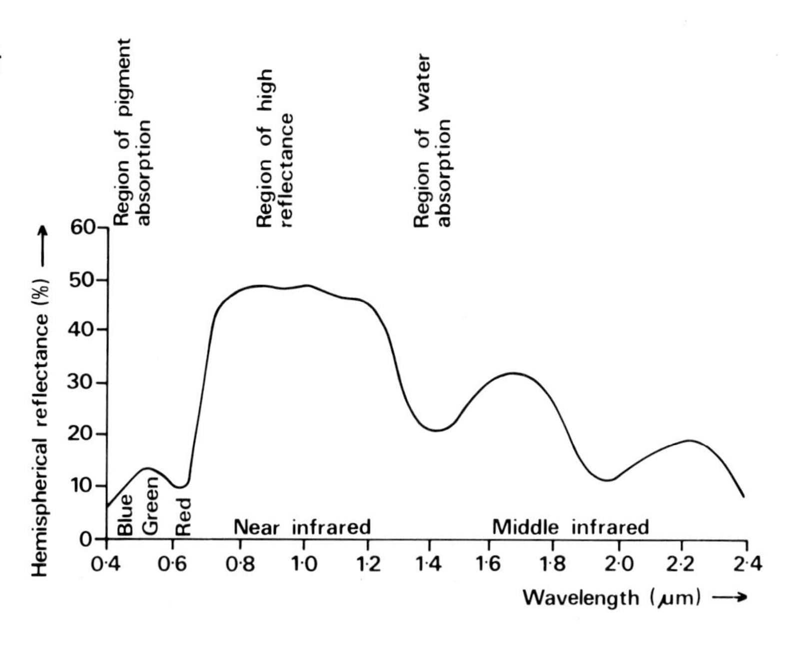


**Fig. 2.11**
The hemispherical reflectance of a Rhododendron leaf.

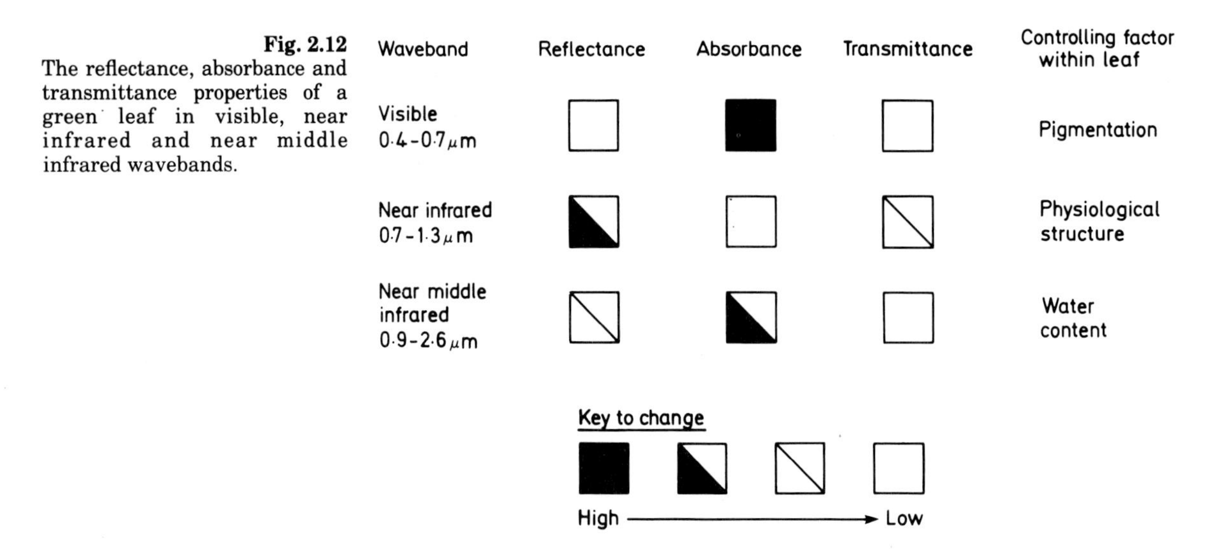


**Fig. 2.12** The reflectance, absorbance and transmittance properties of a green leaf in visible, near infrared and near middle infrared wavebands.

### 2.5.2.1 Pigment absorption in visible wavelengths of electromagnetic radiation

Higher plants contain four primary pigments, chlorophyll a, chlorophyll b, β carotene and xanthophyll, all of which absorb visible light for photosynthesis. Chlorophyll a and b, which are the more important pigments, absorb portions of blue and red light; chlorophyll a absorbs at wavelengths of 0.43 μm and 0.66 μm and chlorophyll b at wavelengths of 0.45 μm and 0.65 μm. The carotenoid pigments, carotene and xanthophyll, both absorb blue to green light at a number of wavelengths (Whittingham 1974; Curran and Milton 1983).

### 2.5.2.2 Physiological structure and reflectance in near infrared wavelengths of electromagnetic radiation

The discontinuities in the refractive indices within a leaf determine its near infrared reflectance. These discontinuities occur between membranes and cytoplasm within the upper half of the leaf and more importantly between individual cells and air spaces of the spongy mesophyll within the lower half of the leaf (Gausman 1974).

The combined effects of leaf pigments and physiological structure give all healthy green leaves their characteristic reflectance properties: low reflectance of red and blue light, medium reflectance of green light and high reflectance of near infrared radiation (Fig. 2.12). The major differences in leaf reflectance between species, are dependent upon leaf thickness which affects both pigment content and physiological structure. For example, a thick wheat flag leaf will tend to transmit little and absorb much radiation whereas a flimsy lettuce leaf will transmit much and absorb little radiation (Fig. 2.13), (Curran 1980a).

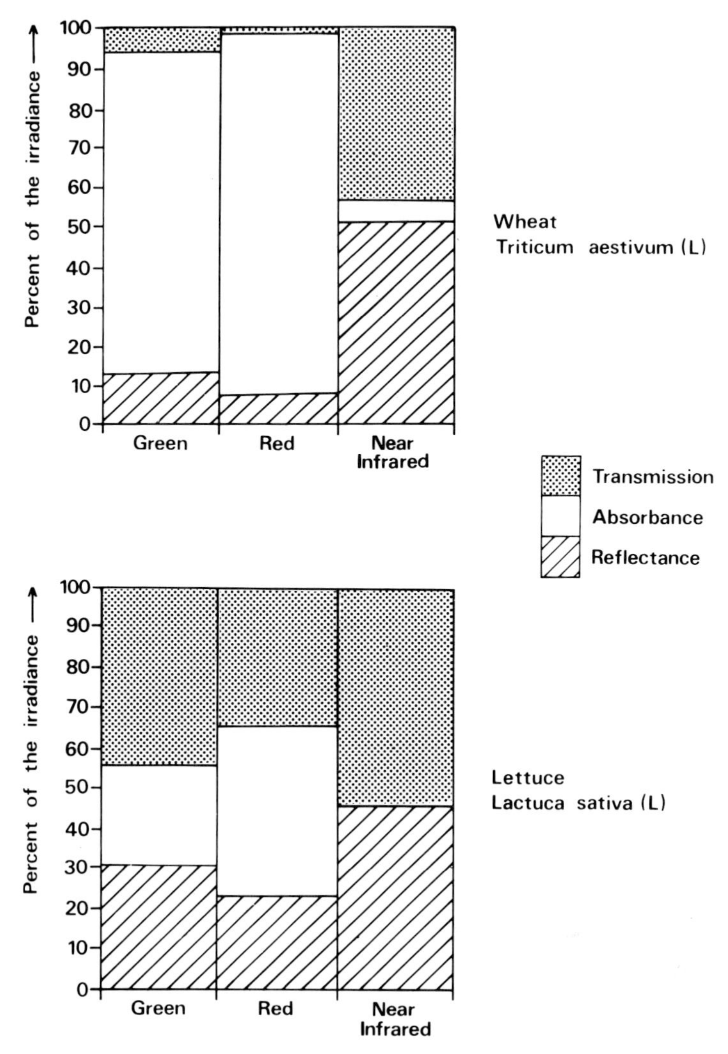


**Fig. 2.13**
The reflectance, absorbance and transmittance properties of wheat and lettuce leaves.

## 2.5.2.3 Water absorption in near and middle infrared wavelengths of electromagnetic radiation

Leaf reflectance is reduced as a result of absorption by three major water absorption bands that occur near wavelengths of 1.4 $\mu$m, 1.9 $\mu$m and 2.7 $\mu$m and two minor water absorption bands that occur near wavelengths of 0.96 $\mu$m and 1.1 $\mu$m (Fig. 2.11). The reflectance of the leaf within these water absorption bands is negatively related to both the amount of water in the leaf and the thickness of the leaf. However, water in the atmosphere also absorbs radiation in these water absorption bands and therefore the majority of sensors are limited to three atmospheric 'windows' that are free of water absorption at wavelengths of: 0.3 to 1.3 $\mu$m; 1.5 to 1.8 $\mu$m; and 2.0 to 2.6 $\mu$m. Fortunately within these wavebands, electromagnetic radiation is still sensitive to leaf moisture (Hoffer 1978), as can be seen from Fig. 2.14.

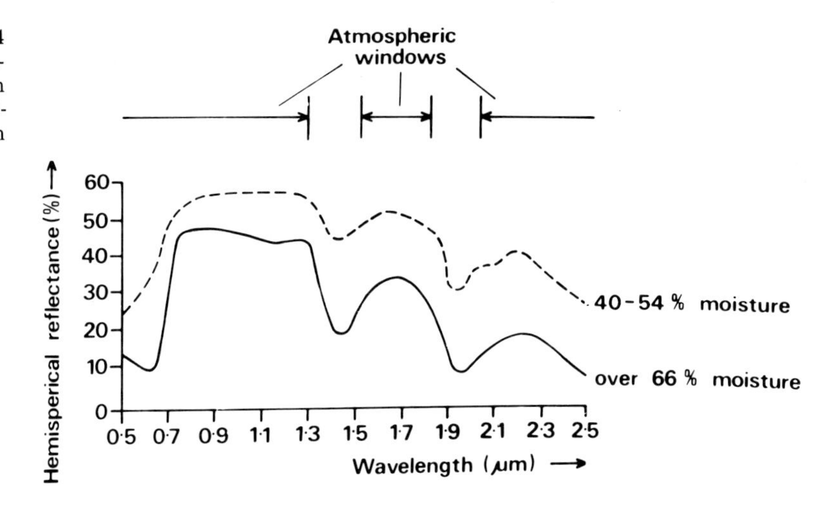


**Fig. 2.14**
The effect of moisture content in the reflectance of corn (*Zea mays* L.) leaves. (Modified from Hoffer and Johannsen 1969)

The hemispherical reflectance of an individual leaf is insufficient to describe the remotely sensed bidirectional reflectance of a vegetation canopy. This is because a vegetation canopy is not a large leaf but is composed of a mosaic of leaves, other plant structures, background and shadow. The bidirectional reflectance, recorded by a sensor is therefore primarily related to the area of leaves within the canopy, rather than to the hemispherical reflectance of the component leaves (Colwell 1974).

The area of leaves within the canopy can be recorded by the leaf area index or LAI – the unit leaf area per unit area of ground. For canopies with a high LAI, as is often the case for grasslands, then the bidirectional reflectance properties of the canopy are similar to the hemispherical properties of the leaf. In these cases the canopy will have *low blue* and *low red* bidirectional reflectance, *medium green* bidirectional reflectance and *high near infrared* and *high middle infrared* bidirectional reflectance. While these reflectance properties hold good for a healthy grass crop on a medium toned soil they show considerable variation with the environment. With constant hemispherical reflectance of the individual leaves the bidirectional reflectance could vary appreciably due to the effect of the soil background, the presence of senescent vegetation, the angular elevation of Sun and sensor, the canopy geometry and certain episodic and phenological canopy changes (Curran 1981e).

## 2.5.2.4 The effect of the soil background

The bidirectional reflectance of the soil has a considerable effect on bidirectional reflectance of the vegetation canopy (Colwell 1974; Curran 1981f). This is summarised in Fig. 2.15, where simulated green, red and near infrared bidirectional reflectance is plotted against LAI, for a light and a dark soil. The soil / waveband combinations that are unsuitable for the remote sensing of vegetation can be identified. For example, on dark toned soils with low red bidirectional reflectance there is little change in the red bidirectional reflectance of the canopy with an increase in the canopy LAI as the leaves have

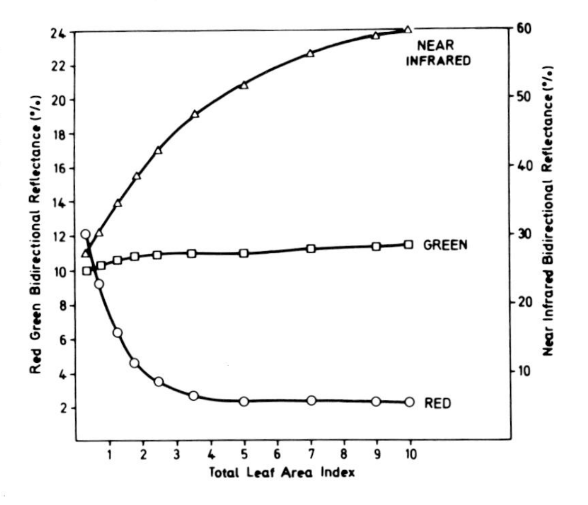


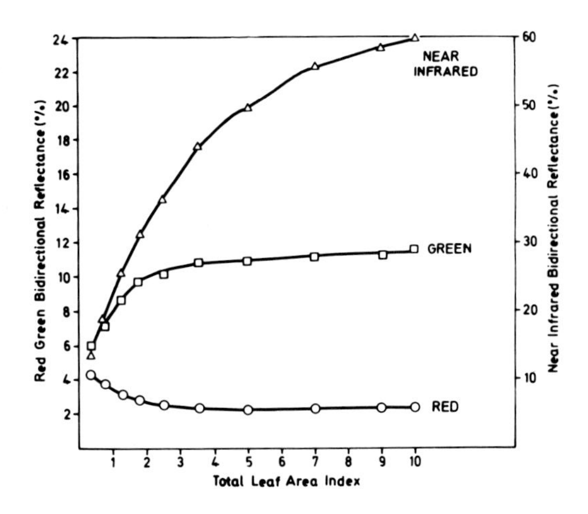


**Fig. 2.15**
The bidirectional reflectance of simulated vegetation canopies on a lght soil (top graph) and a dark soil (bottom graph) in green, red, and near infrared wavebands. (Modified from Colwell 1974)

similar reflectance properties to the soil. On a light toned soil with a high bidirectional reflectance, the relationship between near infrared bidirectional reflectance and LAI is weaker than on a dark soil, as on a dark soil the contrast between leaves and soil is high in near infrared wavelengths (Curran 1983a).

**Table 2.10**
The relationship between multispectral bidirectional reflectance and LAI: the effect of vegetation senescence and soil colour.

| | | *Waveband* | | | | |
|---|---|---|---|---|---|---|
| *Vegetation* | *Soil colour* | *Blue* | *Green* | *Red* | *Near infrared* | *Ratio Near infrared/Red* |
| Live | Light | −ve | 0 | −ve | +ve | +ve |
| | Dark | −ve | +ve | −ve | +ve | +ve |
| Live/Dead mixture | Light | 0 | 0 | 0 | +ve | +ve |
| | Dark | 0 | +ve | 0 | +ve | +ve |
| Dead | Light | 0 | 0 | 0 | +ve | +ve |
| | Dark | +ve | +ve | +ve | +ve | 0 |

where +ve = positive relationship probable
−ve = negative relationship probable
0 = no, or poor, relationship probable

### 2.5.2.5 The effect of vegetation senescence

As vegetation senesces due to ageing and the crop begins to ripen, the near infrared reflectance of the leaf does not significantly decrease. However, the breakdown of the plant pigments, result in a rise in the reflection of blue and red wavelengths. As a result there is a positive relationship between bidirectional reflectance, at each wavelength, and the LAI of senescent vegetation (Table 2.10).

### 2.5.2.6 The effect of solar and sensor elevation

As vegetation does not reflect radiation equally in all directions the elevation of the Sun and the sensor in relation to a vegetation canopy will affect the bidirectional reflectance. Two interrelated factors contribute to the effect of solar elevation on the reflectance of a vegetation canopy. The first is the degree to which solar radiation can penetrate the canopy and this is negatively related to solar elevation. When the Sun is high in the sky radiation will penetrate deep into the canopy and reflectance will be low and when the Sun is low in the sky radiation will only penetrate the canopy to a shallow depth and so reflectance will be high (Ahmad and Lockwood 1979). Solar elevation is also related to the amount of canopy shadow which tends to decrease the early morning and late evening reflectance. This is most noticeable in wavebands where leaf transmittance is low and the shadow is dark; namely in the visible region up to wavelengths of 0.7 μm. In near infrared wavelengths (0.7 μm–1.0 μm) where leaf reflectance is higher, shadow has little effect and so the majority of vegetation canopies have a negative relationship between near infrared reflectance and solar elevation (Duggin 1977; Jackson *et al.* 1979; Curran 1983b).

The angular elevation of the sensor determines the amount of soil that is visible; for as the angle of elevation moves from the vertical, less soil and more vegetation is seen. This effect is particularly severe when the angle of elevation varies by only a few degrees around the vertical. For example, the red and near infrared bidirectional reflectance of an 80% cover of the grasses *Lolium perenne* and *Poa* spp. were recorded at a solar elevation of 31 ° and an altitude of ten m (Fig. 2.16). A move of only 10 ° from the vertical is seen to increase the difference in bidirectional reflectance and the red and near infrared wavebands from 28% to 49% with no change in the LAI of the canopy or the hemispherical reflectance of the leaves. For this reason it is preferable to take measurements of bidirectional reflectance at one angle of elevation wherever possible (Suits 1972). Many remotely sensed measurements are vertical but during the 1980s there will be an increased use of sensors which collect data non-vertically. For example the Landsat Thematic Mapper sensor views up to 9 ° off-vertical (sect. 5.4.2.2) and the SPOT HRV sensor (sect. 5.4.6) views up to 27 ° off-vertical.

### 2.5.2.7 The effect of solar and sensor azimuth

The bidirectional reflectance of a canopy is usually higher if the sensor is looking into, as opposed to away from, the Sun. For the majority of

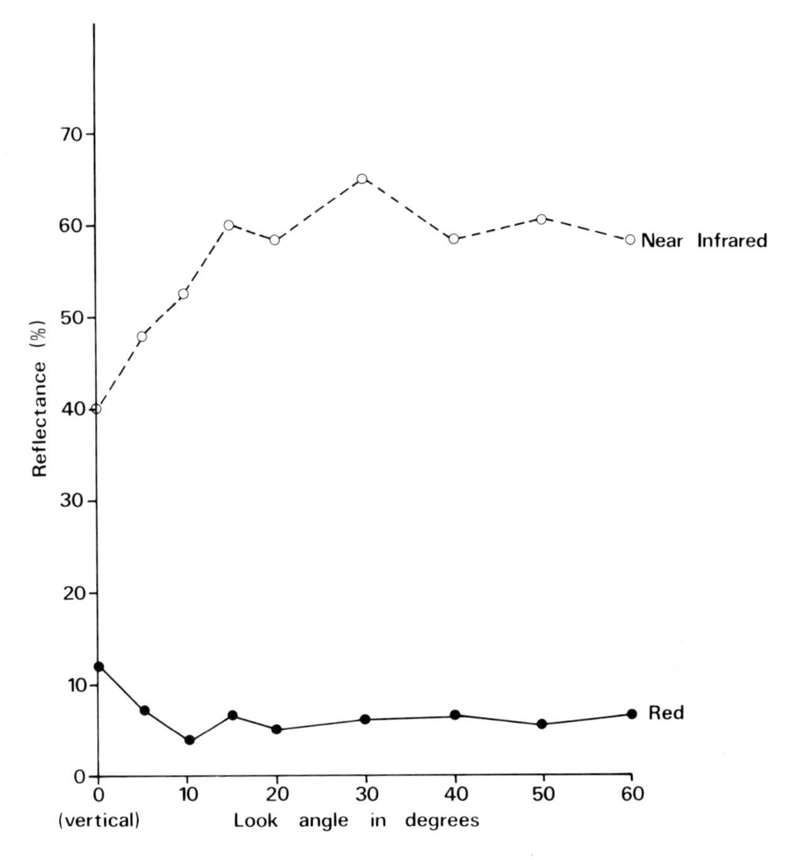


**Fig. 2.16**
Red and near infrared bidirectional reflectance for a grass canopy with 80% ground cover, recorded using a radiometer at a number of look angles from a height of 10 m.

remote sensing applications where the sensor look angle is vertical the effect of the solar azimuth on bidirectional reflectance increases with a decrease in solar angle and an increase in the roughness of the vegetation canopy. At high solar angles there is little shadow and at low solar angles shadow is increased by an amount dependent upon the roughness of the vegetation canopy. To take an extreme example: imagine that this page represents a field of wheat planted in rows with the paper representing the soil and each letter representing the plan view of a head of wheat. When the Sun is low in the sky and orientated parallel to the rows (to the left or right of the page) the sensor will sense a greater proportion of soil reflectance and a lesser proportion of shadow reflectance than if the Sun were at right angles to the row (to the top or bottom of the page). For further details refer to Bauer *et al.* (1980) and Curran (1983c).

### 2.5.2.8 The effect of canopy geometry

The geometry of a vegetation canopy will determine the amount of shadow seen by the sensor and will therefore influence the sensitivity of bidirectional reflectance measurements to angular variations in Sun and sensor. For example the reflectance of a rough tree canopy unlike a smoother grassland canopy is greatly dependent upon the solar angle.

### 2.5.2.9 The effect of episodic events

There are several episodic events that significantly increase or decrease the bidirectional reflectance of the canopy while causing little or no change in LAI. For example, Suits (1972) reported that for a canopy over a medium to dark toned soil the change from a predominantly horizontal to a predominantly vertical leaf orientation (wilting), resulted in an increased red and a decreased near infrared bidirectional reflectance without any change in the LAI of the canopy.

When studying a small area of the Earth's surface it is possible to restrict observation of canopy bidirectional reflectance to times when such short term canopy changes are minimal, however, this is not possible when using satellite sensors for the observation of large areas.

### 2.5.2.10 The effect of phenology

The seasonal change in canopy bidirectional reflectance as recorded by aerial photography, is well documented (Steiner 1970; Curran 1980b). From quantitative studies it is known that for a non-deciduous canopy (e.g. grassland) red bidirectional reflectance is maximised in the autumn and minimised in the spring, and near infrared bidirectional reflectance is maximised in the summer and minimised in the winter.

**Fig. 2.17**
Four hysteresis loops for the seasonal development of vegetation: wheat (modified from Kauth and Thomas 1976); rice (modified from Tanaka *et al.* (1977); corn (modified from Tucker *et al.* 1979) and a nature reserve.

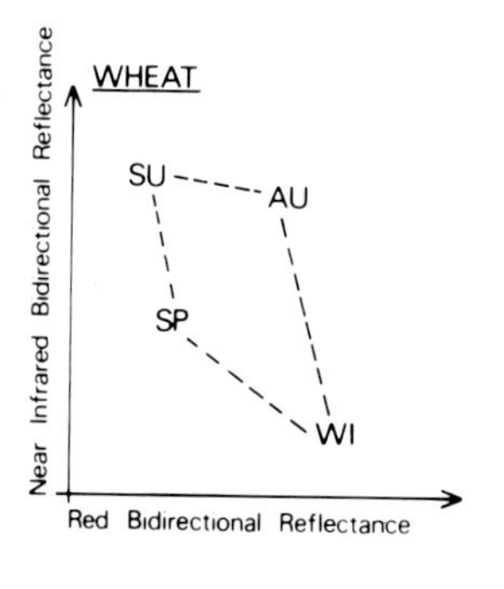


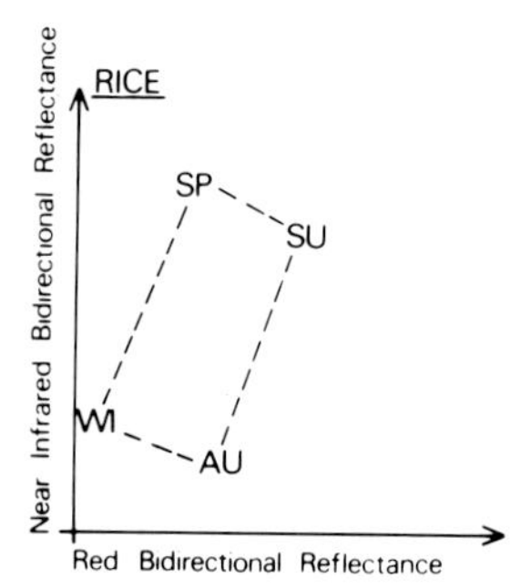


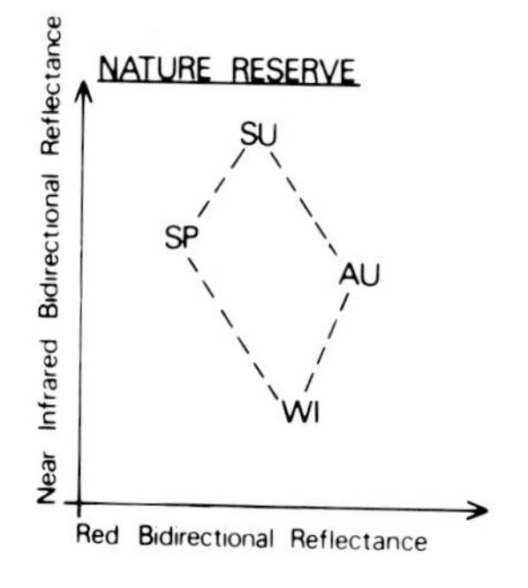


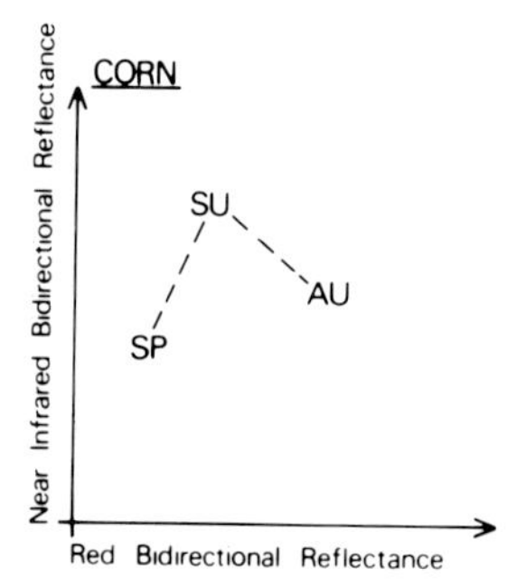


Key
SP Spring
SU Summer
AU Autumn
WI Winter

These relationships can be presented as hysteresis loops of bidirectional reflectance. Four examples of such loops are given in Fig. 2.17: wheat in the USA, rice in Japan, a wooded nature reserve in the UK and corn in the USA. Each hysteresis plot contains the expected seasonal pattern, with minor variations for the vegetation of the nature reserve and corn crop and major variations for the wheat and rice crop. The wheat crop has a lower than expected red bidirectional reflectance in the summer, probably due to high productivity and a higher than expected near infrared bidirectional reflectance in the autumn, probably as a result of senescent stubble left in the fields. Irrigation as well as vegetation LAI determines the bidirectional reflectance of the rice crop: for example, in the summer the wet soil background reduces the otherwise high near infrared bidirectional reflectance from the fields. These are examples where multitemporal, multispectral remote sensing of vegetation is unlikely to be successful without adequate knowledge of the vegetation and sensor.

### 2.5.3 The interaction of visible, near infrared and middle infrared wavelengths of electromagnetic radiation with soil

The majority of the flux incident on a soil surface is either reflected or absorbed and little is transmitted. The reflectance properties of the majority of soils are similar, with a positive relationship between reflectance and wavelength, as can be seen in Fig. 2.18. The five characteristics of a soil that determine its reflectance properties are, in order of importance: its moisture content, organic content, texture, structure and iron oxide content (Fig. 2.19), (Hoffer 1978; Stoner and Baumgardner 1981). These factors are all interrelated, for example the texture (the proportion of sand, silt and clay particles) is related to both the structure (the arrangement of sand, silt and clay particles into aggregates) and the ability of the soil to hold moisture. Therefore texture, structure and soil moisture will be discussed together.

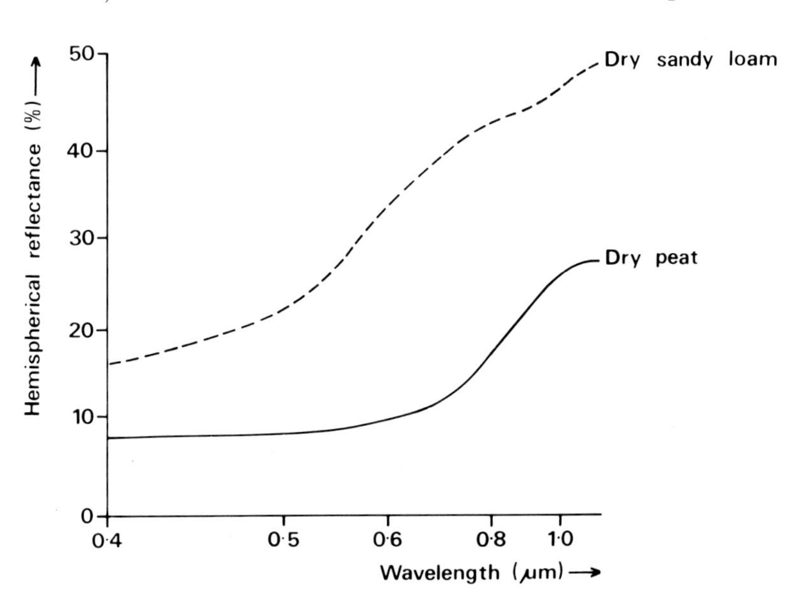


**Fig. 2.18**
The hemispherical reflectance of a dry sandy loam soil and a dry peat soil.

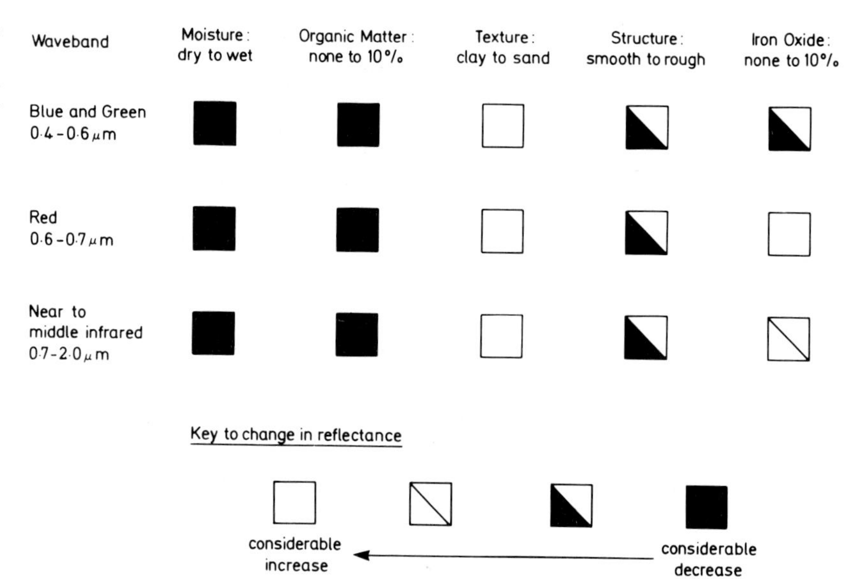


**Fig. 2.19**
The relationship between soil properties and soil reflectance in blue/green, red and near to middle infrared wavebands.

## 2.5.3.1 Texture, structure and soil moisture

The relationships between texture, structure and soil moisture can best be described by reference to two contrasting soil types. A clay soil tends to have a strong structure which leads to a rough surface on ploughing; clay soils also tend to have a high moisture content and as a result have a fairly low diffuse reflectance. In contrast a sandy soil

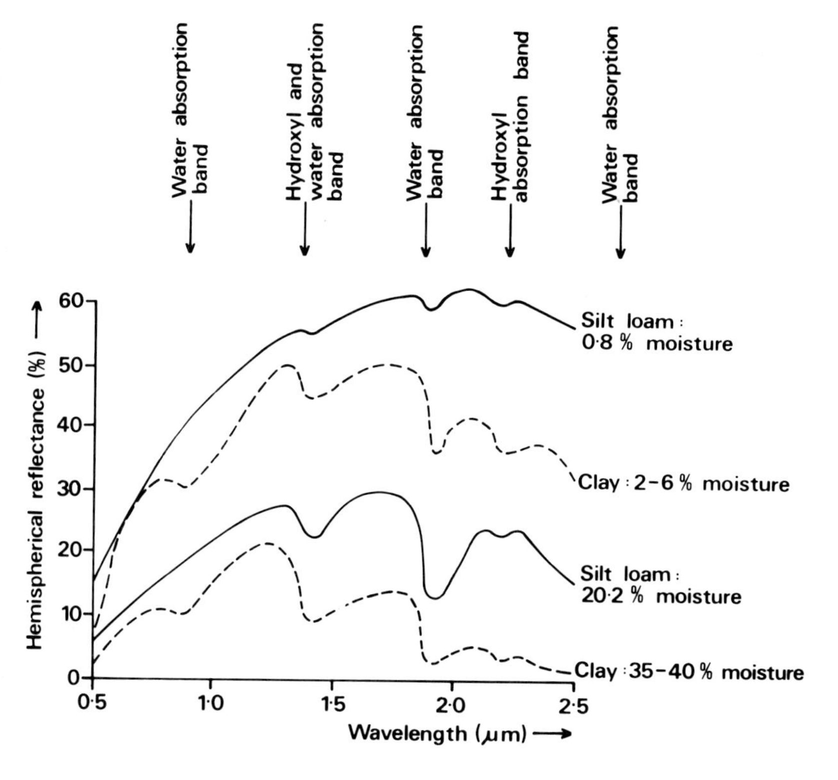


**Fig. 2.20**
Spectral reflectance curves for wet and dry silt loam and wet and dry clay. (Modified from Bowers and Hanks 1965; Hoffer and Johannsen 1969)

tends to have a weak structure which leads to a fairly smooth surface on ploughing; sandy soils also tend to have a low moisture content and as a result have fairly high and often specular reflectance properties (Bowers and Hanks 1965). In visible wavelengths the presence of soil moisture considerably reduces the surface reflectance of soil (Jensen 1983), (wet some soil yourself and see). This occurs until the soil is saturated, at which point further additions of moisture have no effect on reflectance.

Reflectance in near and middle infrared wavelengths is also negatively related to soil moisture; an increase in soil moisture will result in a particularly rapid decrease in reflectance in the water ($H_2O$) and hydroxyl (HO) absorbing wavebands that absorb at wavelengths centred at approximately 0.9 $\mu$m, 1.4 $\mu$m, 1.9 $\mu$m, 2.2 $\mu$m and 2.7 $\mu$m. The effect of water and hydroxyl absorption is more noticeable in clay soils for these soils have much bound water and very strong hydroxl absorption properties, as can be seen in Fig. 2.20.

The surface roughness (determined by the texture and structure) and the moisture content of soil also affect the way in which the reflected visible and near infrared radiation is polarised. This is because when polarised sunlight is specularly reflected from a smooth wet surface it will become weakly polarised to a degree that is positively related to the smoothness and the wetness of that surface (sect. 3.3.3). This effect has been used to estimate soil surface moisture from aircraft-borne sensors at altitudes of up to 300 metres (Curran 1978b, 1981a), (Fig. 2.21).

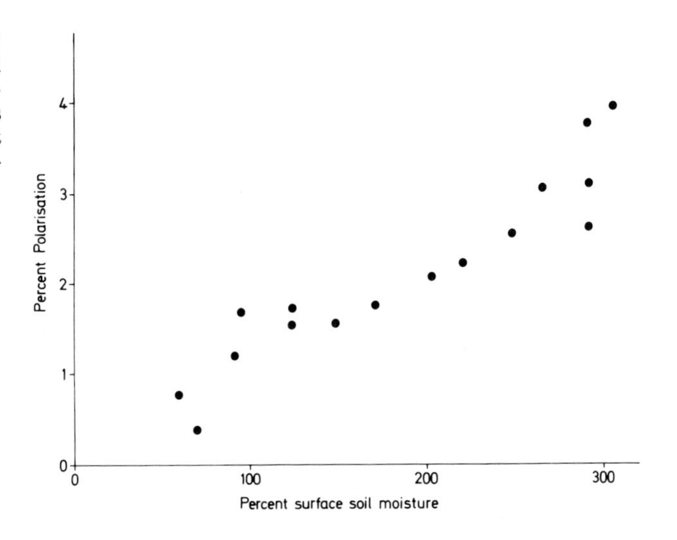


**Fig. 2.21**
Percent polarisation recorded for a range of surface soil moisture contents. The peat soil is recorded from a light aircraft using a camera with a polarising filter.

### 2.5.3.2 Organic matter

Soil organic matter is dark and its presence will decrease the reflectance from the soil up to an organic matter content of around 4–5 per cent (Fig. 2.22). When the organic matter content of the soil is greater than 5 per cent, the soil is black and any further increases in organic matter will have little effect on reflectance.

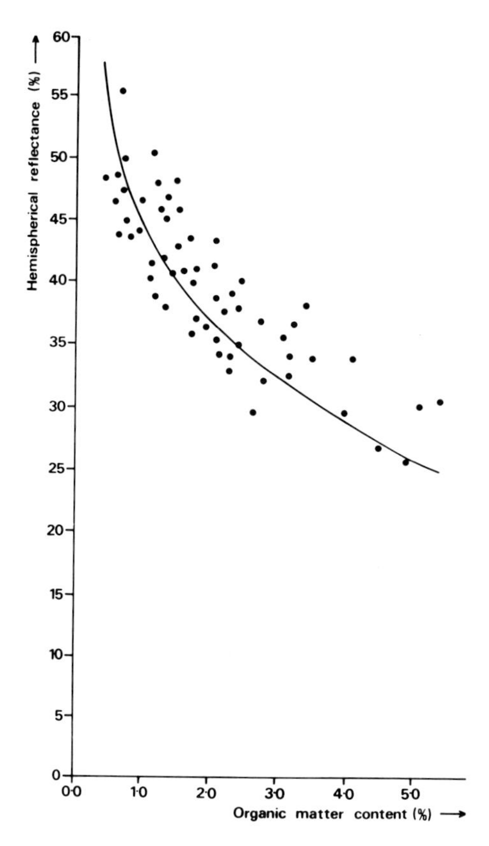


**Fig. 2.22**
The relationship between organic matter and hemispherical reflectance in visible wavelengths. (Modified from Page 1974)

### 2.5.3.3 Iron Oxide

Iron oxide gives many soils their 'rusty' red colouration by coating or staining individual soil particles. Iron oxide selectively reflects red light (0.6–0.7 $\mu$m) and absorbs green light (0.5–0.6 $\mu$m) (Obukhov and Orlov 1964). This effect is so marked that workers have been able to use a ratio of red to green bidirectional reflectance to locate iron ore deposits from satellite altitudes (Vincent 1973).

## 2.5.4 The interaction of visible near infrared and middle infrared wavelengths of electromagnetic radiation with water

Unlike vegetation or soil, the majority of the radiant flux incident upon water is not reflected but is either absorbed or transmitted, as indicated in Fig. 2.23. In visible wavelengths of electromagnetic radiation little light is absorbed, a small amount, usually under 5%, is reflected and the majority is transmitted. Water absorbs near infrared and middle infrared wavelengths strongly, (Fig. 2.24) leaving little radiation to be either reflected or transmitted. This results in a sharp contrast between any water and land boundaries, as can be seen in a

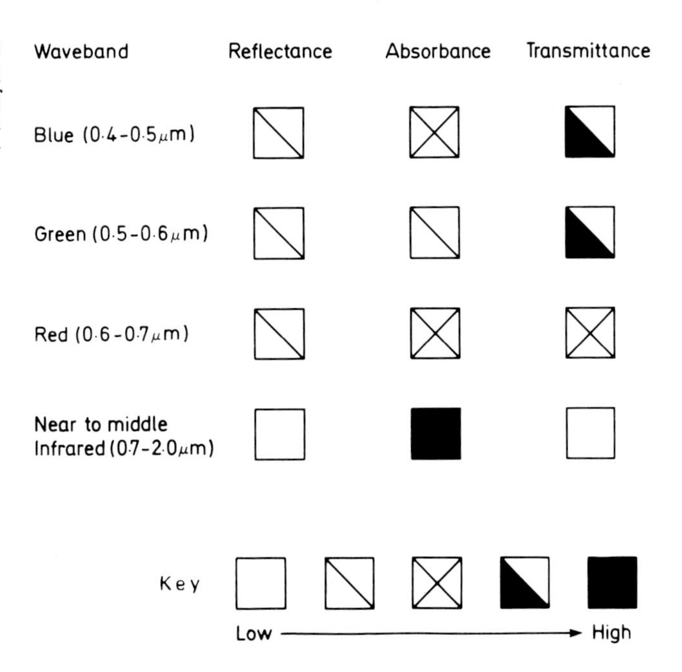


**Fig. 2.23**
The reflection, absorption and transmission characteristics of water in blue, green, red and near to middle infrared wavebands.

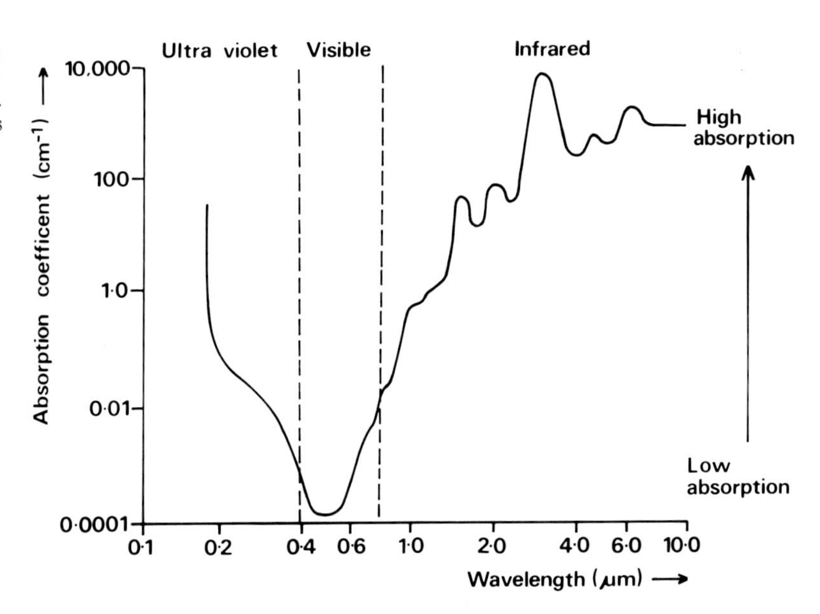


**Fig. 2.24**
The absorption of electromagnetic radiation by sea water. (Modified from Wolfe and Zissis 1978)

near infrared image recorded by the Landsat multispectral scanner in Fig. 5.9.

The factors that affect the spatial variability in the reflectance of a body of water are usually determined by the environment. The three most important factors are, the depth of the water, the materials within the water and the surface roughness of the water. In shallow water some of the radiation is reflected not by the water itself but from the bottom of the waterbody. Therefore, in shallow pools and streams it is often the underlying material that determines the waterbody's reflectance properties and colour. The three most common materials suspended in water are, non-organic sediments, tannin and chlorophyll

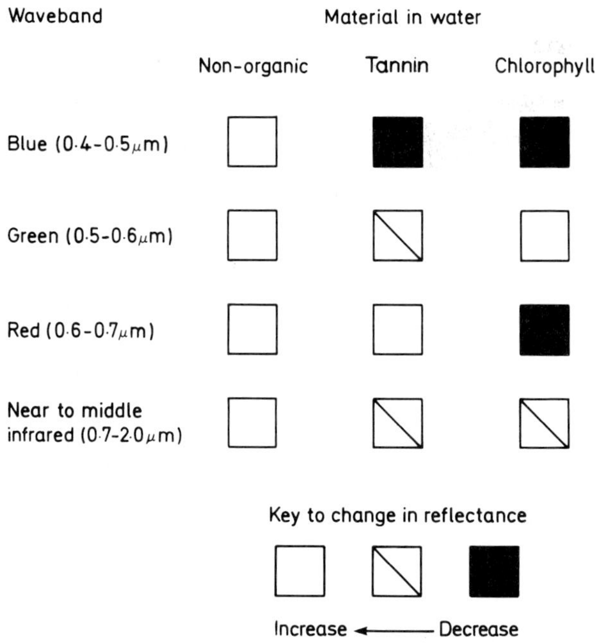

**Fig. 2.25**
The relationship between the presence of materials within a water body and its reflectance in blue, green, red and near to middle infrared wavebands.

Waveband

Material in water

Non-organic Tannin Chlorophyll

Blue (0·4–0·5 μm)

Green (0·5–0·6 μm)

Red (0·6–0·7 μm)

Near to middle infrared (0·7–2·0 μm)

Key to change in reflectance

Increase ← Decrease

(Fig. 2.25). The effect of non-organic silts and clays, is to increase the scatter and reflectance, in visible wavelengths (Weisblatt *et al.* 1973), as can be seen for a coastal region in Fig. 5.15. In agricultural scenes the main colouring agent is tannin produced by decomposing humus; this is yellowish to brown in colour and results in decreased blue and increased red reflectance. Waterbodies that contain chlorophyll have reflectance properties that resemble, at least in part, those of vegetation with increased green and decreased blue and red reflectance. However, the chlorophyll content must be very high before these changes in reflectance can be detected (Piech *et al.* 1978).

The roughness of the water surface can also affect its reflectance properties. If the surface is smooth then light is reflected specularly from its surface, giving very high or very low reflectance, dependent upon the location of the sensor. If the surface is very rough then there will be increased scattering at the surface, which in turn will increase the reflectance.

## 2.5.5 The interaction of thermal infrared wavelengths of electromagnetic radiation with the Earth's surface

Any object with a temperature above absolute zero (0 K) will emit energy as electromagnetic radiation (sect. 2.4.1). The Earth, which has a mean temperature of around 300 K (27 °C) radiates its energy with a peak of radiant exitance ($M$) in the thermal infrared region of the electromagnetic spectrum, at wavelengths of between 3–50 μm. Remote sensing devices sensitive to these thermal infrared wavelengths can be used to record some of this energy and thus measure the radiating temperature $T_{rad}$ of objects on the Earth's surface. The

$T_{rad}$ of an object can be used in two ways: to differentiate it from other objects and to determine certain characteristics of that object. However, the $T_{rad}$ of the object is dependent upon four factors: (i) emissivity, (ii) kinetic temperature, (iii) thermal properties, and (iv) rate of heating.

### 2.5.5.1 Emissivity

In the discussion of emission in section 2.4.1, it was assumed that objects emitted all of the energy they received – they acted as blackbodies. In fact the environment contains no true blackbodies, but rather comprises 'grey bodies', which emit only a proportion of the energy they receive. The proportion that is emitted is expressed by the object's emissivity ($\varepsilon$). A high emissivity (near to the maximum of 1) indicates an object that absorbs and radiates a large proportion of the incident energy and a low emissivity (considerably less than the maximum of 1) indicates an object that absorbs and radiates a small proportion of the incident energy. For example Table 2.11 gives a range of object emissivities from agricultural and urban scenes.

**Table 2.11**
The average emissivity (at 20°C) of objects in a rural and an urban scene.

| | *Emissivity* $\varepsilon$ *(where 1.00 is the emissivity of a blackbody)* |
|---|---|
| **1. Rural scene** | |
| Vegetation with a closed canopy | 0.99 |
| Water | 0.98 |
| Vegetation with an open canopy | 0.96 |
| Wet loamy soil | 0.95 |
| Dry loamy soil | 0.92 |
| Sandy soil | 0.90 |
| Organic soil | 0.89 |
| **2. Urban scene** | |
| Tar/stone | 0.97 |
| Plastic and paint | 0.96 |
| Building bricks | 0.93 |
| Wood | 0.90 |
| Stainless steel | 0.16 |

*Sources*: Hatfield 1979; Sabins 1978.

Knowledge of an object's emissivity is essential if the actual kinetic temperature $T_{kin}$ (i.e. that measured with a thermometer) of the object is to be estimated using the $T_{rad}$ recorded by a remote sensor. However, this is easier said than done as emissivity has great spatial variability and wavelength variability. The effect of spatial variability is perhaps the greater problem because very small changes in emissivity result in considerable changes in $T_{rad}$. Wavelength variability, although of lesser importance is still a problem as the majority of natural objects, excluding water, are selective radiators – their emissivity is wavelength dependent. This would pose serious problems for remote sensing if very wide thermal infrared wavebands were sensed, however, it is usual to consider only two rather narrow thermal infrared wavebands of 3–5 $\mu$m and 8–14 $\mu$m, and within these wavebands the majority of objects have a fairly stable emissivity.

### 2.5.5.2 Kinetic temperature

Kinetic temperature, $T_{kin}$, is the temperature of an object as recorded by a thermometer or any other direct thermal sensing device. This is positively related to the remotely sensed radiant temperature, $T_{rad}$, because hot objects with a high $T_{kin}$ emit strongly in the thermal infrared region of the electromagnetic spectrum and therefore have a high $T_{rad}$. The $T_{kin}$ of an object will not be the same as the $T_{rad}$ recorded by a remote sensing device unless the object is a blackbody. For a grey body $T_{rad}$ is positively related to the object's emissivity and the formula [2.6] applies:

$$T_{rad} = \varepsilon^{\frac{1}{4}} T_{kin} \qquad [2.6]$$

If the effect of emissivity is not accounted for when analysing remotely sensed $T_{rad}$ then $T_{kin}$ will be underestimated.

At present our rather crude estimates of emissivity are a limiting factor in the application of remotely sensed thermal infrared data, as emissivity to an accuracy of 0.02 is required before $T_{kin}$ can be estimated to an accuracy of 1 °C (Slater 1980). Fortunately however, field methods are now being developed to increase the accuracy of emissivity measurement (Vlcek 1982).

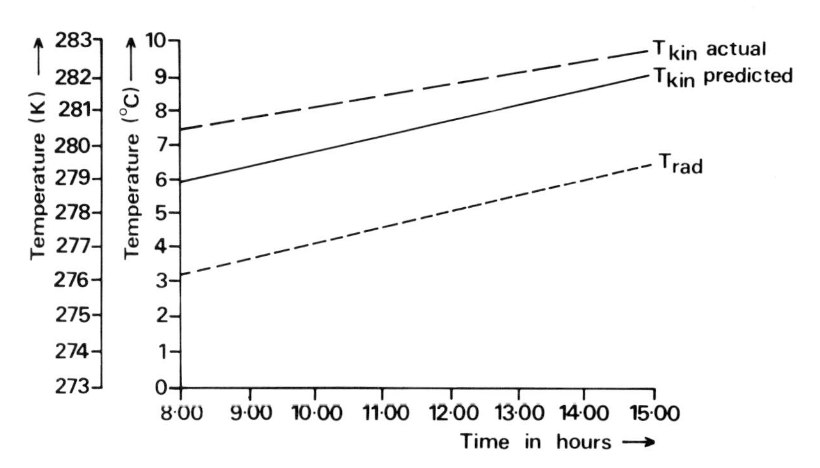


**Fig. 2.26**
The change over time of radiant temperature ($T_{rad}$, predicted), kinetic temperature ($T_{kin}$, predicted), and the actual ($T_{kin}$, actual) for a freshly dug moist soil on a warm summer's day.

Even when emissivity has been accounted for, it is at best only possible to predict the probable $T_{kin}$ from the remotely sensed $T_{rad}$ because $T_{rad}$ is the detected radiation from the surface of an object while $T_{kin}$ is the internal temperature of an object. This is illustrated in Fig. 2.26 where the $T_{rad}$ recorded by a thermal infrared radiometer was used to predict the $T_{kin}$ for a freshly dug soil on a warm summer's day. In the morning the soil surface is moist and due to evaporative heat loss the predicted $T_{kin}$ is 1.6 ° cooler than the actual $T_{kin}$, which is recorded 1 cm below the soil surface. By early afternoon the soil surface is no longer moist and has a predicted temperature $T_{kin}$ which is only 0.7 °C cooler than the actual $T_{kin}$ temperature, recorded 1 cm below the soil surface. Remote sensing can only estimate the magnitude of temperature differences within the environment and any attempt to measure actual temperatures tend to be fraught with difficulty (sect. 4.3.6).

### 2.5.5.3 Thermal properties

The thermal properties of an object on the Earth's surface determine how heat is distributed within the object and how the temperature of the object varies with time and depth.

The four thermal properties that are of importance are, thermal capacity ($c$), thermal conductivity ($k$), thermal diffusivity ($K$) and thermal inertia ($P$), the formulae for these properties are given in Table 2.6.

Thermal capacity ($c$), or specific heat, as it is sometimes called, is a measure of an object's ability to store heat. For example, water has a high thermal capacity and can hold more heat than vegetation or soil (Fig. 2.27).

Thermal conductivity ($k$) is a measure of the rate at which heat can pass through a material. Urban areas are good conductors of heat, but by contrast natural materials are relatively poor conductors of heat (Fig. 2.27). Therefore, the diurnal temperature change in rural areas is primarily a function of the upper layer of the soil, vegetation or water rather than the whole of its depth.

Thermal diffusivity ($K$) is a measure of the rate of temperature change within a volume of material. In general, dry surfaces diffuse temperature changes downwards at a slower rate than do wet surfaces (Fig. 2.27).

Thermal inertia ($P$) is a measure of the thermal response (or resistance) of a material to temperature changes. Materials with a low thermal inertia, for example dry sandy soils, reach high temperatures

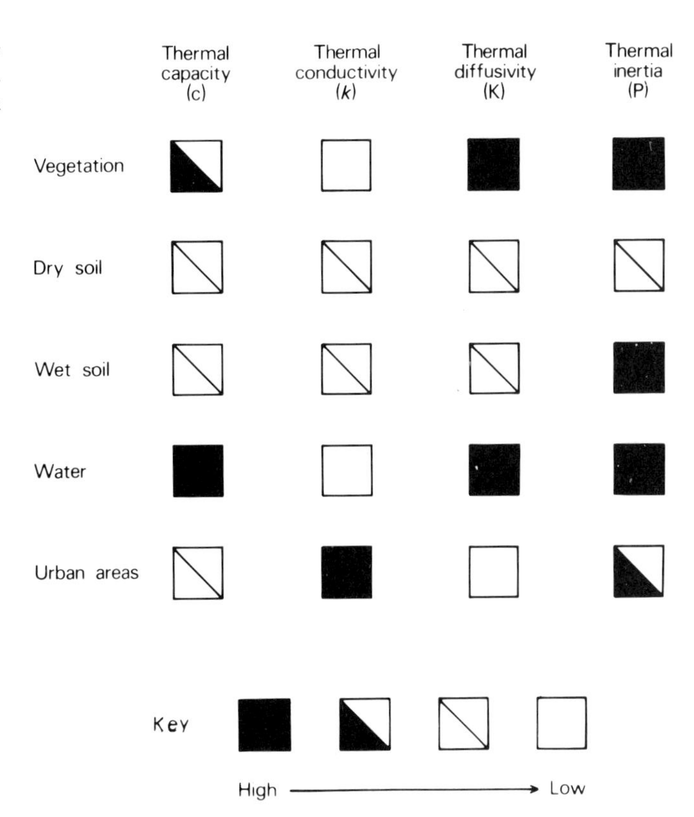


**Fig. 2.27** The thermal properties of vegetation, soil, water and urban areas.

during the day and cool to low temperatures at night. By contrast, materials with a higher thermal inertia, for example, wet clay soils, are more resistant to change and will have a smaller diurnal temperature range. However, just before sunset and about two hours after sunrise the majority of objects have a similar temperature regardless of their thermal inertia, as is illustrated in (Fig. 2.28). For further details refer to Sabins (1978), Kahle (1980), Kahle *et al.* (1981) and Colcord (1981).

#### 2.5.5.4 Rate of heating

The rate of heating of a particular area of ground is primarily the result of the intensity and rate of absorption of solar insolation. The intensity of solar insolation reaching the ground is reduced by such obstructions as trees, clouds and buildings and is also considerably affected by slope and aspect; for example solar insolation has less effect on north facing slopes in the Northern Hemisphere. These shadowed, north facing areas have a lower $T_{\mathbf{kin}}$ and $T_{\mathbf{rad}}$ than similar but unshadowed or south facing areas (Fig. 4.7). The albedo of the surface also determines its rate of absorption; for example, dark soils with a low albedo have a higher overall temperature than light soils with a high albedo.

The spatially variable rate of solar heating can be confusing when interpreting remotely sensed thermal infrared images and makes it unlikely that thermal infrared images will be successfully used for automated scene classification on an operational basis in the near future (Price 1981).

For further details refer to Lillesand and Kiefer (1979) and Kahle (1980).

### 2.5.6 The interaction of thermal infrared wavelengths of electromagnetic radiation with vegetation

The thermal properties of a vegetation canopy are more complex than the thermal properties of water, soil or urban areas because vegetation absorbs a greater amount of solar energy, especially at visible wavelengths (sect. 2.5.2) which it then re-emits at thermal infrared wavelengths to maintain its energy balance. As a result, the diurnal range of a vegetation canopy is small and therefore the canopy is at a different temperature to the surrounding air for the majority of the day. For example, on a warm summer's day in the UK, leaf temperatures in a cereal crop can be 10–15 °C below air temperature at midday and 5 °C above air temperature at midnight. During the warming and cooling parts of the day the leaf temperature will follow a cycle between these two extremes (Fig. 2.28), the range of which is ultimately dependent upon the thermal inertia of the vegetation.

As was discussed in section 2.5.5, the thermal inertia of a vegetation canopy is similar to the thermal inertia of soil and yet their diurnal

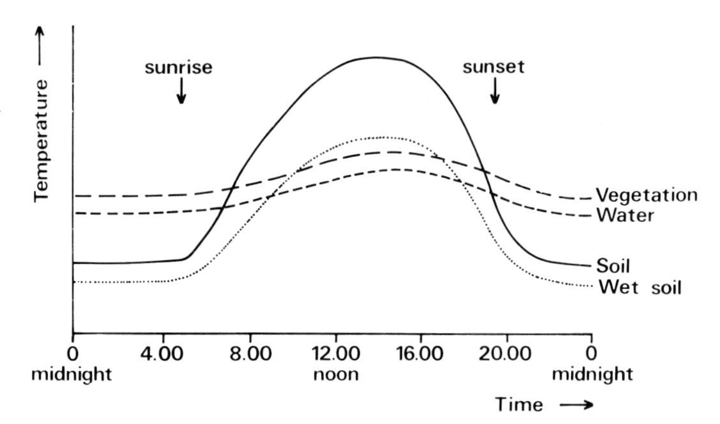


**Fig. 2.28**
Diagrammatic diurnal temperature curves for four components of a rural scene. (Modified from Sabins 1978)

temperature ranges are very different. This is because the temperature of a vegetation canopy is modified by its large surface area and high transpiration rate. The large surface area of a vegetation canopy results in very high radiation rates which can be around 30 times higher than the radiation rates of the surrounding soil (Geiger 1965). This, as a result, reduces canopy temperatures during both day and night.

Transpiration is the evaporative water loss of a plant that keeps it cool during the heat of the day; the actual rate of transpiration is controlled by the characteristics of the vegetation and the environment of the vegetation. The characteristics of the vegetation that are of importance include the root to shoot ratio, leaf area and leaf structure, which vary little within a crop canopy. Of greater importance to the user of remotely sensed data are the environmental factors that are usually short-term in nature; for example, atmospheric humidity is negatively related to transpiration while light availability, temperature, wind speed and soil moisture are all positively related to transpiration. Therefore, on one of the UK's rare sunny, hot, windy days when the soil is moist and the air is fairly dry, transpiration will be high and the vegetation canopy will be much cooler than the surrounding air.

One of the prime aims of quantitative thermal infrared sensing is the use of $T_{\mathbf{rad}}$ to estimate $T_{\mathbf{kin}}$. To date, thermal estimation has met with only partial success when working with vegetation canopies because of the structural and dynamic nature of vegetation. The four main problems have proved to be (i) the soil temperature, (ii) sensor angle, (iii) vegetation cover, and (iv) vegetation moisture content (Slater 1980).

### 2.5.6.1 Soil temperature

The $T_{\mathbf{rad}}$ of a vegetation canopy is a mixture of soil and leaf temperatures. In the middle of the day the $T_{\mathbf{rad}}$ of a vegetation canopy would be a mixture of high soil temperatures and relatively low leaf temperatures and the $T_{\mathbf{rad}}$ for the canopy would be much higher than the $T_{\mathbf{kin}}$ of the individual leaves. Therefore, in order to determine the $T_{\mathbf{kin}}$ of leaves the relative area and temperature of the soil is required.

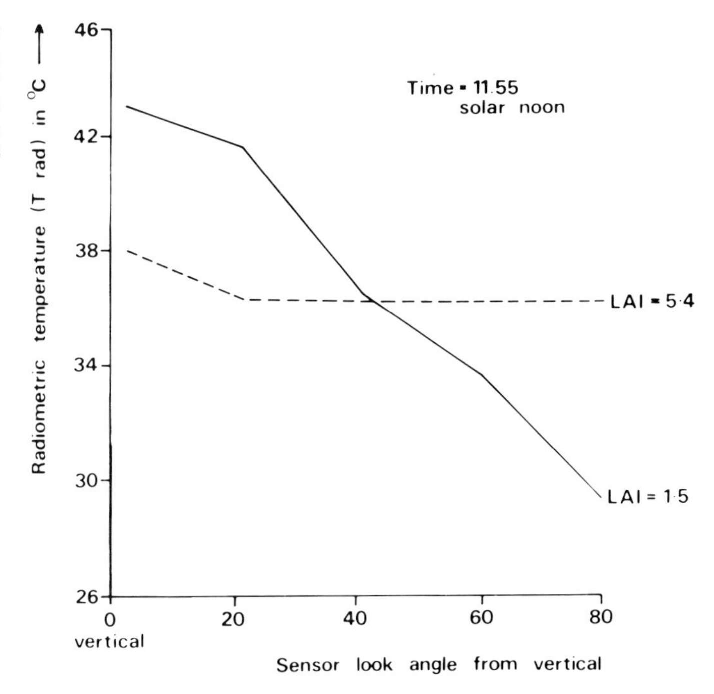


**Fig. 2.29** Radiometric temperature as a function of sensor look angle for two wheat canopies, one with a high LAI and the other with a low LAI. (Modified from Kimes *et al.* 1980)

### 2.5.6.2 Sensor angle

The $T_{rad}$ of a canopy is often related to the sensor angle, as a vertical view will sense a greater area of soil than will an oblique view (Kimes 1983). This is illustrated in Fig. 2.29 where for a noon measurement the canopy is cooler (has a lower $T_{kin}$) than the soil. With a vertical view both soil and vegetation are seen and the recorded canopy temperature $T_{rad}$ appears high, while with an oblique view only vegetation is seen and the recorded canopy temperature $T_{rad}$ appears low.

### 2.5.6.3 Vegetation cover

Incomplete vegetation canopies have a very variable $T_{rad}$ and as a consequence many experiments on the relationship between $T_{rad}$ and $T_{kin}$ have been undertaken on the simpler closed canopies. For further details refer to Millard *et al.* (1980).

### 2.5.6.4 Vegetation moisture content

The moisture content of a leaf determines its emissivity, as dry leaves can have an emissivity as low as 0.96 while moist leaves can have an emissivity as high as 0.99. As the moisture content of a canopy varies seasonally, diurnally and spatially so does both the emissivity and the

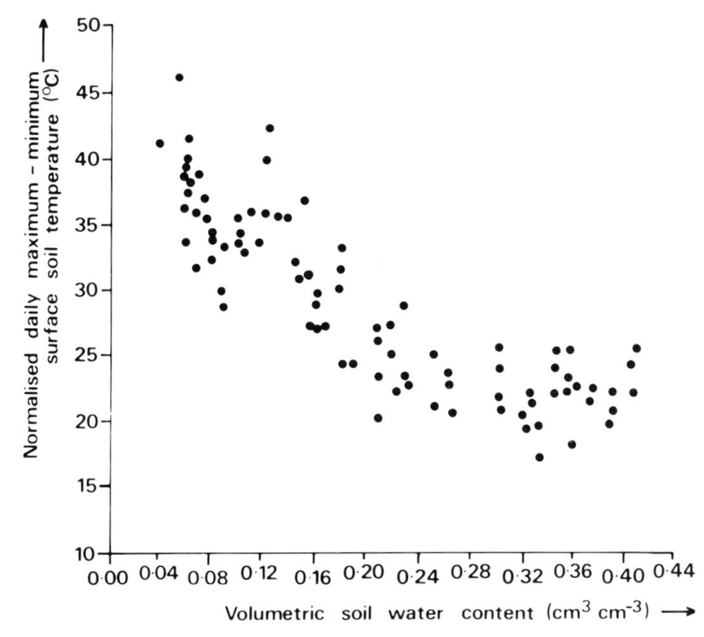


**Fig. 2.30**
The maximum-minimum temperature differential of a smooth bare loam soil, normalised by the air temperature. This is plotted against the average daily volumetric water content of the uppermost 2 cm. Data from thermocouples, ground based and airborne thermal infrared radiometers. (Modified from Idso *et al.* 1976)

$T_{rad}$ recorded by the remote sensor (Hatfield 1979). For further details refer to Kimes *et al.* (1981) and Smith *et al.* (1981).

### 2.5.7 The interaction of thermal infrared wavelengths of electromagnetic radiation with soil

The $T_{rad}$ of a soil is primarily determined by its moisture content, (Fig. 2.28). The wetter the soil is, the cooler it will be during the day and the warmer it will be at night (Fig. 2.30). For example, Watson (1971) demonstrated that an increase in the soil moisture content from 0 to 100% resulted in a two-fold increase in thermal inertia and a considerable decrease in the thermal range. The depth in the soil at which soil moisture ceases to have an effect on $T_{rad}$ varies with soil type from a few millimetres to tens of centimetres (Hatfield 1979).

### 2.5.8 The interaction of thermal infrared wavelengths of electromagnetic radiation with water

Water has a similar thermal inertia to soil and yet it has a much smaller diurnal thermal range. This is due to internal convection which maintains a fairly constant water surface temperature (Fig. 2.28), unlike soil which oscillates in large heating and cooling cycles. For further details refer to Sabins (1978).

## 2.5.9 The interaction of microwave wavelengths of electromagnetic radiation with the Earth's surface

Radar images like those shown in Fig. 4.20 are a visual representation of microwave scattering by the components of the environment. To adequately interpret these images the interpreter needs to be familiar with the means by which these very long microwaves are scattered. When a microwave is transmitted towards the Earth from the radar it is either: (i) scattered from the surface of an object, as is often the case with soil and urban areas, or (ii) scattered within the volume of an object, as is often the case with vegetation, or (iii) specularly reflected at the angle of incidence, as is often the case with water and urban areas. If the wave is scattered from a surface it will not have penetrated to any great depth and there will be some depolarisation and backscatter to the sensor. If the wave is scattered by the volume of the surface it will have penetrated the surface and depolarisation and backscatter to the sensor will occur from a number of depths. If the wave is scattered specularly from a smooth surface, there will be little depolarisation or backscatter to the sensor, especially at high angles of incidence (Janza 1975).

The fate of this wave as it sets off towards the Earth is not only dependent upon the radar system but is also dependent on the characteristics of the terrain. The features of the radar system that are of importance are the characteristics of the radar signal, which are wavelength, angle of incidence and polarisation, and these have already been introduced in section 2.4.2. The two characteristics of the terrain that are of importance are roughness and conductivity and these will be discussed in general before the characteristics of vegetation, soil and water are discussed in detail.

### 2.5.9.1 Surface roughness

Surface roughness and the microwave wavelength are interrelated, as at very long wavelengths the majority of surfaces appear smooth and at very short wavelengths the majority of surfaces appear rough (Wang *et al.* 1983). Therefore when using long microwave wavelengths, the variability of surface heights within a pasture or fallow field may be so small that the waves will be specularly reflected and unless the angle of incidence is very low most of the signal will be reflected away from the sensor. At shorter, microwave wavelengths the variability of surface heights within a pasture or fallow field may be almost as large as the wavelength and this will result in a certain amount of backscatter towards the sensor. This interrelationship between roughness and wavelength is summarised in Fig. 2.31.

Surface roughness at the long microwave wavelengths is often a function of surface morphology (Bryan 1975) and as a result radar imagery has been used to map urban morphology and geomorphology. For example, using data from the long wavelength (L band) radar onboard the satellite Seasat, Henderson and Wharton (1980) differentiated the smooth commercial and residential areas of towns from the rough (in the textural sense!) urban fringe, and Sabins *et al.* (1980)

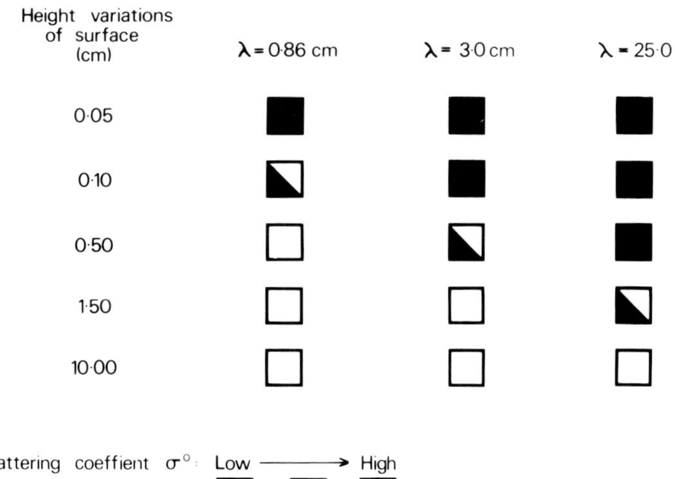


**Fig. 2.31**
The scattering coefficient (σ°) for different values of vertical relief (roughness) with a sensor depression angle of 45°. (Modified from Sabins 1978)

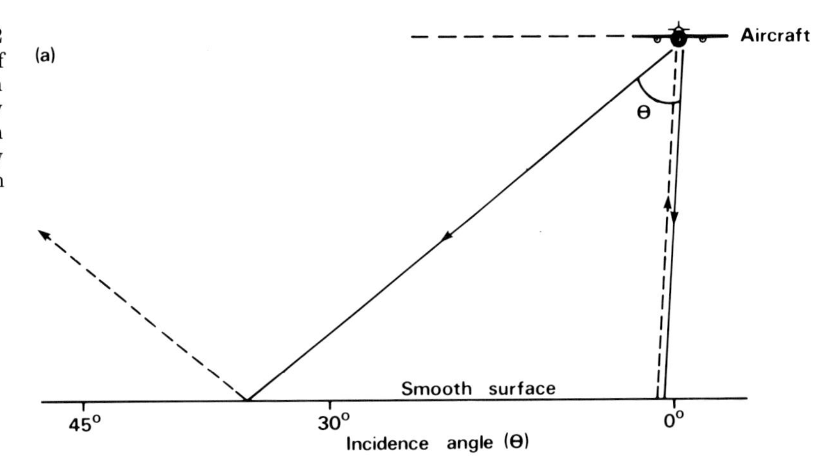


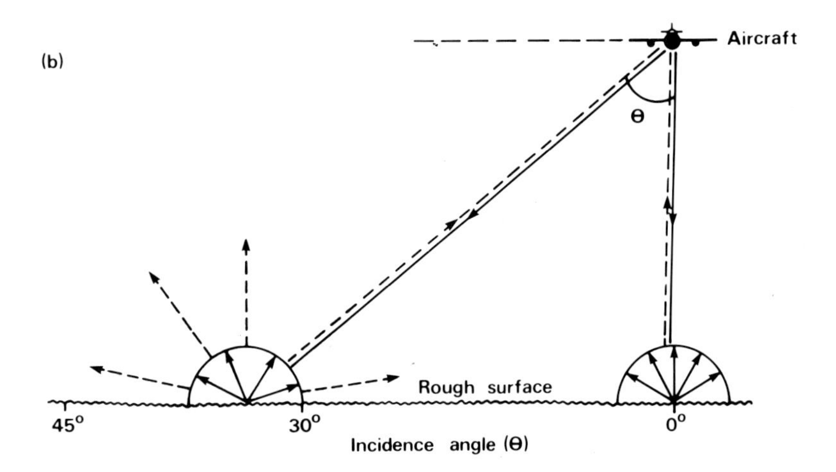


**Fig. 2.32**
Radar return as a function of depression angle for (a) an idealised smooth, specularly reflecting surface and (b) an idealised rough, diffusely reflecting surface. (Modified from Sabins 1978)

differentiated smooth from rough terrains. The angle of incidence of the microwave also has an effect upon the radar return by an amount dependent upon the surface roughness. The angle of incidence not only varies between images but also varies within each image as a radar will scan a swath of countryside with a low angle of incidence near to the aircraft and a higher angle of incidence further away from the aircraft. A smooth surface at unusually low incidence angles will reflect the radar signal specularly straight back towards the sensor, resulting in a very high return and yet a smooth surface at high incidence angles will reflect the radar signal specularly away from the sensor, resulting in a very low return (Fig. 2.32). A rough surface scatters almost equally in all directions, therefore the radar return from a rough surface is moderate and similar regardless of the angle of incidence (Fig. 2.32). This effect can be seen in Fig. 2.33 where a very rough maize canopy produces a similar backscatter at all incidence angles and the very smooth lucerne canopy produces a high backscatter at low angles of incidence and a low backscatter at high angles of incidence. The angle of incidence also determines the scattering mechanism. If the angle of incidence is low and the wavelength is long then the microwave can penetrate through both the vegetation canopy and the surface soil and volume scattering can occur from sub-canopy and even sub-surface features. If the angle of incidence is high and the wavelength is short then the microwave will be scattered from the surface of the vegetation canopy or soil and its scatter will be dependent upon the roughness of the surface (Ulaby *et al.* 1983).

Many radar missions now involve the acquisition of two images of the Earth's surface. One using like or plane polarisation (HH) and the other using cross polarisation (HV) as an area can have a different radar return in these two images to a degree that is dependent upon its surface roughness. If a surface is very rough at radar wavelengths,

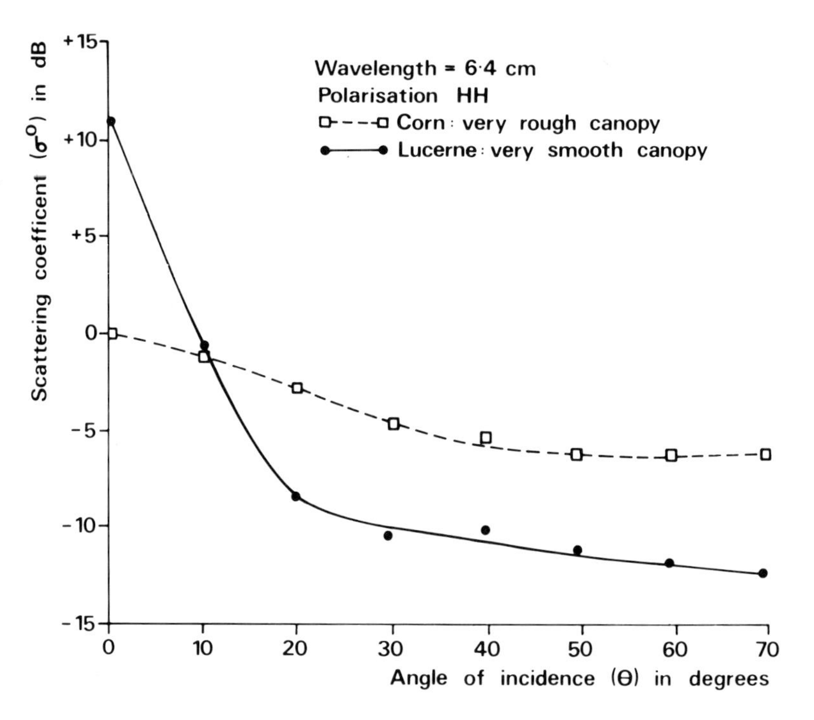


**Fig. 2.33**
Scattering coefficient for a rough and diffusely scattering canopy (corn) and for a smooth and specularly scattering canopy (lucerne) at a range of incidence angles. (Modified from Ulaby 1975)

as is often the case for woodland, then the polarised radar wave will be depolarised and scattered in all directions resulting in a similar scattering coefficient ($\sigma°$) for both the HH and HV images. If, however, a surface were very smooth at radar wavelengths as is often the case for pasture and urban areas, then the polarised wave will suffer little depolarisation and this will result in a moderate scattering coefficient in the HH image and a low scattering coefficient in the HV image (Fig. 2.34). However, a word of caution as the intensity of the cross polarised return is lower than the like polarised return, the HV antenna is often set to near maximum gain in order to increase its sensitivity and as a result of this, rough surfaces can have a lighter image tone.

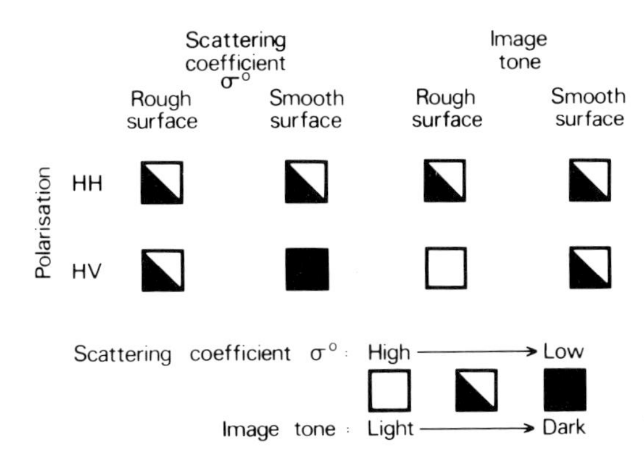


**Fig. 2.34**
The effect of surface roughness and microwave polarisation upon the scattering coefficient ($\sigma°$) and image tone of radar images.

## 2.5.9.2 Surface conductivity

The conductivity of the terrain has a considerable effect on radar backscatter, as highly conductive surfaces, for example, metal or water, will have a greater reflectivity than non-conducting surfaces, for example dry soil or sand. Conductivity, which is dependent upon the electrical properties of the surface, is commonly measured by the complex dielectric constant. Dielectric refers to the free electrical charges contained by a substance with a poor conductor having few free electrical charges. The complex dielectric constant is positively related to the conductivity and positively related to the reflectivity of the surface. For example, the complex dielectric constant for water is between 60 to 80 and yet the complex dielectric constant for most dry and naturally occurring materials is between three and eight (MacDonald and Waite 1973). In the majority of cases, the amount of water within the vegetation or soil will affect the radar return to a greater extent than the morphology of an individual vegetation canopy or the texture of an individual soil. In practice, the dielectric constant is so difficult to measure and varies so much with factors like temperature and salinity that image interpretation is usually undertaken on the basis of changes in the surface roughness and moisture content within the scene.

The discussion on the effect of surface roughness and conductivity on radar return is summarised, in very general terms, in Fig. 2.35. The table only refers to the natural surfaces within an agricultural scene and does not include urban areas that have a very high and variable radar return, as a result of many specular and corner reflections (Fig.

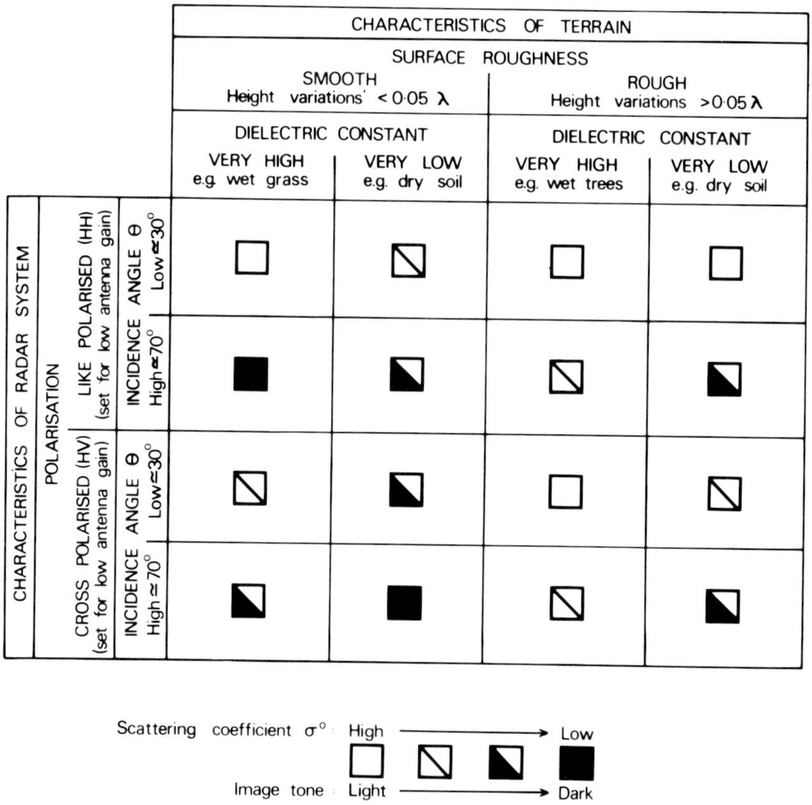

**Fig. 2.35**
The effect of terrain and sensor characteristics on the scattering coefficient ($\sigma^\circ$) and tone of radar images.

| CHARACTERISTICS OF RADAR SYSTEM | | CHARACTERISTICS OF TERRAIN: SURFACE ROUGHNESS — SMOOTH (Height variations < 0·05 λ) — DIELECTRIC CONSTANT: VERY HIGH e.g. wet grass | SMOOTH — VERY LOW e.g. dry soil | ROUGH (Height variations > 0·05 λ) — DIELECTRIC CONSTANT: VERY HIGH e.g. wet trees | ROUGH — VERY LOW e.g. dry soil |
|---|---|---|---|---|---|
| POLARISATION: LIKE POLARISED (HH) (set for low antenna gain) | INCIDENCE ANGLE θ Low ≈ 30° | □ | ⧄ | □ | □ |
| | INCIDENCE ANGLE θ High ≈ 70° | ■ | ◪ | ⧄ | ◪ |
| POLARISATION: CROSS POLARISED (HV) (set for low antenna gain) | INCIDENCE ANGLE θ Low ≈ 30° | ⧄ | ◪ | □ | ⧄ |
| | INCIDENCE ANGLE θ High ≈ 70° | ◪ | ■ | ⧄ | ◪ |

Scattering coefficient $\sigma^\circ$: High → Low
□ ⧄ ◪ ■
Image tone: Light → Dark

4.22), (Moore and Wellar 1969). Natural surfaces with a high radar return can be identified as vegetation covered surfaces and some rough soil surfaces when recorded at low angles of incidence. The areas that are likely to have a low radar return are likely to be smooth vegetation canopies or soils or man-made features like roads, which are recorded at high angles of incidence (Janza 1975).

As the interaction of a microwave with a non-urban scene is more complicated than the simplified view given in Fig. 2.35 would suggest, the remainder of this section will, therefore, be devoted to a more detailed examination of the interaction of a microwave with vegetation, soil and water. For further details refer to Easams (1972), Janza (1975), Guyenne and Levy (1981) and Ulaby *et al.* (1982).

## 2.5.10 The interaction of microwave wavelengths of electromagnetic radiation with vegetation

The radar return from vegetation is primarily determined by the roughness and the conductivity of both the vegetation canopy and underlying soil (Ulaby 1975; Lang and Sidhu 1983).

### 2.5.10.1 Surface roughness

The roughness of a vegetation canopy is an important factor in influencing the radar return. Roughness is dependent upon the size, shape, orientation and number of leaves in relation to the wavelength, angle of incidence and polarisation of the radar signal. The effect of canopy roughness is particularly evident in radar images produced by short wavelength microwaves that have been generated and recorded at a high angle of incidence. The roughness of a surface also varies over space and time; for example a downpour of rain that results in leaf droop and soil compaction will cause the radar return of the canopy to increase and the radar return of the soil to decrease (Ulaby *et al.* 1975). The geometry of a vegetation canopy also affects its apparent roughness at microwave wavelengths. This is noticeable for row crops where a short radar signal is scattered to a greater extent if it is transmitted across as opposed to down the rows (Ulaby and Bare 1979). Depolarisation by a vegetation canopy is similar regardless of the canopy orientation and therefore, cross polarised HV images are insensitive to row direction (Fig. 2.36).

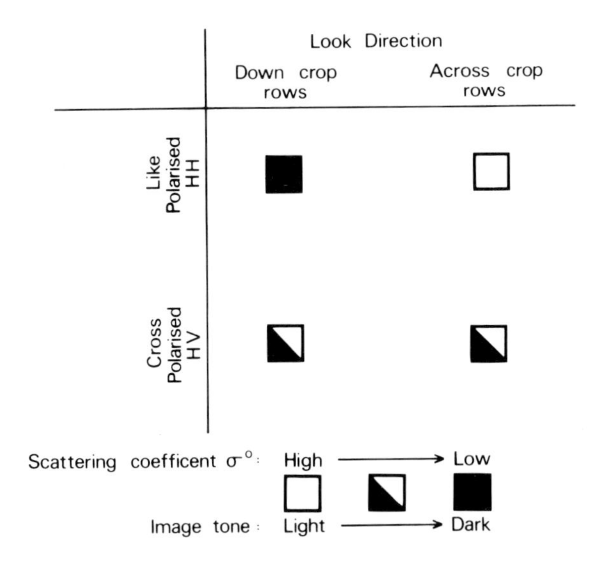


**Fig. 2.36**
The effect of the look direction upon the cross (HV) and like (HH) polarised scattering coefficient ($\sigma^{\circ}$) and radar image tone for row crops.

## 2.5.10.2 Surface conductivity

The complex dielectric constant of a complete vegetation canopy is high and similar for most types of vegetation and increases with moisture content during growth and decreases with moisture content during senescence (Brakke *et al.* 1981). This seasonal change in conductivity can be seen in Fig. 2.37 in which the like polarised X band radar return parameter ($\gamma$) is plotted during a period of rapid growth for three structurally and physiologically dissimilar crops. The radar return decreases as the senescing wheat screens the moist soil. The radar return increases as the rapidly growing sugar beet increases its leaf area and therefore both its above ground moisture content and complex dielectric constant. The lucerne is intermediate in its response as its initial growth serves to cover the moist soil and decrease the radar return, however by mid-July the moisture content, complex dielectric constant and the radar return are increasing. As can be seen, some of the changes in the radar return from a crop are attributable

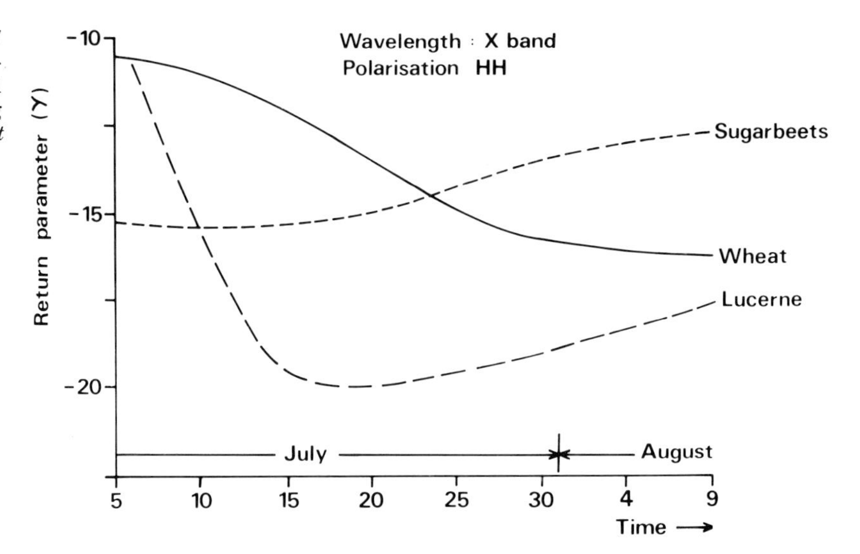


**Fig. 2.37** Variation of the return parameter (γ) for three agricultural crops through the growing season. (Modified from de Loor *et al.* 1974)

to changes in the moisture content of the soil (de Loor *et al.* 1974). Figure 2.38 illustrates this point, as when using HH polarised radar with a wavelength of 5.1 cm an increase in soil moisture is seen to increase the radar return from the crop (Ulaby *et al.* 1979).

The depth in the soil to which radar is sensitive to moisture is dependent on the wavelength, angle of incidence and the moisture content of the soil. When the wavelength is long, the angle of incidence low and the moisture content low, then soil moisture conditions down to the depth of a wavelength can be detected (Ulaby *et al.* 1979).

As all vegetation tends to have a similar roughness and complex dielectric constant the variation in radar return between for example, crops, is not large. The variation usually lies within the range of 8–10 dB (de Loor 1976) increasing to 15 dB if the bare fields and woods are

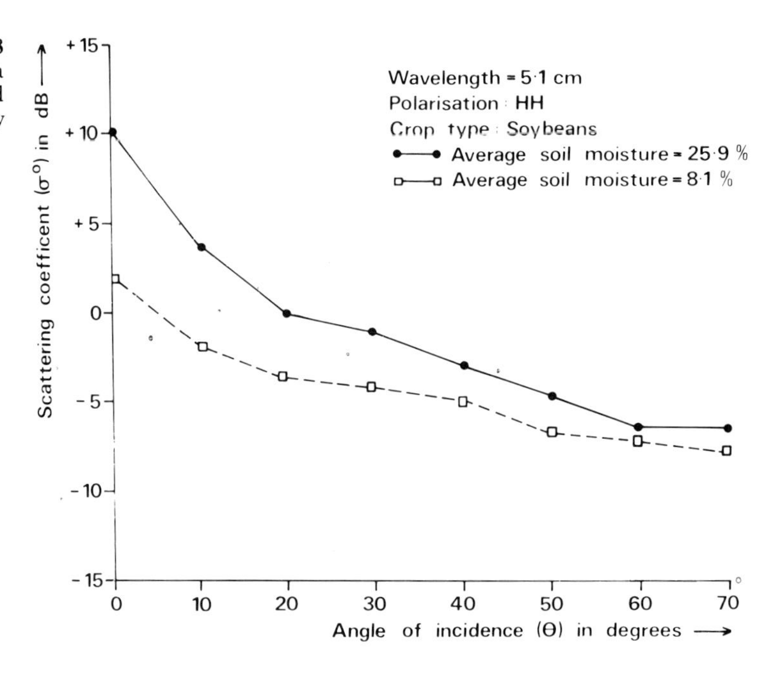


**Fig. 2.38** Scattering coefficient for a soybean canopy over a wet and a dry soil. (Modified from Ulaby 1975)

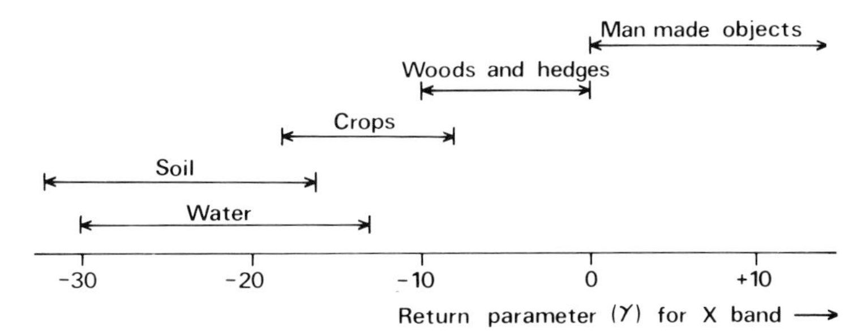


**Fig 2.39**
The range of radar returns for five components of an agricultural scene. (Modified from de Loor 1976).

also included. Not only is the range of radar return for vegetation small but the actual return is similar to the return from other features within a scene; tall moist crops have a return similar to soil and water. This is summarised for X band radar in Fig. 2.39. It is therefore not surprising that differentiation of vegetation and crop type on radar imagery has proved to be very difficult (Haralick *et al.* 1970) and it is only with the advent of calibrated, high resolution, multispectral, multitemporal, dual polarisation radar systems that vegetation classification by radar is at least becoming a possibility (Ulaby *et al.* 1980; Inkster *et al.* 1980; Ulaby *et al.* 1981; Ulaby *et al.* 1982).

## 2.5.11 The interaction of microwave wavelengths of electromagnetic radiation with soil

The radar return of soil is determined by its roughness and complex dielectric constant, as has been discussed in section 2.5.9. Soils in general generate a low radar return and it is only when they are recorded at moderate to low incidence angles that they generate a moderate return and are sensitive to soil moisture variations (Ulaby *et al.* 1978). For example, the radar return for a 4.2 cm radar (Fig. 2.40) has a high sensitivity to soil moisture at incidence angles lower than 20 ° (Ulaby *et al.* 1982).

As radar can penetrate the soil surface it has been used to provide information on subsurface moisture conditions. The depth of penetration is called the 'skin depth' which is defined as the depth at which the amplitude of the energy is reduced to 37% of its value at the surface (Janza 1975). Skin depth is largest at low angles of incidence, long wavelengths and paradoxically low soil moisture contents, as the skin depth of a moist soil to an X band radar at medium angles of incidence are at most a few millimetres (Fig. 2.41). For further details refer to Easams (1973) and Ulaby *et al.* (1982).

## 2.5.12 The interaction of microwave wavelengths of electromagnetic radiation with water

Water specularly reflects microwaves away from the radar antenna. A waterbody is therefore usually an area of low radar return and appears black on radar images (Fig. 4.20). The exception to this occurs if there are waves on the water surface that are orientated at right

**Fig. 2.40**
The sensitivity of the radar return to the moisture content of a bare soil, as a function of incidence angle. (Modified from Ulaby *et al.* 1975)

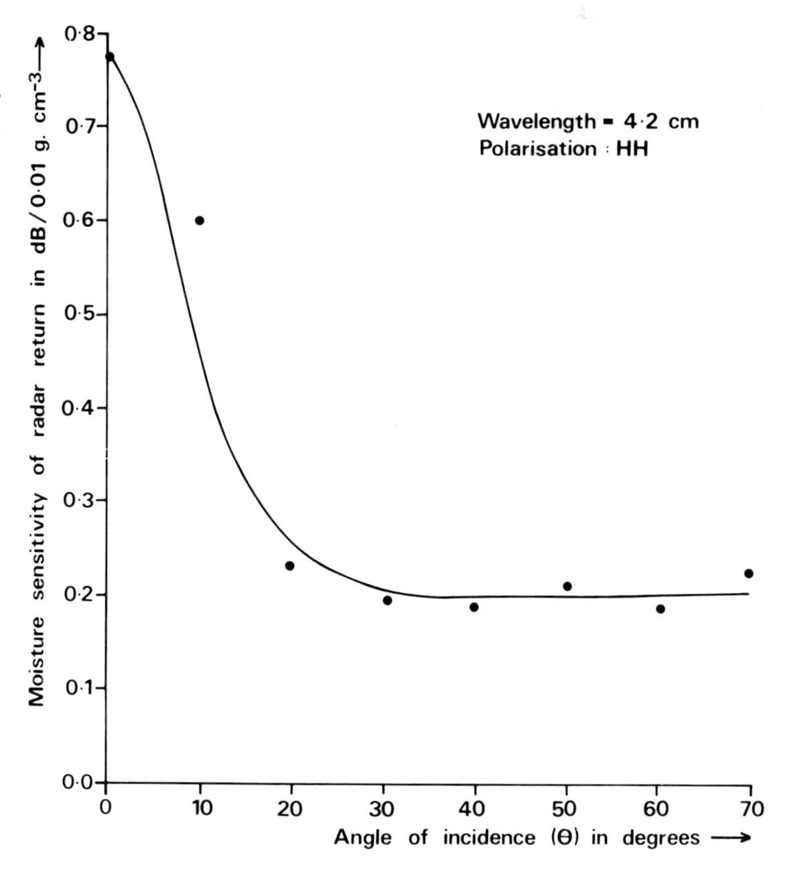


**Fig. 2.41**
The skin depth of sand and clay at two wavelengths of 3 and 23 cm, as a function of volumetric water content. (Modified from Cihlar and Ulaby 1974)

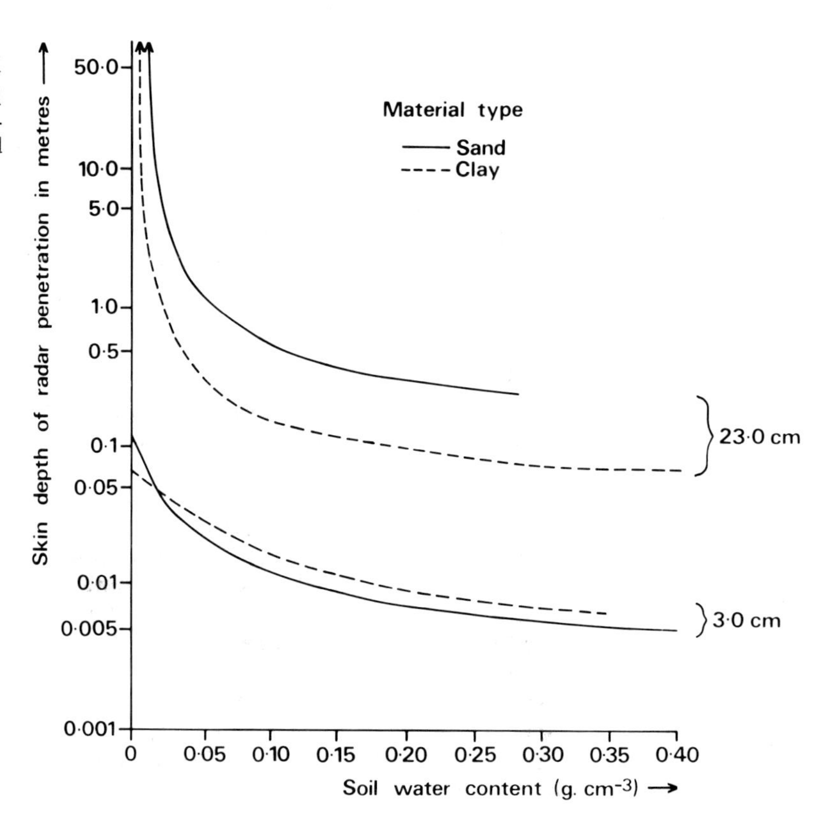

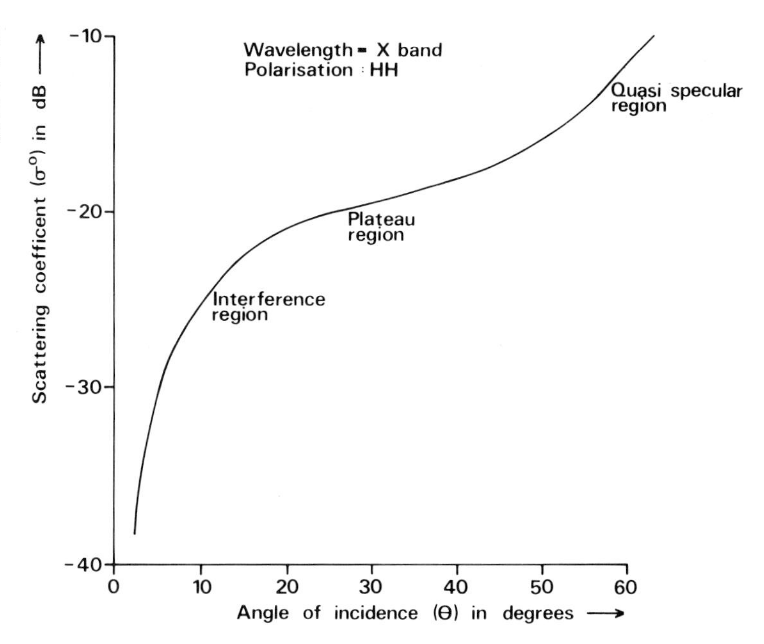


**Fig. 2.42**
The dependence of the scattering coefficient on the incidence angle for wind generated waves in a water tank. (Modified from Skolnik 1970)

angles to the radar pulse as this will result in backscattering and a very speckled image. The backscattering is dependent upon the angle of incidence with high backscattering at low angles of incidence, as indicated in Fig. 2.42. If the angle of incidence is low then water penetration is possible but only to a depth of a few millimeters. For further details refer to MacDonald (1980).

## 2.6 Atmospheric interactions with electromagnetic radiation

All electromagnetic radiation before and after it has interacted with the Earth's surface, has to pass through the atmosphere prior to its detection by a remote sensor. This passage will alter the radiation's speed, frequency, intensity, spectral distribution and direction as a result of atmospheric scattering, absorption and refraction. These effects are most severe in visible and infrared wavelengths. In fact, the atmosphere has such a minor effect on electromagnetic radiation at microwave wavelengths that it is of no practical importance and will not be considered here; for further details refer to Easams (1972).

Atmospheric scattering primarily affects the direction of visible radiation but can also alter the spectral distribution of visible and near visible wavelengths. The four types of scattering in order of importance are given in Table 2.12. By far the most common type of scatter is Rayleigh scatter which affects short visible wavelengths and results in haze; therefore, in order to reduce the effect of haze, aerial photography is usually taken with a minus-blue filter.

Unlike scatter, absorption affects wavelengths that are both shorter and longer than those of visible light. Absorption occurs when an atmospheric atom or molecule is excited by the absorption of electro-

| *Type of scatter* | *Size of effective atmospheric particles* | *Type of effective atmospheric particles* | *Scatter* | *Effect of scatter on visible and near visible wavelengths* |
|---|---|---|---|---|
| Rayleigh | Smaller than the wavelength of radiation. Usually $< 0.1\ \lambda$ | Gas molecules | Molecule absorbs high energy radiation and re-emits. Scatter is inversely proportional to fourth power of wavelength | Affects short visible wavelengths, resulting in haze in photography, skylight and blue skies |
| Mie | Same size as the wavelength of radiation. | Spherical particles of water vapour, fumes and dust | Physical scattering under overcast skies | Affects long visible wavelengths |
| Non-selective | Larger than the wavelength of radiation | Water droplets and dust | Physical scattering by fog and clouds | Affects all visible wavelengths equally, resulting in white fog and clouds |
| Raman | Any | Any | Photon has elastic collision with molecule resulting in a loss or a gain in energy; this can decrease or increase wavelength. | Variable |

**Table 2.12**
Types of atmospheric scatter, in order of importance.

magnetic energy; instead of re-emitting radiation at the wavelength at which it was absorbed, as occurs during scatter, it uses the energy in heat motion and eventually releases it at much longer wavelengths. Water vapour, carbon dioxide and ozone are the primary absorbers of electromagnetic energy. The regions of the electromagnetic spectrum in which atmospheric absorption is low are called atmospheric windows and it is through these 'windows' that remote sensing of the Earth's surface takes place. For example, photography occurs through the visible window at 0.4–0.9 $\mu$m and thermal infrared sensing occurs through the two atmospheric windows at 3–5 $\mu$m and 8–14 $\mu$m. These atmospheric windows are not totally free from atmospheric absorption as gases and suspended particles in the atmosphere absorb the radiation from ground objects, resulting in a decrease in radiation at the sensor, and also emit radiation of their own, thus adding to the radiation at the sensor. These effects are very variable in space and time and often necessitate the atmospheric correction of remotely sensed data (sect. 6.3.5.3).

Refraction occurs when electromagnetic radiation passes from one medium to another and occurs as electromagnetic radiation passes through a stratified atmosphere. This is not a problem when the atmosphere is stable but when it is turbulent the bending of waves and therefore the effect on the geometric accuracy of remotely sensed images is unpredictable. A full discussion of the way in which electromagnetic radiation interacts with the atmosphere is beyond the scope of this book. For further details refer to Fraser (1975), Fraser and Curran (1976) and Chahine (1983).

## 2.7 Recommended reading

**American Society of Photogrammetry** (1978) American Society of Photogrammetry, usage of the international system of units, *Photogrammetric Engineering and Remote Sensing*, **44**: 923–38.

**Chahine, M.T.** (1983) Interaction mechanisms within the atmosphere: *In* Colwell, R.N. (ed.) *Manual of Remote Sensing* (2nd edn). American Society of Photogrammetry, Falls Church, Virginia, pp. 165–230.

**Curran, P. J.** (1980) Multispectral remote sensing of vegetation amount, *Progress in Physical Geography*, **4**: 315–41.

**Hoffer, R. M.** (1978) Biological and physical considerations in applying computer-aided analysis techniques to remote sensor data: *In* Swain, P. H. and Davis, S. M. (eds), *Remote Sensing: The Quantitative Approach*. McGraw Hill, pp. 227–89.

**Kahle, A. B.** (1980) Surface thermal properties: *In* Siegal, B. S. and Gillespie, A. R. (eds) *Remote Sensing in Geology*. John Wiley, New York, pp. 257–73.

**Smith, J.A.** (1983) Matter-energy interaction in the optical region: *In* Colwell, R.N. (ed.) *Manual of Remote Sensing* (2nd edn). American Society of Photogrammetry, Falls Church, Virginia, pp. 61–113.

**Ulaby, F. T., Moore, R. K.** and **Fung, A. K.** (1982) *Microwave Remote Sensing Active and Passive, Volume II, Radar Remote Sensing and Surface Scattering and Emission Theory*. Addison-Wesley, Reading, Massachusetts; London.

# Chapter 3 Aerial photography

*'Photography . . . a means of fixing time within a framework of space.'* (de Latil 1961)

## 3.1 Introduction

Aerial photography was the first method of remote sensing (Fig. 3.1) and even today in the age of the satellite and electronic scanner, aerial photographs still remain the most widely used type remotely sensed data. The six characteristics of aerial photography that make it so popular are its availability, economy, synoptic viewpoint, time freezing ability, spectral and spatial resolution and three dimensional perspective.

1. **Availability**: aerial photographs are readily available at a range of scales for much of the world (Appendix A).
2. **Economy**: aerial photographs are cheaper than field surveys and are often cheaper and more accurate than maps for many countries of the world.
3. **Synoptic viewpoint**: aerial photographs enable the detection of small scale features and spatial relationships that would not be evident on the ground.
4. **Time freezing ability**: an aerial photograph is a record of the Earth's surface at one point in time and can therefore be used as an historical record.
5. **Spectral and spatial resolution**: aerial photographs are sensitive to radiation in wavelengths that are outside of the spectral sensitivity range of the human eye, as they can sense both ultra-violet (0.3–0.4 $\mu$m) and near infrared (0.7–0.9 $\mu$m) radiation. They can also be sensitive to objects outside the spatial resolving power of the human eye.

For example, the aerial photographs taken by the S190 B camera from the satellite Skylab, at a height of 435 km, were used to detect objects 50 m in length (Fig. 5.5) and this would be equivalent to the human eye detecting a pedestrian ant from the top of St Paul's Cathedral!

6. **Three dimensional perspective**: a stereoscopic, view of the Earth's surface can be created and measured both horizontally and

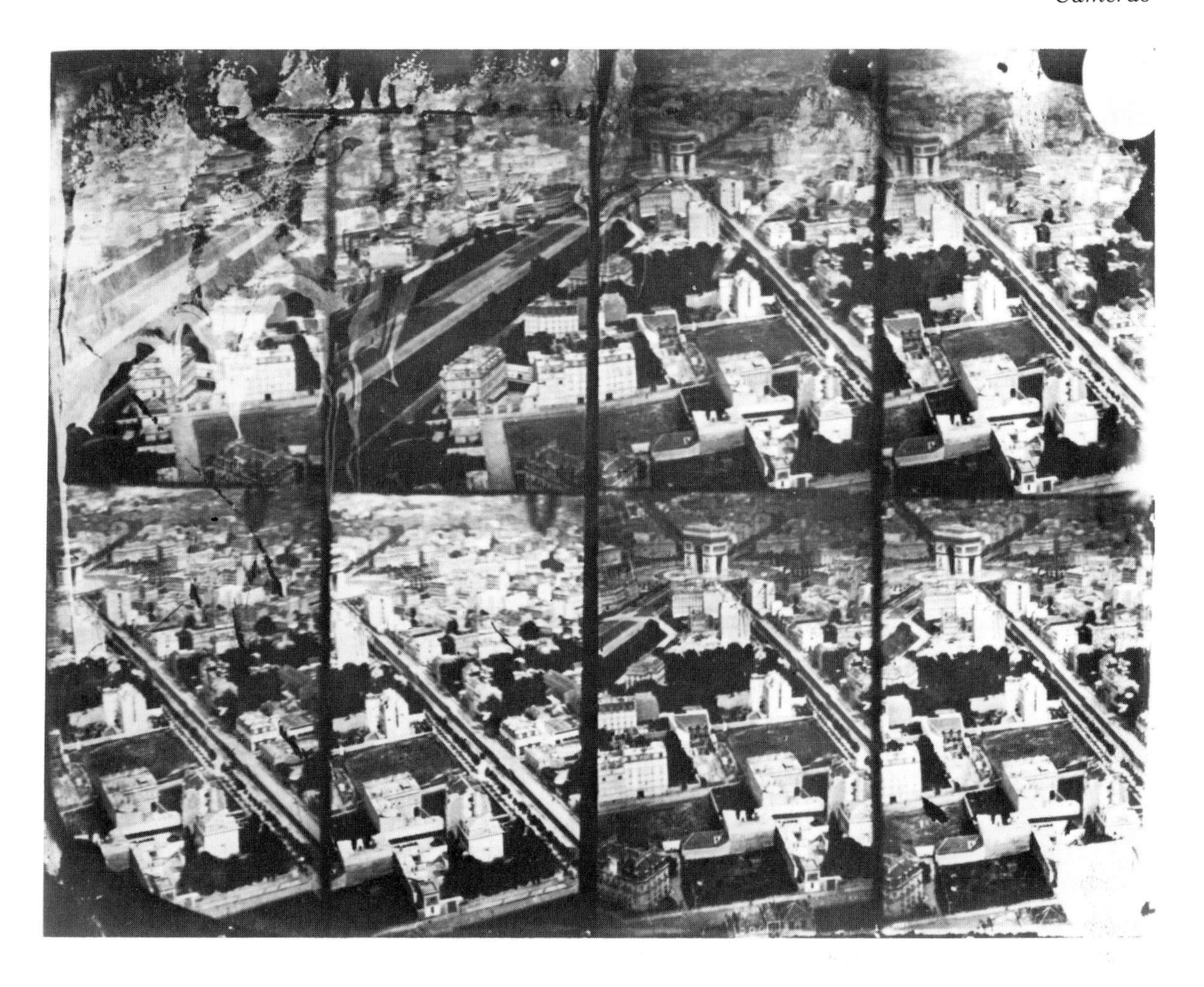

**Fig. 3.1**
Some of the first aerial photographs. This strip of aerial photography of the Arc de Triomphe and the Place de l'Etoile in Paris was taken using a plate camera by F. Nadar whilst in the nude, from a hot-air balloon in 1868 (Gosling 1976). (Courtesy, CNMHS/Spadem)

vertically (sect. 3.4.4); a characteristic that is lacking for the majority of remotely sensed images.

Surprisingly aerial photographs are not considered as a 'tool-of-the-trade' by all environmental scientists. The reasons given for this are the problems of obtaining suitable aerial photographs, the need to interpret as opposed to read aerial photographs, an uncertainty as to what equipment is required to interpret and take measurements from aerial photographs and more commonly a lack of knowledge of what aerial photographs have to offer. Hopefully after reading this chapter these will not be seen as limitations.

## 3.2 Cameras

There is nothing magical about the cameras used for taking aerial photographs. The aerial photograph in Fig. 3.20 was taken with a small amateur camera, which is large in comparison with some models (Fig. 3.2).

**Fig. 3.2**
One of the smaller models of aerial camera, dated 1907. (Courtesy, Deutsches Museum, West Germany)

For the aerial study of large areas, high geometric and radiometric accuracy are required and these can only be obtained by using cameras that are purpose built. These cameras usually have a medium to large format, a high quality lens, a large film magazine, a mount to hold the camera in a vertical position and a motor drive.

## 3.2.1 Aerial cameras

There are six types of aerial camera. These are: the mapping camera, reconnaissance camera, strip camera, panoramic camera, multilens camera and the multicamera array.

### 3.2.1.1 Mapping camera

The majority of aerial photographs are taken with high quality mapping cameras (Fig. 3.3). These cameras are relatively simple in design and comprise a low distortion lens, a shockproof lens cone and a large magazine capable of holding the film during exposure and

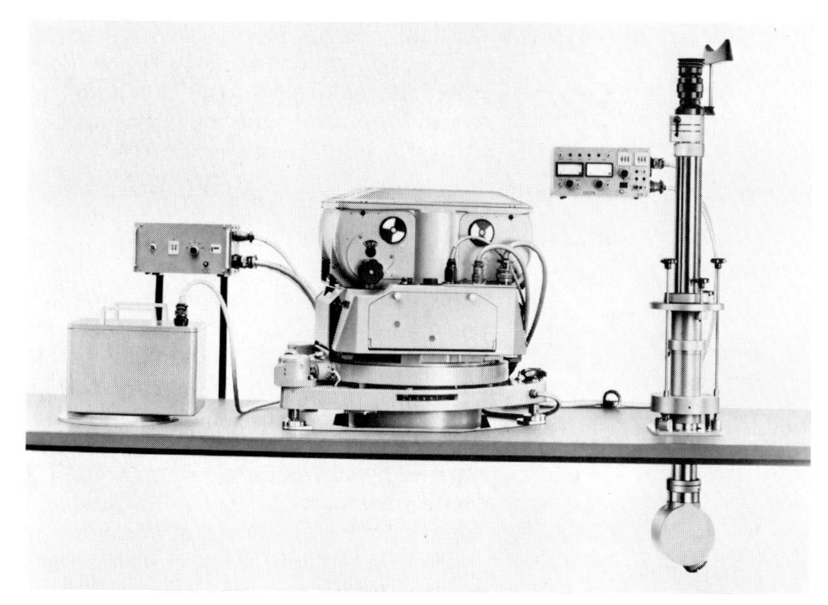

**Fig. 3.3**
An aerial mapping camera. This is a Carl Zeiss RMK/A15/23 with automatic levelling and exposure control. It is mounted on a suspension mount, between the remote control unit on the left and its navigation telescope on the right. (Courtesy, Carl Zeiss, Oberkochen Ltd, West Germany)

storing the film after exposure. Attached to this camera are a motor drive, which moves the film during exposure, to compensate for image motion, and winds the film after exposure, a viewfinder to sight the camera, an exposure meter, to take regular measurements of light intensity, and an intervalometer, to set the speed of the motor drive (Slater 1975, 1983).

### 3.2.1.2 Reconnaissance camera

These cameras are cheaper than mapping cameras to both buy and operate. Their two main disadvantages are their relatively high levels of geometric distortion, which prevents their use for mapping, and their lens design, which prevents the use of colour film in the majority of cases (Slater 1975, 1983).

### 3.2.1.3 Strip camera

The lens of this camera focuses light onto an adjustable slit, under which the film moves at a speed that is proportional to the ground speed of the aircraft. This mechanism compensates for image motion and results in long strips of unblurred photography (Slater 1975, 1983). The camera was originally designed for military purposes and later found limited application in transport studies, but it is little used today.

### 3.2.1.4 Panoramic camera

The lens in this camera oscillates, scanning from horizon to horizon while focusing the light onto a cylinder upon which a photographic film is held. In some of the modern versions of this camera the lens remains stationary and a prism scans the landscape. The result is a distorted image that cannot be used for mapping even when corrected (ASP 1966). Its sole advantage lay in its ability to cover large areas of terrain, an advantage that has now been lost due to the ready availability of satellite sensor data.

#### 3.2.1.5 Multiband aerial cameras

Objects on the Earth's surface vary in their spectral response within the photographically sensitive part of the electromagnetic spectrum. To take advantage of this to differentiate between objects, it is possible to photograph a scene through different filters. This 'multiband' photography can be taken using either a multilens camera or a multicamera array. The multilens camera uses one film which is exposed by radiation passing through several filters and lenses. The multicamera array comprise a number of small format cameras, each camera having its own film and filter (Fig. 3.4).

There are many multilens cameras, the main difference between them being the number of lenses that they contain which have ranged from four to nine. The photograph in Fig. 3.14 was taken with a four lens multilens camera, the film was black and white near infrared and the four filters were blue, green, red and near infrared.

Multicamera arrays began as a relatively cheap and flexible means of obtaining multiband photography without having the expense of a special multilens camera. A simple multicamera array can be two 35 mm format cameras held together in a frame (Curran 1981c) but the most common multicamera array comprises four 70 mm format cameras mounted in a frame and sighted to take images of the same area at the same time. This system is very flexible as it offers a very wide range of film and filter combinations (Yost and Wenderoth 1967).

## 3.3 Film

Photographic film is a sandwich of thin polyester and gelatin. The thin polyester gives the film rigidity and the gelatin binds the light sensitive silver halide crystals together. If the camera shutter is opened, electromagnetic radiation can reach this film, causing one of the silver halide crystals to release an electron. This electron can move off but it usually gets caught up within the lattice of the silver halide crystal and turns a silver ion to a silver atom. If this happens twice within about a second then it will be the trigger that converts the whole silver halide crystal into metallic silver (Sturge 1977; James 1977). At this stage the film has a concealed image that can be turned into a visible image by the chemical process of development, and once developed it will be the negative image from which positive prints are made. An aerial photograph is a very compact method of storing information about the Earth's surface, as every silver halide crystal, whether exposed or not, contains some details of the scene. It is interesting to note that a photograph has around $25 \times 10^9$ units of information per square centimetre, which is $5 \times 10^4$ times more than can be held on the computer tapes discussed in section 6.3 (Sabins 1978).

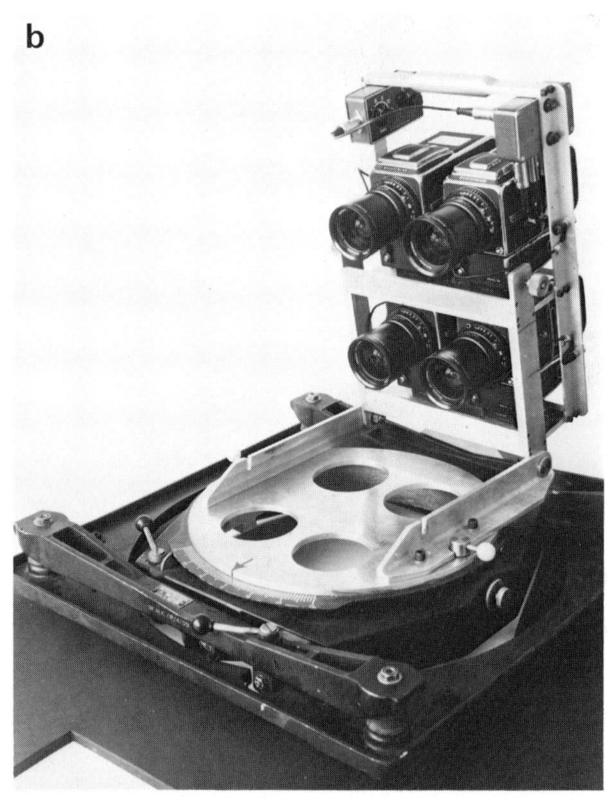


**Fig. 3.4**
Multiband aerial cameras: (a) A multilens camera system NAC MB 490B suitable for use in a light aircraft. (Courtesy, IIMC Ltd, Oxfordshire, UK); (b) A multicamera array comprising four 70 mm format Hasselblad cameras in a frame.

### 3.3.1 The characteristic curve of a film

The tone of an object on an aerial photograph is directly related to the radiance of that object on the Earth's surface and an understanding of the relationship between image tone and object radiance is necessary for the successful interpretation and measurement of tones on aerial photographs.

As the shutter on an aerial camera is opened the different radiance levels within the scene expose the film for the same length of time. Light toned objects with a high radiance, strongly expose the negative and produce a dark image and dark toned objects with a low radiance, weakly expose the negative and produce a light image. The degree of darkness can be measured using a transmission densitometer which passes a beam of light through the developed film negative. There are many densitometers on the market and these can be grouped into macrodensitometers, that record the light transmission over an image spot of around 1 mm in diameter, and microdensitometers, that record the light transmission over an image spot of around 10 μm in diameter.

These densitometers can be used to measure the degree of film darkness in units of either transmittance ($T$), opacity ($O$) or density ($D$), (formulae [3.1], [3.2] and [3.3]).

$$\text{Transmittance (T)} = \frac{\text{light passing through the film}}{\text{light impinging on the film}} \quad [3.1]$$

$$\text{Opacity } (O) = \frac{1}{T} \quad [3.2]$$

$$\text{Density } (D) = \log 0 \quad [3.3]$$

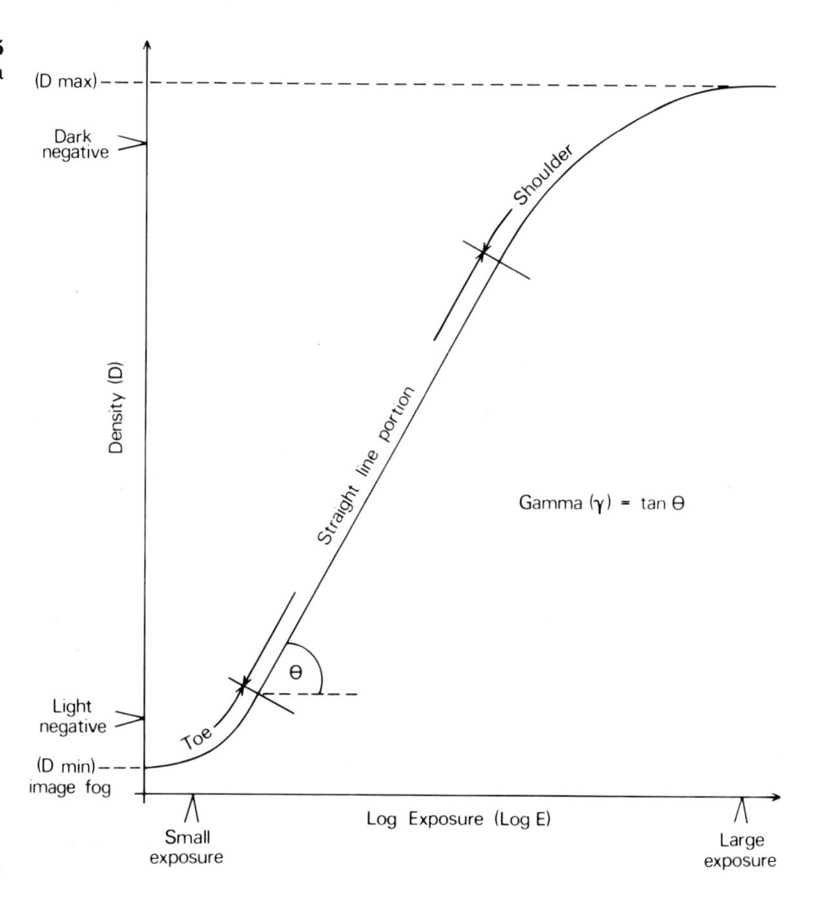


**Fig. 3.5**
The characteristic curve of a film.

Density ($D$) is the most popular scale as it has a positive relationship to the logarithm of film exposure (log $E$). This relationship is termed the characteristic curve or $D$log$E$ curve of a film (Fig. 3.5) and is unique for each batch of film and method of processing (Slater 1980). This curve enables density ($D$) to be predicted for a given level of exposure ($E$), but more importantly, it also enables exposure ($E$) to be predicted for a given level of density ($D$), (Lillesand and Kiefer 1979). As its name would suggest the characteristic curve provides a considerable amount of information on the characteristics of a particular film. To illustrate this the characteristic curves of two very different films (Fig. 3.6) are discussed.

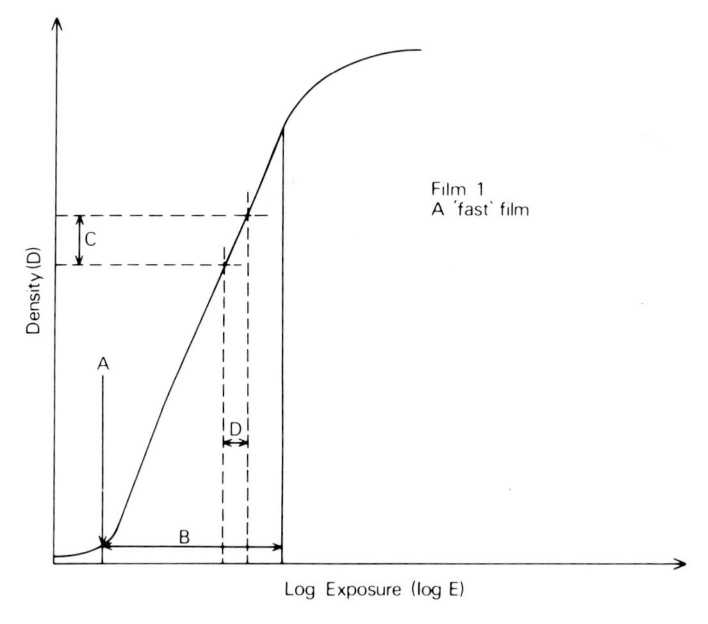


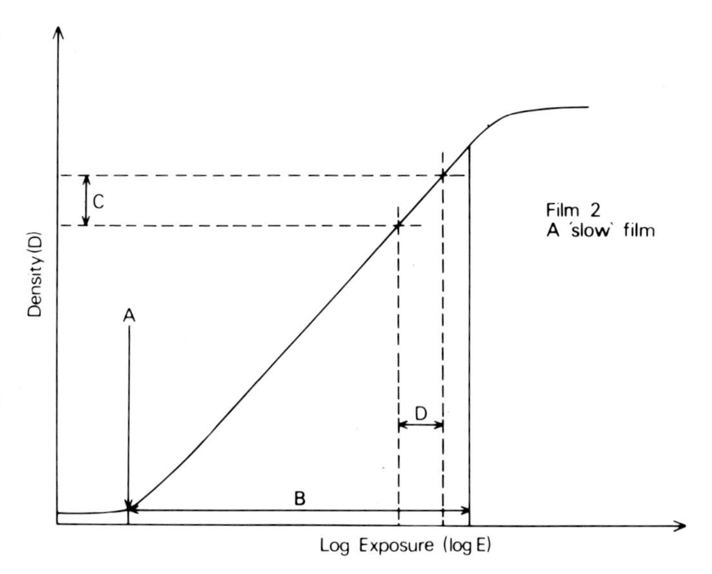


**Fig. 3.6**
A comparison between the characteristic curves of two films. *Film 1* is a fast film with a high gamma, high contrast, high granularity and a narrow exposure latitude. As the film is sensitive to low levels of exposure it can be used with fast shutter speeds to minimise the effect of image blurr. The radiometric resolution of the film is poor but the spatial resolution of the film is good. *Film 2* is a slow film with a low gamma, low contrast, low granularity and a wide exposure latitude. As the film is not sensitive to low levels of exposure it cannot be used with fast shutter speeds to minimise the effect of image blurr. The radiometric resolution of the film is poor but the spatial resolution of the film is good.
A = Minimum level of exposure to which the film will respond. Film 1 is a fast film because it will respond to lower levels of exposure than will film 2.
B = Exposure range (or latitude) over which log exposure is linearly related to image density. Film 2 can record exposure over a much larger exposure range than can film 1.
C = The resolution attainable by a given densitometer, this is dependent upon the densitometer and not the film.
D = The exposure resolution with which exposure can be determined by a given densitometer. The fast film (film 1) has the finer exposure resolution.

### 3.3.1.1 Gamma of the characteristic curve

The angle that the straight line portion of the characteristic curve makes with the log $E$ axis is termed gamma ($\gamma$), (formula [3.4]).

$$\gamma = \tan \theta \qquad [3.4]$$

For example, a characteristic curve with a straight line position at 45° to the log $E$ axis has a gamma of 1. A film with a gamma greater than 1 (film 1 in Fig. 3.6) has a high contrast and likewise a film with a gamma less than 1 (film 2 in Fig. 3.6) has a low contrast. A high contrast film distributes the radiance within the scene over a large range of density values using the full range from white to black. A low contrast film distributes the radiance within the scene over a smaller range of density values from light to dark grey. As can be seen in Fig. 3.24 the difference between a high and a low contrast photograph can make a considerable difference to interpretation.

### 3.3.1.2 Film speed, granularity, spatial resolution and the characteristic curve

The speed of a film is a measure of the minimum level of exposure to which it will respond. Film 1 (Fig. 3.6) which is a fast film, will accommodate low levels of exposure and will require less exposure than film 2 (Fig. 3.6), which is a slow film. Film speeds are provided by the manufacturer in either American Standard Association (ASA) units, if the films are for use on the ground, or Aerial Film Speed (AFS) units if the films are for use in the air (Horder 1976). An aerial photographer in temperate latitudes often wishes to use the fastest possible film as this will ensure that he can use short exposure times to reduce the blurring effect of image motion. Unfortunately, very fast films result in photographs that look 'grainy' at high magnification and have a low spatial resolution. This is because the very large silver halide crystals that are needed to intercept radiation at low levels of exposure form into clearly visible clumps or 'grains' (Slater 1980).

To determine the spatial resolution of different films, manufacturers take aerial photographs of test charts under near perfect conditions. These test charts comprise groups of parallel lines separated by spaces equal to the width of the lines. The resolving power of the film under these optimum conditions is expressed as the number of lines per millimetre that can be distinguished under a microscope (Jensen 1968; Sabins 1978). As the visibility of the lines is dependent upon the contrast between the lines and their background, manufacturers usually report spatial resolution for a range of contrasts (Eastman Kodak 1976). For the environmental scientist the spatial resolution of the film at a contrast of 1.6 : 1 is the most useful as this is similar to the contrast present in a typical aerial photograph. These figures in lines per millimetre at a contrast of 1.6 : 1 can be converted to spatial resolution in centimetres using formula [3.5].

$$\text{Spatial resolution on the ground in cm} = \frac{\text{reciprocal of image scale}}{\text{film resolution in lines per mm at a contrast of 1.6 : 1}} \times 10 \qquad [3.5]$$

| Kodak film number | Film type | Film speed (AFS units) | Granularity (ranking where 1 = low, 4 = high) | Film spatial resolution in lines per mm at a contrast of 1.6 : 1 | Likely ground resolution of film in cm at an image scale of 1 : 15,000 |
|---|---|---|---|---|---|
| 3414 | Reconnaissance | 8 (very slow) | 1 | 250 | 6 |
| 3410 | Reconnaissance | 40 | 2 | 80 | 19 |
| 2402 | Mapping | 200 | 3 | 50 | 30 |
| 2403 | Mapping | 640 (very fast) | 4 | 25 | 60 |

*Source*: Modified from Eastman Kodak 1976.

**Table 3.1**
The relationship between film speed, granularity and spatial resolution for four aerial photographic films.

For example Kodak 2402 mapping film has an image resolution of 50 lines per mm at a contrast of 1.6 : 1 (Table 3.1). Therefore at a scale of 1 : 15,000 an aerial photograph will have a spatial resolution, on the ground of 30 cm.

$$\frac{15{,}000}{50 \times 10} = 30 \text{ cm}$$

Calculation of spatial resolution allows an easy comparison between different film types and gives an indication of the likely levels of resolution under perfect weather conditions. Unfortunately aerial photography cannot always be taken under perfect weather conditions due to the effects of poor visibility, low cloud cover, inadequate illumination, rainfall and snow. In the UK these effects conspire to restrict the taking of panchromatic aerial photographs with a high spatial resolution to one day in four over urban areas and one day in two over rural areas.

For colour and false colour near infrared aerial photography the conditions are even more critical. Winter illumination levels are often insufficient and the requirement of greater visibility restricts the taking of colour the false colour near infrared aerial photography with a high spatial resolution to less than one day in two or three (Evans 1974). It is therefore not surprising that much aerial photography in the UK is taken in sub-optimal conditions with a resultant decrease in its spatial resolution.

### 3.3.1.3 Exposure latitude and the characteristic curve

The exposure latitude determines the exposure range that can be recorded by a film; film 1 (Fig. 3.6) is a fast film with a narrow exposure latitude and film 2 (Fig. 3.6) is a slow film with a wide exposure latitude. Ideally, an aerial photographer would like to spread the exposure range of the scene along the straight line portion of the characteristic curve but this is rather difficult, if not impossible for some fast films. For example false colour near infrared film (sect. 3.3.6), which is a fast film has a very small exposure latitude and if you look carefully at examples of such photography in this book (Fig. 3.21) and elsewhere, you will note that very light and very dark objects tend to be off of the characteristic curve.

### 3.3.1.4 Radiometric resolution and the characteristic curve

Radiometric resolution is the smallest difference in exposure that can be detected by a densitometer. A densitometer will be able to record exposure with a relatively high radiometric resolution on a fast film (film 1 in Fig. 3.6) but relatively low radiometric resolution on a slow film (film 2 in Fig. 3.6). It is therefore prudent to use a fast film for studies where radiometric data are being derived from aerial photography (Curran 1982a).

### 3.3.1.5 Calculating surface radiance from the characteristic curve

It is possible to derive relative measurements of object radiance from aerial photography by using the characteristic curve. The method is as follows: first, densitometry using a densitometer (sect. 3.3.1) or a calibrated analogue image processor (sect. 6.2.5), to measure the tone of the film negative. Second, the translation of image tone into relative exposure values via the characteristic curve of the film (Fig. 3.5) and third, the linking of scene radiance to a known standard or standards (usually reflectance targets on the ground, as in Plate 5), to suppress the temporal variability in the irradiance of the Sun and sky, atmospheric transmission, film exposure and the optical properties of the camera lens and filter (Lillesand and Kiefer 1979; Curran 1980d; Curran *et al.* 1981).

## 3.3.2 Types of film

The two types of film are one layer black and white and three layer colour, both of which can be sensitised to either visible only or visible and near infrared wavelengths. The four resultant film types are black and white, black and white near infrared, colour and false colour near infrared.

## 3.3.3 Black and white film

Black and white film that is sensitive to a broad waveband of visible light is called panchromatic film. The two types of panchromatic black

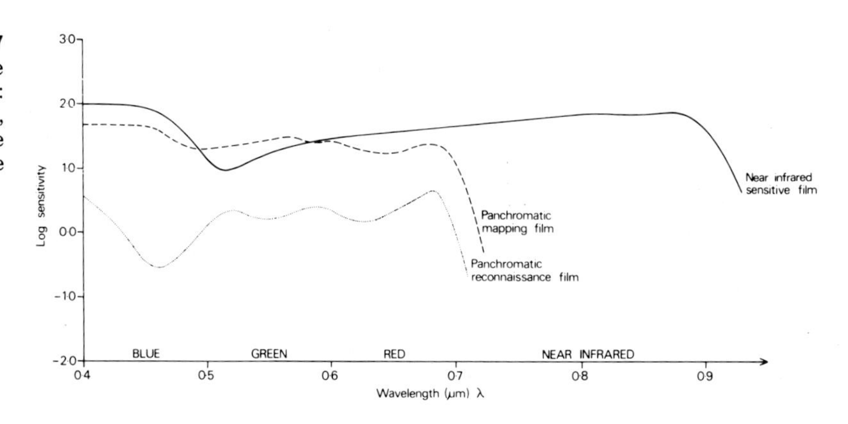


**Fig. 3.7**
The spectral sensitivity of three types of black and white film: panchromatic mapping film, panchromatic reconnaissance film and near infrared sensitive film.

and white film are mapping film which has equal sensitivity to all visible wavelengths, and reconnaissance film, which has reduced sensitivity to blue wavelengths to minimise the effect of atmospheric scatter (Fig. 3.7). The majority of panchromatic black and white aerial photographs are taken using mapping film (Fig. 3.8 and 3.9). This is the most popular film for use in the environmental sciences. The reasons for this are first, black and white aerial photographs are readily available for most of the world as they have been taken for topographic mapping and other uses and then made available to the public through commercial firms, planning authorities and national distribution centres (Appendix A). Second, they are geometrically stable and are ideal for mapping purposes. Third, they are cheaper to take, process, print and purchase and have a higher spatial

**Fig. 3.8**
Black and white panchromatic aerial photograph of an arid area of southern Asia, showing a heavily dissected pitching anticline in sedimentary rocks. (Courtesy, Clyde Surveys Ltd, UK)

resolution than colour or false colour near infrared films (Wolf 1974; Paine 1981).

### 3.3.3.1 Applications of black and white aerial photography

Black and white aerial photographs have been used for many applications in the Earth sciences, for geological mapping (Fig. 3.8), (Allum 1980; Press *et al.* 1980), hydrogeological investigations (La Riccia and Rauch 1977), and terrain analysis (Fig. 3.9 and 3.10), (Ball *et al.* 1971; Sauer 1981). For example, they have been used for the mapping of glacial deposits (Welch and Howarth 1968), desert dunes (Finkel 1961; Davis and Neal 1963) and coastal formations (Kidson and Manton 1973; Welsted 1979). They are used in agriculture for the identification of crop types (Smit 1978; Philipson and Liang 1981), crop diseases (Bell 1974) and soil erosion (Stephens *et al.* 1982). In soil surveys they

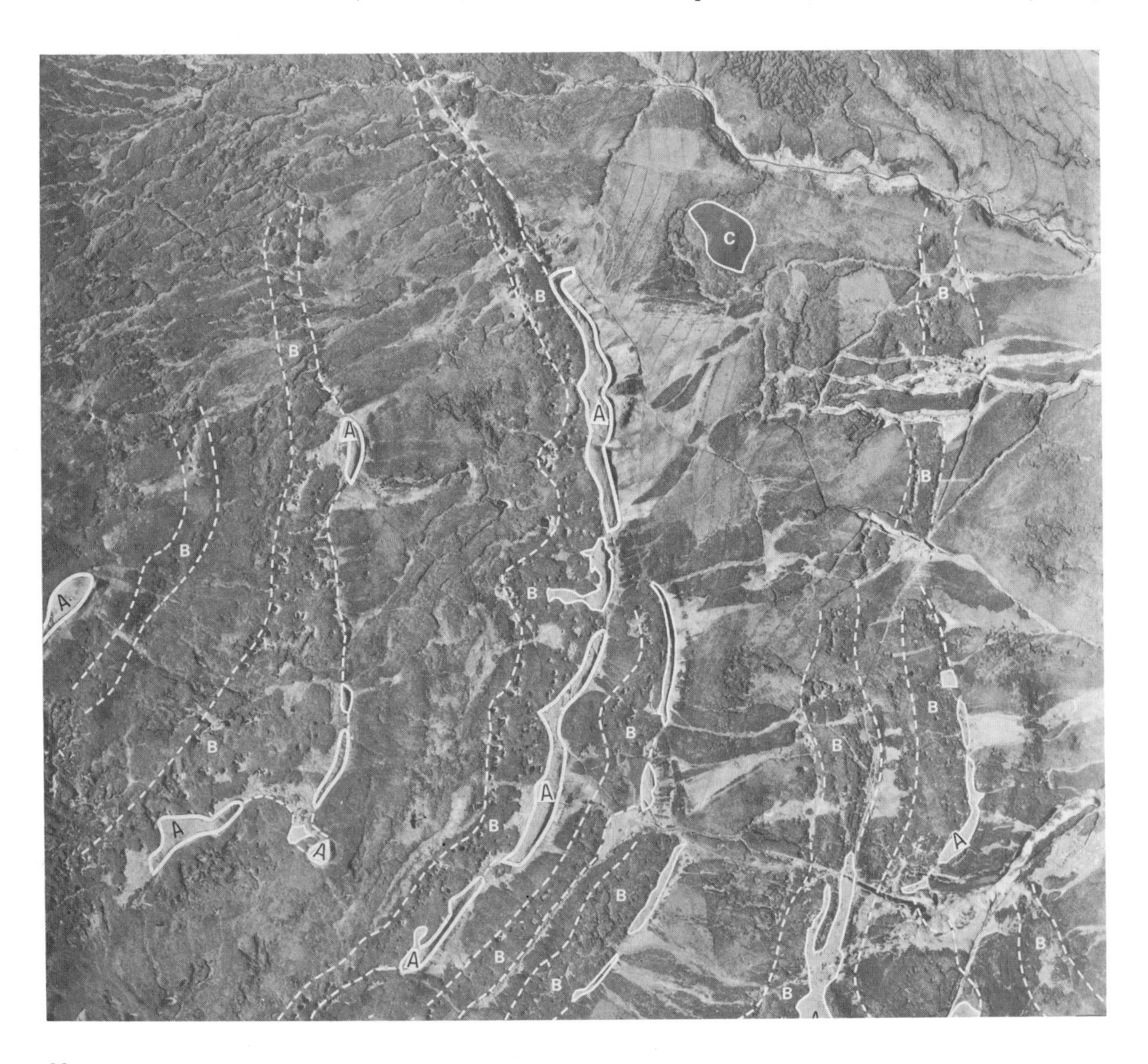


**Fig. 3.9** Panchromatic aerial photograph of a section of the upper Teesdale National Nature Reserve, UK. A, represents soil complexes developed over limestone. B, represents belts of swallow holes where the limestone outcrops and C, represents peaty rankers over sandstone (Ball *et al.* 1971). (Courtesy, Meridian Airmaps Ltd, UK)

**Fig. 3.10**
A vertical panchromatic aerial photograph of a moorland area in the North York Moors, UK. This photograph formed part of the information used for the construction of a number of maps, extracts of five of these are reproduced below. In this case there is a strong link between lithology and the potential and actual use of the land. (Image courtesy, Meridian Airmaps Ltd, UK)

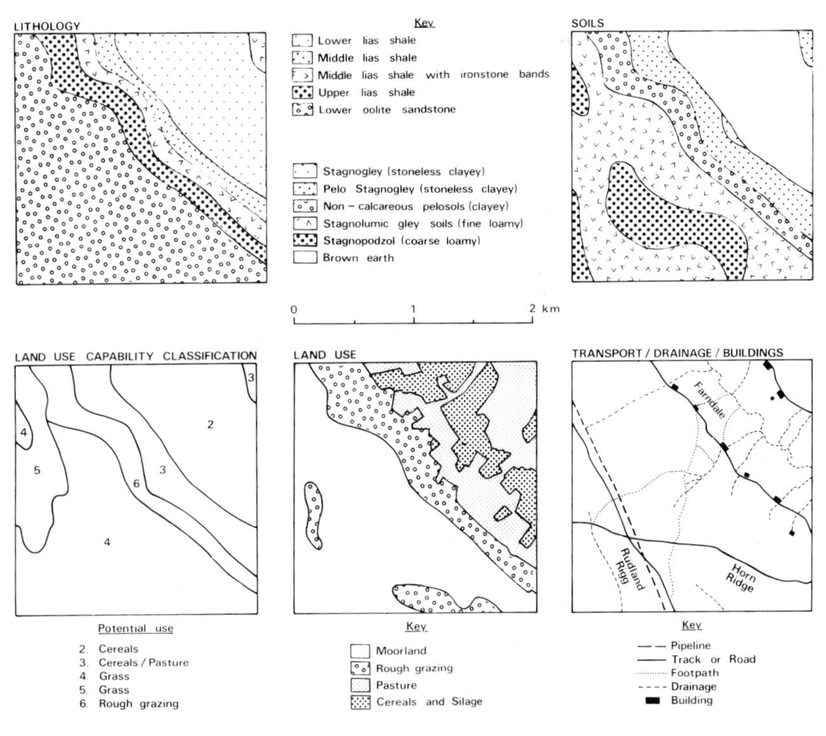

have been used for reconnaissance and field mapping (Evans 1972; White 1977; Antrop 1979), as is also the case for vegetation surveys (Goodier and Grimes 1970; Stove and Hulme 1980; Grainger 1981), (Fig. 3.10).

The majority of planning authorities throughout the world hold black and white aerial photographs of their region (Appendix A) which have been used for regional planning (Avery 1965), urban planning (Fig. 3.24), (Collins and El-Beck 1971) and for more specific studies, for example, the census of population (Hsu 1971; Lo and Chan 1980; Olorunfemi 1982), the census of derelict land (Bush and Collins 1974; Erb *et al.* 1981) and the monitoring of urban growth (Ellefsen and Davidson 1980).

Not all of the black and white photography used on a regular basis is taken vertically (Curran 1981c); oblique photography is often used for both military reconnaissance (Gustafson 1980), aerial archaeology (Edwards and Partridge 1978), and low cost environmental monitoring (Curran 1979b). This film can also be effectively filtered for example, a polarising filter that allows only polarised visible light to reach the film has been used for studies of both soil (Curran, 1978b, 1981a) and vegetation (Curran, 1981b, 1982b).

For further recent examples of the application of black and white aerial photography refer to recent issues of the major remote sensing journals and symposia (Appendix B).

## 3.3.4 Black and white near infrared film

Black and white near infrared film has similar characteristics to panchromatic black and white film (sect. 3.3.3). The main difference is its spectral sensitivity which extends beyond visible wavelengths to a wavelength of around 1.0 μm, in the near infrared region of the electromagnetic spectrum (Fig. 3.7). This film can be used with a near infrared filter to record a near infrared waveband or with both visible and near infrared filters to record visible and near infrared wavebands (Paine 1981).

### 3.3.4.1 Applications of black and white near infrared aerial photography

When this film is used with a near infrared filter in aerial photography it has proved to have value in a number of applications; for example in studies of vegetation it has been used for the location of crop diseases (Fig. 3.11), crop type (Fig. 4.10), and the mapping of forest and semi-natural vegetation (Marshall and Meyer 1978; Aldrich 1979). The sensitivity of the film to surface soil moisture has been usefully employed to map the soil moisture status of agricultural fields (Curran 1981f), to locate geological boundaries where they coincide with a change in soil depth and moisture content (Hunter and Bird 1970), to monitor the spread of dry sand dunes over moist soils (Stembridge 1978), to locate areas of active soil erosion (Garland 1982) and to map archaeological sites (Fig. 3.12).

**Fig. 3.11**
Vertical, black and white near infrared aerial photograph of three fields in Norfolk, UK, in 1964. There is a potato crop at top left, a sugar beet crop at top right and a dwarf bean crop at the bottom of the aerial photograph. The potatoes are infected with an airborne infestation of blight and the dwarf beans are infected by soil borne halo blight, which is spread by rain splash or the movements of implements through the crop. (Courtesy, Aerial Photography Unit, Ministry of Agriculture, Fisheries and Food, UK)

**Fig. 3.12**
Oblique, black and white near infrared aerial photograph of the Lox Yeo River in Somerset, UK taken looking westwards in April 1980. The outline of the Roman enclosure adjoining the river can be seen. (Courtesy, West Air Photography, Weston-super-Mare, UK)

### 3.3.4.2 Applications of black and white near infrared film in multiband aerial photography

Black and white near infrared film is used for multiband aerial photography where a scene is photographed through a number of either broad band filters of blue, green, red and near infrared or narrow band filters that have been chosen to match the spectral response of the scene. The likely tone of objects on these images can be predicted by reference to section 2.5 and Fig. 3.13. For example, vegetation photographed through broad band filters would display its low to medium blue reflectance, its medium to high green reflectance, its low to medium red reflectance and its high near infrared reflectance. Broad band multiband photography has been used for many applications in the environmental sciences, including crop identification (Curtis and Mayer (1974), soil mapping (Stephens *et al.* 1981) and the monitoring of land cover (Estes and Simonett, 1975), (Fig. 3.14).

Narrow band spectral filters like those listed by Eastman Kodak (1965) are less popular than broad band filters, because although they offer greater spectral detail (Yost and Wenderoth 1967), they often result in images with a low spatial resolution. This is because a narrow waveband will reduce the level of radiation reaching the film, necessitating the use of a high speed film which will result in an increase in granularity and a decrease in spatial resolution (sect. 3.3.1.2). Despite this problem narrow band filters are used where they provide

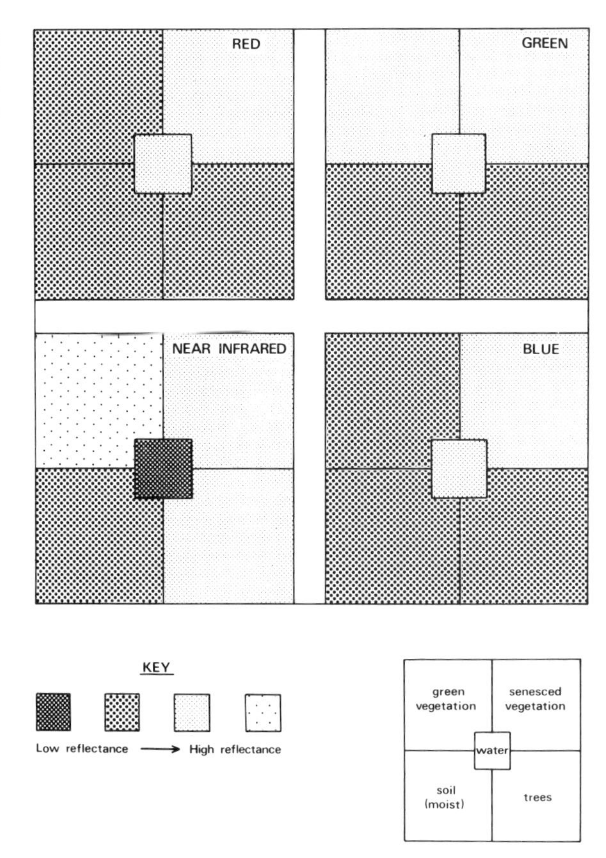


**Fig. 3.13**
The probable image tone of five scene components when photographed with a 4-lens multiband camera.

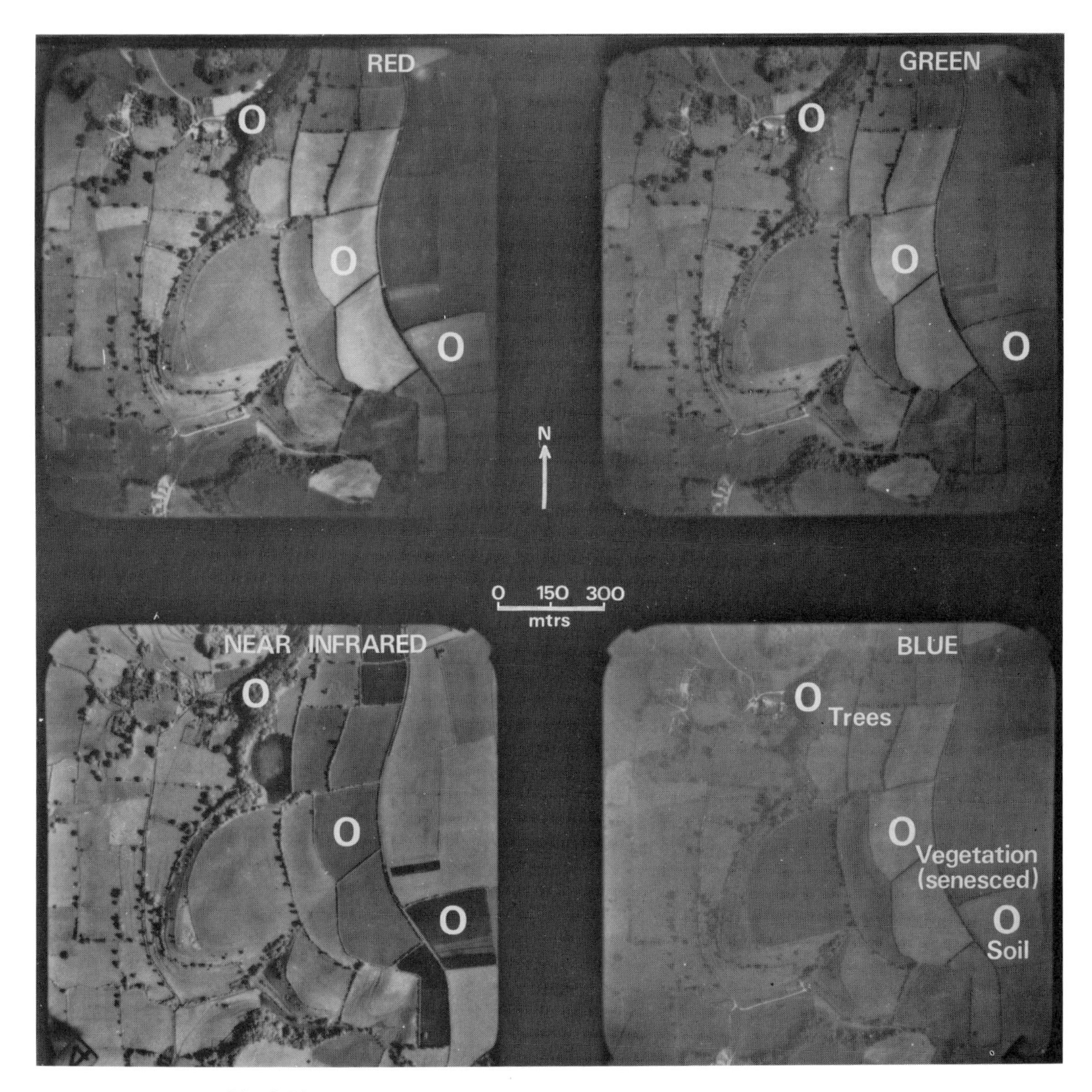


**Fig. 3.14**
Multiband aerial photograph taken on 6.7.1972, of a wooded scarp in South Wales. Above the scarp some of the cereal crops have been harvested, exposing the soil, while below the scarp the soil is covered by cereal and grass crops. This aerial photograph was taken on black and white near infrared film through broad band blue, green, red and near infrared filters. (Courtesy, Hunting Technical Services Ltd, UK)

greater discrimination of features on the Earth's surface. For example, narrow band aerial photography has been used to detect both lithological boundaries and geobotanical anomalies, as an aid to mineral exploration (Gilbertson and Longshaw 1975), sewage outfalls, as an input to environmental planning (Curran 1979a) dye tracers in hydrological studies (Valerio and Llebaria 1982) and land cover types for land use planning (Wenderoth and Yost 1975).

The method used to interpret broad or narrow band images depends upon the application. To identify and map features on the Earth's surface it is often convenient to combine at least three of the images together in a colour composite. This can be done by either using a multi-additive viewer (Fig. 6.4) or a Diazo colour printer (sect. 6.2.2), (Wenderoth and Yost 1975; Townshend 1981b). To obtain quantitative data on scene radiance the image tone is measured for each feature in each waveband using a densitometer (sect. 3.3.1). This can either be used to discriminate features from their surroundings (Curtis and Mayer 1974) or to estimate the characteristics of these features, for example vegetation biomass (Curran 1980, 1983e).

For further recent examples of the application of black and white near infrared aerial photography and multiband photography, refer to recent issues of the major remote sensing journals and symposia (Appendix B).

## 3.3.5 Colour film

As the human eye can perceive more than 20,000 tints and shades of colour but only 200 tones of grey (Ray and Fischer 1957; Yost and Wenderoth 1967), far more information is available in a colour photograph than is available in a black and white photograph. Colour film is designed to mimic the response of the human eye which is broadly sensitive to the three colours of blue, green and red. These colours are called the primary colours because any other colours can be produced by mixing appropriate proportions of each (ASP 1968). For example, to produce a purple paint a child will mix blue and red paint and to produce a purple dress on a television screen small blue and red spots are mixed. This mixing process is called additive colour and has been used in the production of colour films which are spectrally accurate but rather dark (Horder 1976). As our eyes are insensitive to small differences in spectral radiance, a colour film with only approximate mixtures of the primary colours is sufficient to fool our eyes. This can be achieved by the colour subtraction process in which the radiance of an object in the scene, for example, a green field, will on exposure, lead to the production of film colours that will subtract all except green wavelengths. This process of subtracting colours from white light to produce other colours also occurs in nature, for example, the pink skies of late evening are the result of the atmospheric subtraction of some green and much blue light from the white light radiated by the Sun.

In colour photography the subtraction process relies upon the inclusion of three dyes into the film during development; these are yellow, magenta (blue-red) and cyan (blue-green). These dyes are complementary in colour to the primary colours and each absorb one primary colour and transmit two primary colours (Jones 1971). The yellow dye absorbs blue and transmits green and red light, the magenta dye absorbs green and transmits blue and red light and the cyan dye absorbs red and transmits blue and green light.

These dyes which are located in separate emulsion layers within the film (Fig. 3.15) are sandwiched around a filter and are held onto a backing and base. The filter is yellow and is used to absorb the blue light to which the green and red emulsion layers are also sensitive (Fig. 3.15). As our eyes are insensitive to the difference between red and red-green objects, the response of the red sensitive layers to both red and green light is not compensated for by filtration (Slater 1980).

### 3.3.5.1 Processing colour film

The way in which a colour film is processed depends upon whether the

**Fig. 3.15**
The sensitivity of the three emulsion layers in a colour film.

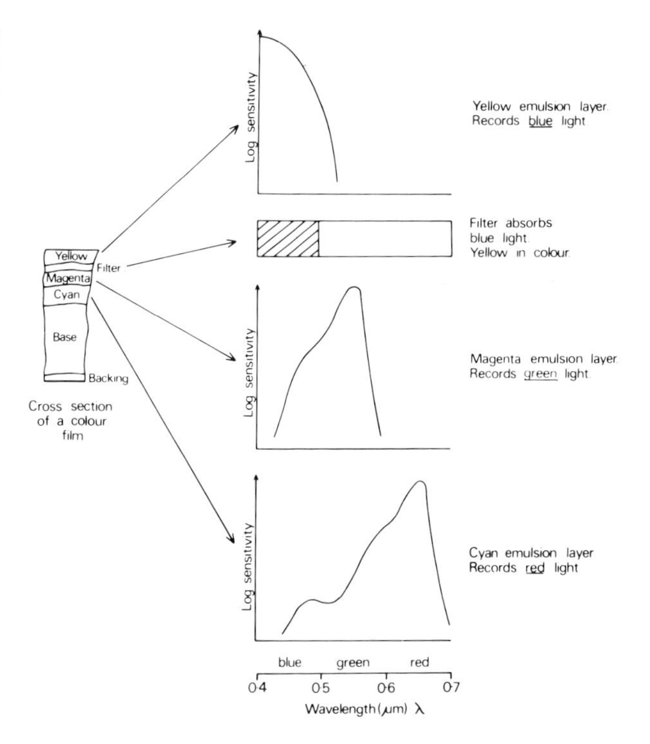


film is a colour negative or a colour positive. A colour negative film produces an image which has reversed scene geometry and brightness and is coloured in the complementary colours of yellow, magenta and cyan. When this image is projected onto photographic paper using white light, a colour print is produced that has similar geometry, brightness and colour to the original scene. A colour positive (or slide) film produces a positive image on the film within the camera and has similar geometry, brightness and colour to the original scene when viewed by projection with white light (Evans *et al.* 1953). After exposure the colour negative film is developed in colour developer to form an image of negative silver and complementary colours. The film is bleached to remove the negative silver image and the yellow filter layer, leaving an image in complementary colours (Fig. 3.16).

After exposure the colour positive film is developed in black and white film developer resulting in a negative silver image in each film emulsion layer. The film is then re-exposed with a white light and developed in a colour developer, with the result that only the emulsion layers exposed by the second exposure are coloured by complementary colours. The film is bleached to remove the exposed silver negative image and yellow filter layer leaving a colour positive image (Fig. 3.16), (Eastman Kodak 1972; Horder 1976).

The choice of whether to use a colour negative or a colour positive film is dependent, first, upon the spatial variability of reflectance within the scene, as a colour negative film has three times the exposure latitude of a colour positive film (Slater 1980) and second, the relative cost and quality of the photographic products (Mott 1966). The highest quality of interpretative product is the first generation colour positive transparency derived from colour positive film (Welch 1968). To produce colour or black and white copies from this film for field

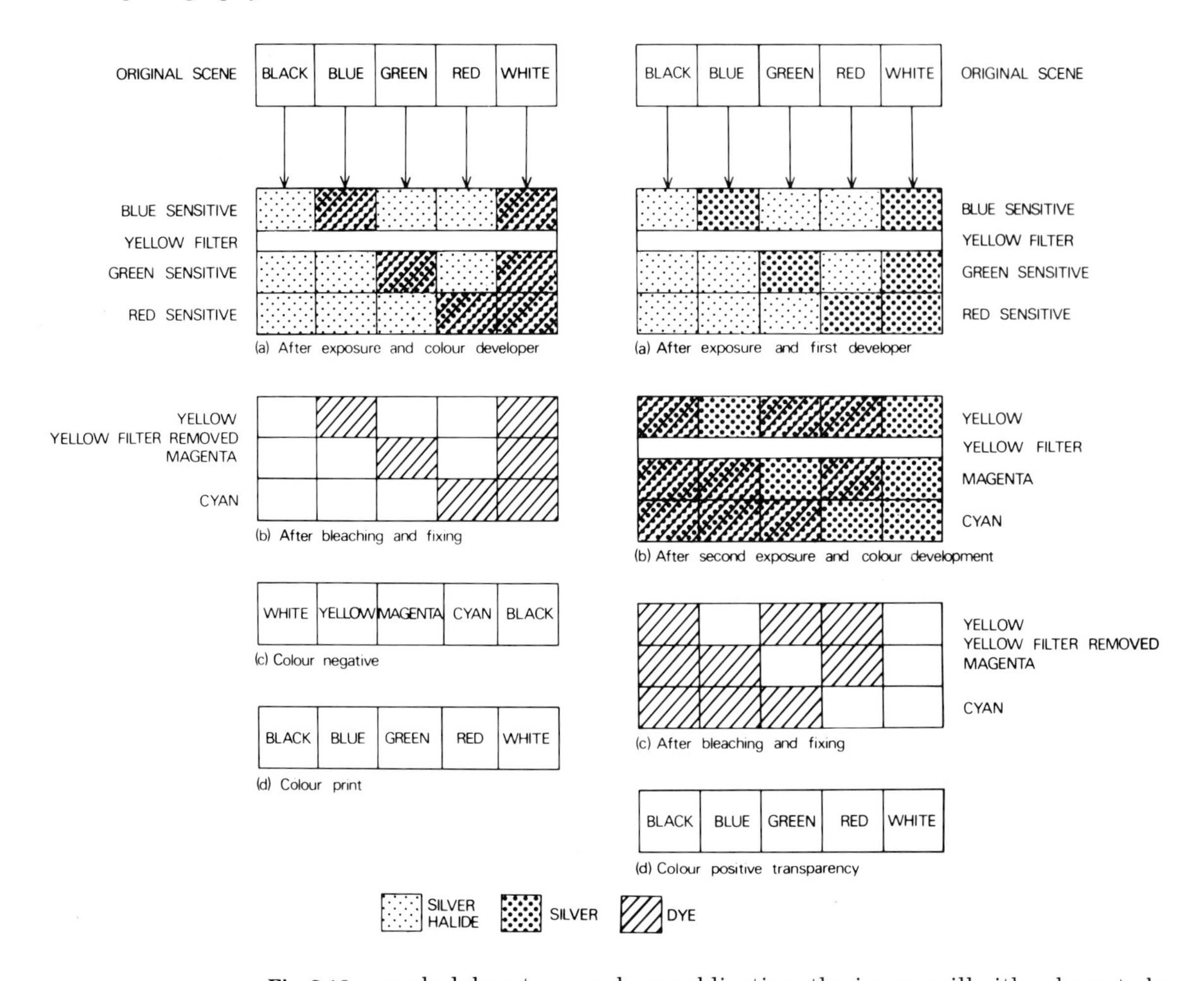


**Fig. 3.16**
The two routes to processing a colour film. (Modified from Slater 1980)

work, laboratory work or publication, the image will either have to be rephotographed to produce an inter-negative and then printed with a resultant loss in image quality or it can be printed directly as a positive print using a material with the trade name of 'Cibachrome' (Horder 1976). The positive colour prints produced by the Cibachrome process are very popular, as they use dyes that are spectrally superior and less liable to fading than those used in conventional colour prints. Their two drawbacks are their high cost, especially if hand printed and their high contrast, which obscures shadow detail. Colour negative film is far more versatile and can be used to produce colour paper prints, colour positive transparencies, black and white paper prints and black and white positive transparencies (Fig. 3.17).

### 3.3.5.2 Applications of colour aerial photography

Although colour aerial photographs are expensive to take and process, colour film has been used for a wide range of mapping and monitoring tasks especially in the fields of agriculture and forestry, ecology, geology and geomorphology, and hydrology and oceanography (ASP 1968). For example, in agriculture and forestry the wide range of colours in colour aerial photography facilitates the identification of

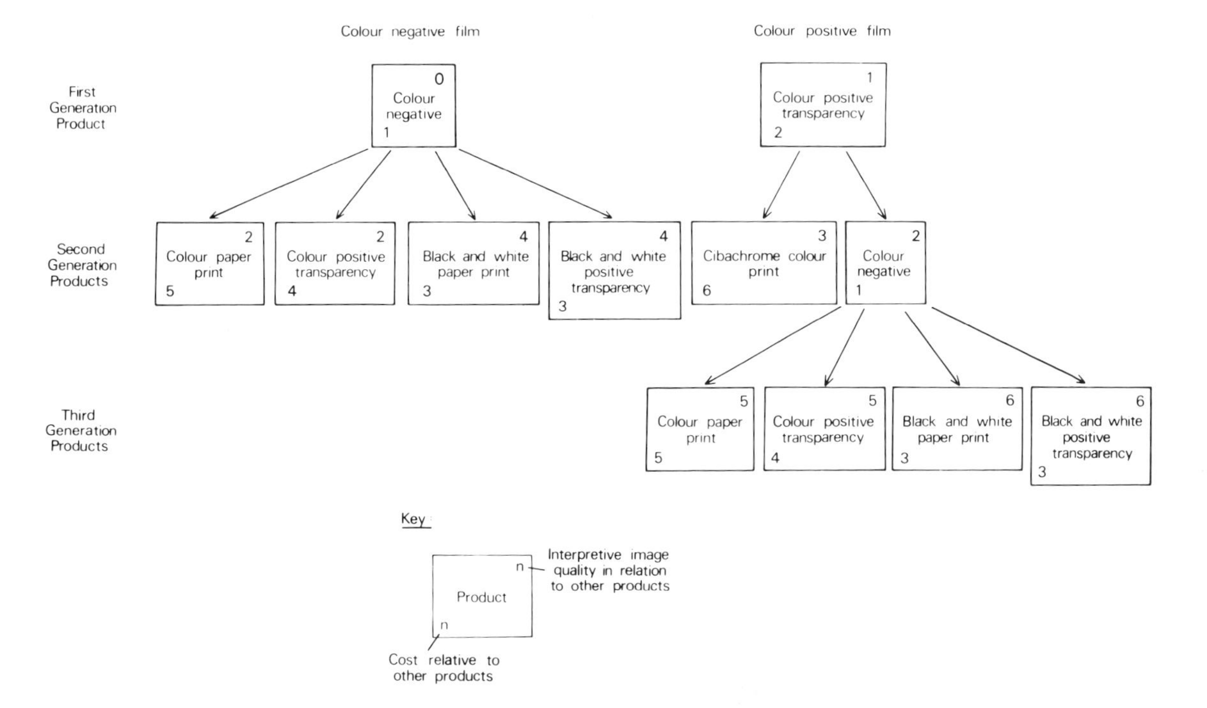


**Fig. 3.17**
Possible photographic products from colour negative and colour positive film.

land cover (Grumstrup *et al.* 1982), (Plate 5), crop condition (Ryerson *et al.* 1980), tree types (Becking 1959), tree stress (Wert *et al.* 1970) and soil type (Gerbermann *et al.* 1971). In ecological studies colour aerial photography has received wide application for a diversity of tasks that vary from vegetation mapping (Perkins 1971; Wallace 1981) to the determination of the sex ratio of wildfowl (Ferguson *et al.* 1981). In geology and geomorphology it has been shown to be superior to panchromatic black and white aerial photography for photogeological mapping (Fischer 1962; Lo 1976) and for the location of geobotanical anomalies (Cole *et al.* 1974). In hydrology and oceanography it has been used successfully in tasks where there is a high contrast between water and land and the sensitivity of the film to sub-surface water has been an advantage. For example, colour aerial photography has been used for monitoring overland flow patterns (Wallace 1973), mapping of flood boundaries (Rodda 1978); definition of coastlines (Specht *et al.* 1973), and estimation of water depth (Helgeson 1970).

Colour aerial photography has also been applied to archaeology (Strandberg 1967) and urban planning (Mumbower and Donoghue 1967; Ford 1979).

For further recent examples of the application of colour aerial photography refer to recent issues of the major remote sensing journals and symposia (Appendix B).

## 3.3.6 False colour near infrared film

Colour and false colour near infrared film have a similar structure, which is built around three emulsion layers, each of which are sensitive to a particular wavelength of electromagnetic radiation. You will recall from section 3.3.5, that colour film records blue radiation in a

**Fig. 3.18**
The sensitivity of the three emulsion layers in a false colour near infrared film.

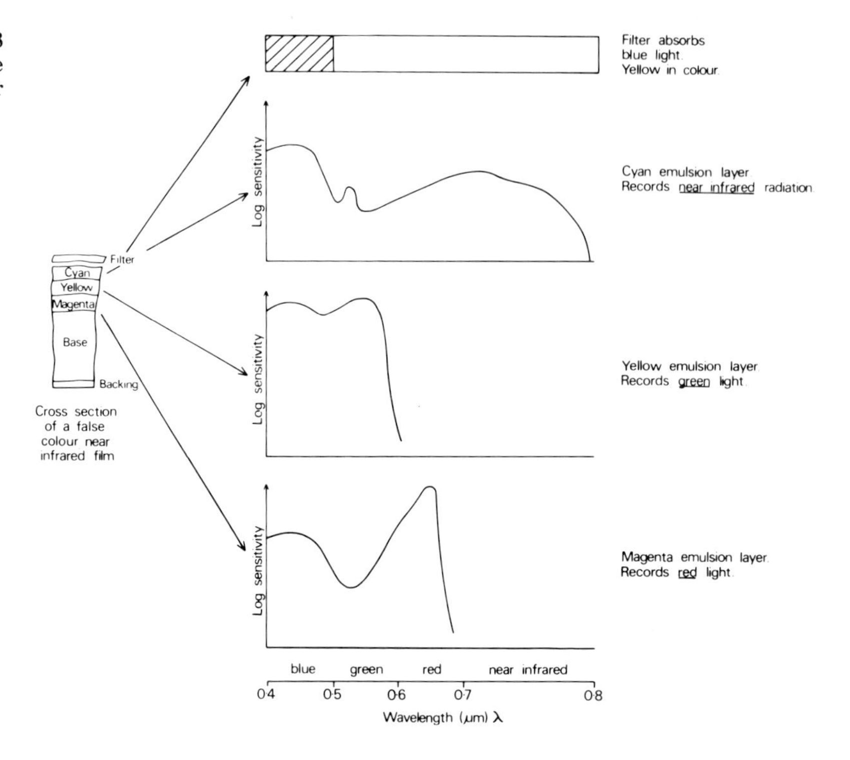


yellow emulsion layer, green radiation in a magenta emulsion layer and red radiation in a cyan emulsion layer. A false colour near infrared film has a different relationship between radiation sensitivity and emulsion colour as it uses its yellow emulsion layer to record green radiation, its magenta emulsion layer to record red radiation and its cyan emulsion layer to record near infrared radiation, although as with colour film there is overlap between the sensitivity of the emulsion layers (Fig. 3.18). No blue light manages to reach the film as a yellow filter which absorbs blue light is placed in front of the lens during exposure (Lillesand and Kiefer 1979). This particular linking of radiation to emulsion colour results in a 'colour shift' (Fig. 3.19). An object on the ground that reflects primarily *blue* radiation is not recorded and appears *black* on the film; an object that reflects primarily *green* radiation appears *blue* on the film; an object that reflects primarily *red* radiation appears *green* on the film and an object that reflects primarily *near infrared* radiation appears *red* on the film (Pease 1969; Curran 1978a; Slater 1980).

An aerial photograph of the Earth's surface will record the radiance of objects that do not have such simple reflectance characteristics. In order to interpret this imagery it is necessary to refer to the reflectance properties of some common scene components like vegetation and soil (Figs. 2.11 and 2.20). Vegetation appears green on colour film as it reflects more green than blue or red radiation, and red on false colour near infrared film as it reflects more near infrared than green or red radiation.

Soil appears green-red on colour film as it reflects more green-red radiation than blue radiation, and blue-green on false colour film because of the effect of colour shift (Plate 1).

**Fig. 3.19**
The stages involved in processing a false colour near infrared film. (Modified from Slater 1980)

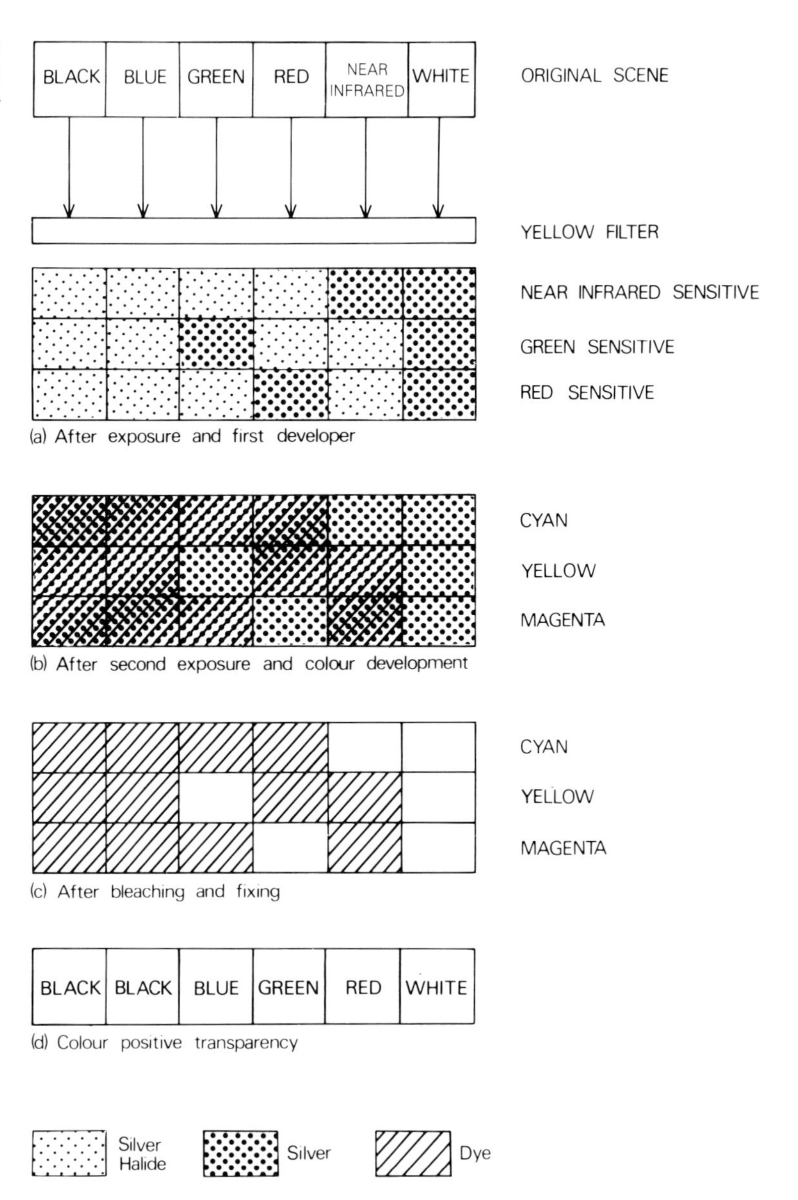


The two characteristics of this film that cause the most confusion are its spectral sensitivity and its colour balance (Curran 1981d). The spectral sensitivity only extends into the near infrared, and as the sensitivity of false colour near infrared film does not extend to thermal infrared wavelengths, it is not sensitive to heat.

The colour balance was originally developed for camouflage detection during World War II, so that green tanks in green woods could be recorded as blue tanks in red woods. As the colours were carefully balanced to achieve this effect from low flying aircraft, aerial photographs taken from either very high or very low altitudes will have a disturbed colour balance that can only be corrected by the careful use of filters (Pease and Bowden 1969).

### 3.3.6.1 Applications of false colour near infrared aerial photography

False colour near infrared film is widely used in a variety of applications which include classification of urban areas (Plate 1), (Lindgren 1971), monitoring of soil moisture (Curran 1979b), mapping of soil (Frazee *et al.* 1973), environmental disaster assessment (Garofalo and Wobber 1974), mapping of flooded land (Rodda 1978) and the census of animals (Heintz *et al.* 1979).

**Fig. 3.20**
Oblique, false colour near infrared aerial photograph, here reproduced in black and white, of an agricultural research station at Long Ashton, in Avon, UK. The photograph, which was taken looking west on 30.6.1975, when the solar elevation was low, emphasises the difference in surface texture between the types of land cover. (Courtesy, Dr T.H. Lee Williams, University of Kansas, USA)

The film is most applicable to studies where vegetation is involved, either in agricultural monitoring (Henneberry *et al.* 1979), (Fig. 3.20), environmental monitoring (Nichol and Collins 1980) or the detection of tree stress in forest (Ashley *et al.* 1978; Talerico *et al.* 1978) and urban environments (Fuhrer *et al.* 1981), (Plate 1). It has also been used for the location of anthropogenically degraded landscapes (Handley 1980), the detection of geobotanical anomalies (Cole *et al.* 1974), the mapping of crop marks on sites of archaeological interest (Curran 1980e) and the mapping of vegetation associations (Hagen and Meyer 1979, (Fig. 3.21). In recent years this kind of film has been applied to mapping the relatively inaccessible wetlands and intertidal zones with great success. (Gammon and Carter 1979; Steward *et al.* 1980; Wallace 1981).

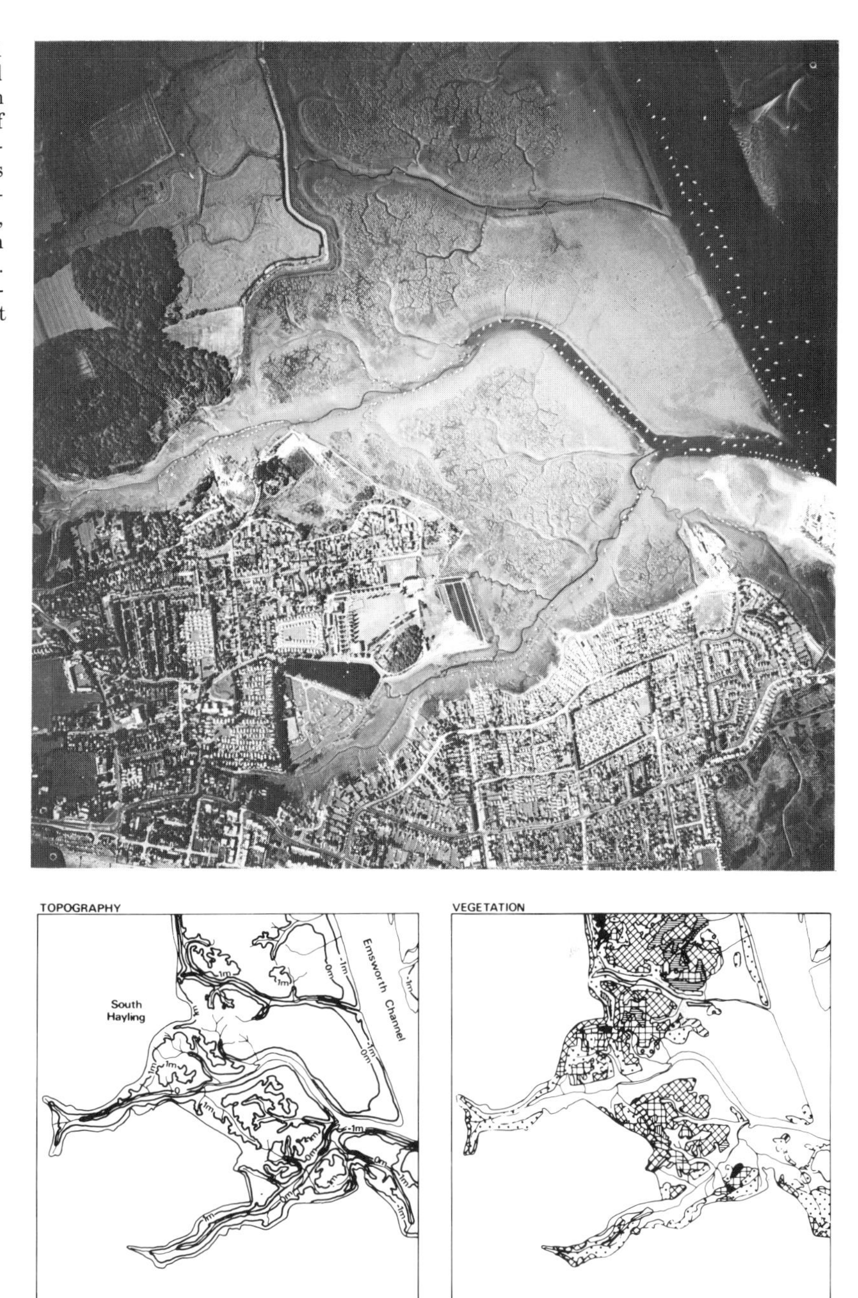


**Fig. 3.21**
False colour near infrared aerial photograph, here reproduced in black and white, of part of Chichester Harbour in Hampshire, UK. This photograph was used in the production of topographic and vegetation maps, extracts of which are shown (Budd and Coulson 1980). (Courtesy, Cambridge University Collection, UK; copyright reserved)

For further recent examples of the application of false colour near infrared aerial photography refer to recent issues of the major remote sensing journals and symposia (Appendix B).

## 3.4 Taking measurements from aerial photographs

To take accurate measurements from an aerial photograph requires knowledge of the geometric properties of the aerial photograph and the properties of viewing and measuring equipment (Estes *et al.* 1983).

### 3.4.1 Geometric properties of aerial photographs

The two most important geometric properties of an aerial photograph are those of angle and scale.

#### 3.4.1.1 Angle of aerial photographs

The angle at which an aerial photograph is taken is used to classify the photograph into one of three types, vertical, high oblique or low oblique (Fig. 3.22). Vertical photography is taken with the camera axis pointing vertically downwards and oblique photography is taken with the camera axis pointing obliquely downwards. High oblique photography incorporates the horizon into the photograph while low oblique photography does not.

Vertical aerial photographs are the most widely used aerial photographs as they have properties that are similar to those of a map with an approximately constant scale over the whole photograph and as a result can readily be used for mapping and measurement. Oblique photographs also have their advantages as they cover many times the area of a vertical aerial photograph, taken from the same height using

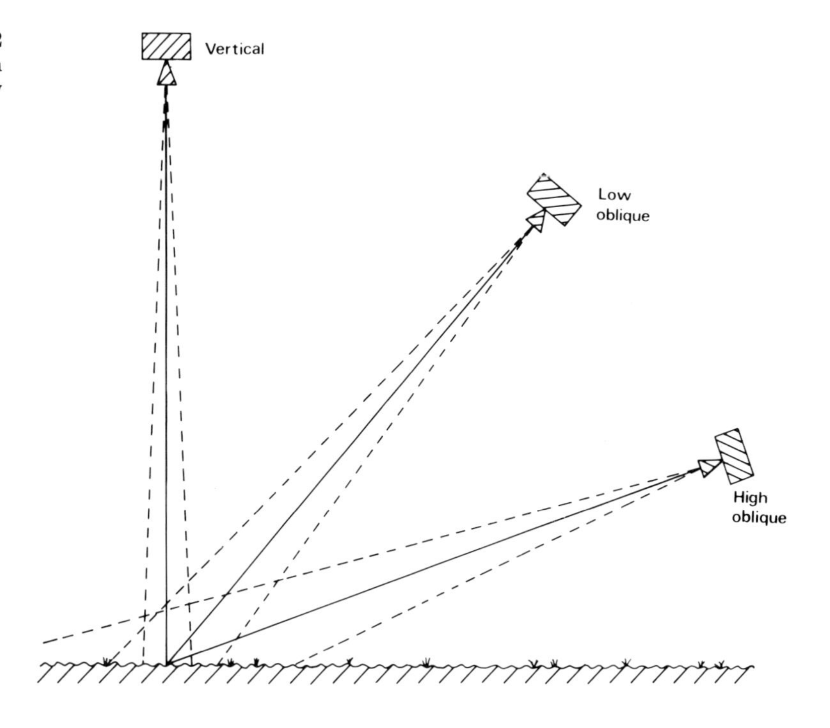


**Fig. 3.22**
The angular difference between a vertical, high oblique and low oblique aerial photographs.

**Fig. 3.23**
Flying pattern used to obtain stereoscopic aerial photography. There is 60% overlap between aerial photographs in a run and 30% overlap between aerial photographs in adjacent runs. Circles mark the photocentres which are located directly beneath the aircraft.

the same focal length lens and in addition present a view that is more natural to the interpreter (Paine 1981).

Vertical aerial photographs are usually taken in sequences along the aircraft's line of flight (Fig. 3.23) in such a way that overlap (called forward lap) is of the order of 60%. This is not for reasons of caution but to enable adjacent prints on a run to be viewed three dimensionally or stereoscopically (sect. 3.4.4.). To ensure that all of an area is covered the aerial photographs also overlap sideways (called sidelap) by around 30%. This means that each point on the ground appears at least two times in a block of aerial photographs.

### 3.4.1.2 Scale of aerial photographs

The scale of an aerial photograph determines its value for particular applications. A small scale aerial photograph (e.g. 1 : 50,000) will provide a synoptic, low spatial resolution overview of a large area and may be of value in structural geology or regional mapping (Fig. 3.8). A large scale aerial photograph (e.g. 1 : 2,000) will provide a detailed and high spatial resolution view of a small area (Murtha 1983) which may be of value in vegetation studies or urban planning, and according to Aldrich (1979) is essential if you wish to map animal droppings! A series of aerial photographs showing the relationships between scale, photograph area and spatial resolution for large, small and intermediate scale aerial photographs are given in Fig. 3.24.

The scale (S) of a photograph is determined by the focal length of the camera and the vertical height of the lens above the ground. The focal length (f) of the camera, is the distance measured from the centre of the camera lens to the film. The vertical height of the lens above the ground (H – h) is the height of the lens above sea level (H), minus the

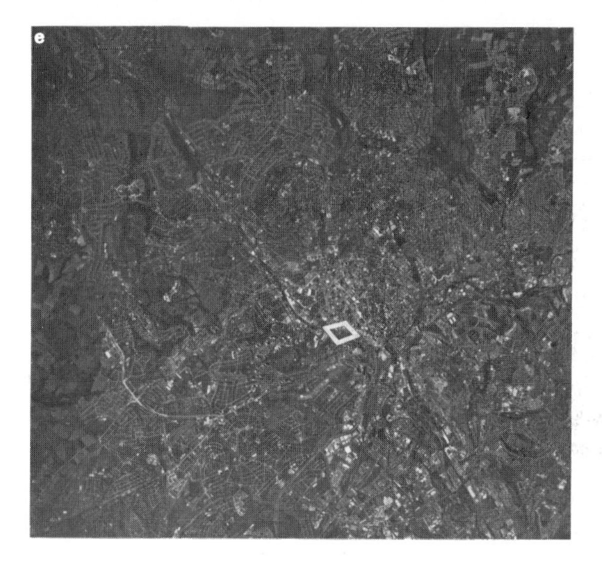

**Fig. 3.24**
The effect of image scale on the area and spatial resolution of aerial photographs is illustrated by means of five aerial photographs of Sheffield, UK. The properties of these aerial photographs prior to printing are given below.

| | Date | Scale | Area ($km^2$) | Approximate spatial resolution (cm) |
|---|---|---|---|---|
| a | 2.7.1969 | : 3,000 | 0.5 | 8 |
| b | 29.10.1969 | : 8,000 | 3.5 | 15 |
| c | 16.8.1966 | : 11,000 | 6.2 | 20 |
| d | 3.7.1971 | : 22,000 | 24.5 | 60 |
| e | 10.4.1981 | : 55,000 | 156 | 150 |

To illustrate further the effect of scale on area and spatial resolution an area round Sheffield railway station has been outlined on each of the aerial photographs. (Courtesy, (a) and (c) Meridian Airmaps Ltd, UK; (b) and (d) Cambridge University Collection, UK; copyright reserved), and (e) British Crown copyright reserved)

**Fig. 3.25**
The measurements required for the calculation of aerial photographic scale.

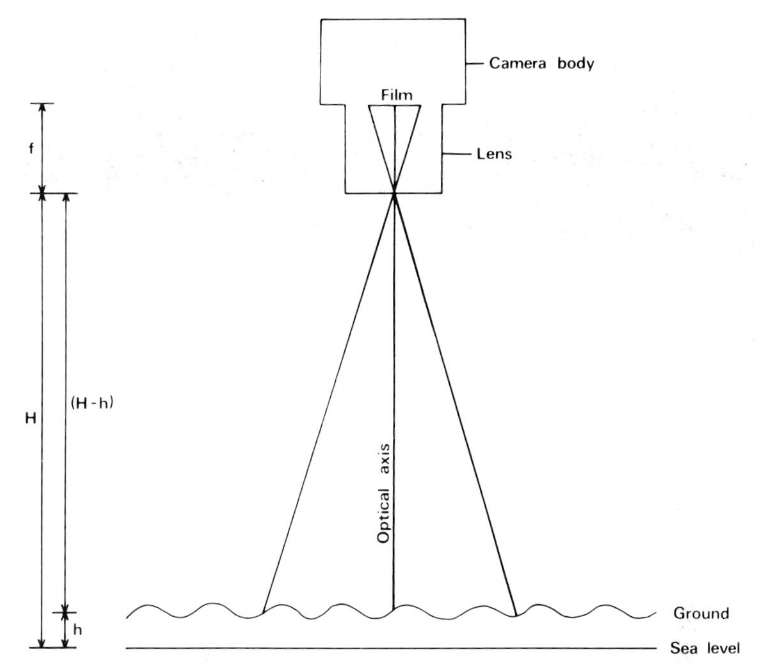


height of the ground above sea level (h), when the optical axis is vertical and the ground is flat (Fig. 3.25). These are related by formula [3.6].

$$S = \frac{f}{H - h} \qquad [3.6]$$

For example if a photograph were taken with a standard 150 mm (6-inch) focal length lens, from an altitude of 2,000 m (6,600 feet), of terrain that was 500 m (1,600 feet) above sea level, then the scale would be:

$$\frac{0.15 \text{ m}}{2000 - 500 \text{ m}} = \frac{1}{10{,}000} \quad \text{or} \quad 1:10{,}000$$

You may find it easier to calculate scale using Imperial as opposed to SI units, as when using a standard 6-inch lens you simply multiply (H-h) by a factor of 2 in this case (6,600-1,600) × 2 = 10,000 as

$$\frac{0.5 \text{ foot} \qquad (\times 2)}{6{,}600 - 1{,}600 \text{ feet}(\times 2)} = \frac{1}{10{,}000} \quad \text{or} \quad 1:10{,}000$$

Under certain circumstances, for example when f, H and h are not known, or the scale for part of an oblique image is required, then photographic scale can be determined by measurement. The ground distance (gd) is measured between two points on a map and this is divided into the photographic distance (pd) which is measured between the same two points on the aerial photograph as in formula [3.7].

$$S = \frac{pd}{gd} \qquad [3.7]$$

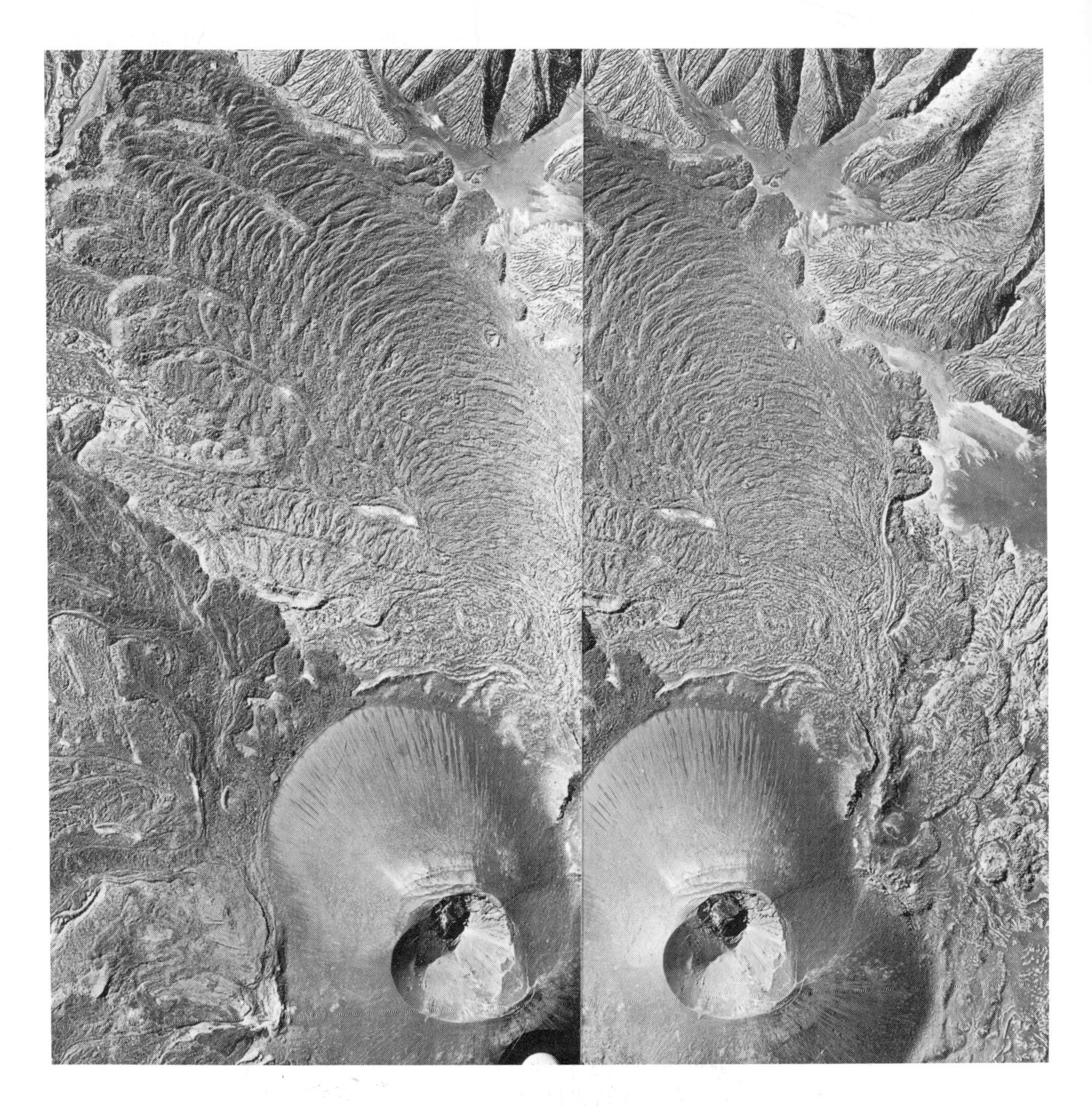

**Fig. 3.26**
The value of three dimensions can be seen in this stereopair of black and white aerial photographs of Paracutin volcano in Mexico. Without the aid of a stereoscope it is very difficult to determine the age of the magma flows but with a stereoscope the relationship between magma age and height are obvious. (Courtesy, Wild Heerbrugg Ltd, UK)

For example, if the distance between two road junctions is 500 m on the ground and 10 cm on the aerial photograph then the photographic scale would be:

$$S = \frac{0.1}{500} = \frac{1}{5,000} \quad \text{or} \quad 1:5,000$$

## 3.4.2 Viewing aerial photographs stereoscopically

One of the advantages of all aerial photographs is that when taken as overlapping pairs they can provide a three dimensional view of the Earth's surface (Fig. 3.26). This three dimensional view is made possible by the effect of parallax. Parallax is common to most of us, as our left and right eyes are continually recording information about objects from two slightly different viewpoints; the brain uses the effect of parallax to give us the perception of depth. We can simulate the view our eyes would see if we were looking out of an aircraft by viewing the left hand photograph of a pair of aerial photographs with the left eye and the right hand photograph with the right eye. With adequate control of our eye muscles it is possible to view a pair of prints stereoscopically without the aid of any equipment but as the effort is considerable, optical devices called stereoscopes are normally used. There are many stereoscopes on the market, the three most popular being the pocket stereoscope, the mirror stereoscope and the scanning stereoscope (Fig. 3.27). The pocket stereoscope is simple and small. It has folding legs, eyebrow pinching adjusters, low magnification and an eyebase that is so small that one of the aerial photographs often has to be bent upwards. The mirror stereoscope is the most widely used stereoscope because it enables the aerial photographs to be seen, without bending them, over a range of magnifications. The scanning stereoscope is similar to the mirror stereoscope but in addition it allows the operator to roam around the stereomodel at low or high magnification without the need to move the aerial photographs. An 'Interpreterscope' is a very sophisticated scanning stereoscope and aerial photographs, whether transparencies or prints, can be simultaneously scanned by two operators using a continuously variable magnification (Fig. 3.27).

### 3.4.3 Equipment required for taking measurements from aerial photographs

Besides interpreting the features shown on aerial photographs environmental scientists often wish to take measurements of distance, area and height or to map boundaries or topographic contours. This section will focus on the principles behind the use of readily available instruments (Fig. 3.28) but will not attempt to outline what the reader will need to know in order to use these instruments. To give incomplete instruction would be like telling you that you can play the flute by blowing through one end and moving your fingers up and down!

### 3.4.4 Measuring distance from aerial photographs

The most simple device for measuring distance from aerial photographs is the interpreter's scale. It is made of transparent plastic and has divisions in both white and black for measuring distances on both dark and light toned areas. For more accurate measurements a fine graticule can be placed in the eyepiece of a stereoscope. Both of these devices are used with the assumption that the ground surface is flat but when this is not the case, these instruments can only be used for very approximate measurements of distance.

### 3.4.5 Measuring area from aerial photographs

Four methods are commonly used to measure area on aerial photographs. These are a transparent overlay, a polar planimeter, a table

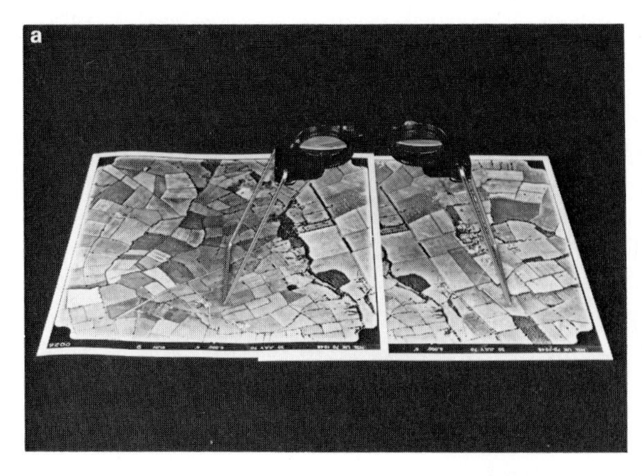

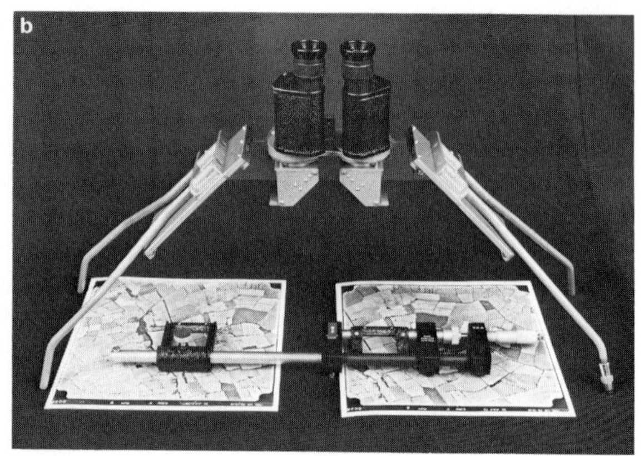

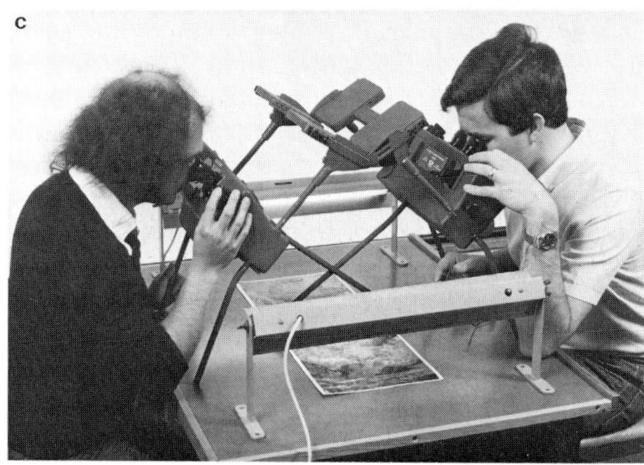

**Fig. 3.27**
Four types of stereoscope. (a) A pocket stereoscope, (b) a mirror stereoscope, with a binocular attachment for increased magnification and a parallax bar for height measurement; (c) two scanning stereoscopes made by Old Delft, The Netherlands and (d) a sophisticated scanning stereoscope the 'Interpreterscope' made by Carl Zeiss, West Germany.

digitiser and an analogue image processor.

The transparent overlay is a fine grid printed onto an acetate sheet; to obtain the area of a region, the number of grid squares covering the region are simply counted up. This method is neither quick nor accurate, but it is cheap.

A polar planimeter is a mechanical device that consists of a fixed arm, and a wheel with a measuring device. The wheel is run around the circumference of the region of interest in a clockwise direction. The circumference of the region can be converted to the area of the region using a look-up table provided by the manufacturer. This is cheap and fiddly, however, with practice, accurate measurements of area can be obtained.

A table digitiser is a quick, expensive and fairly accurate method of measuring area. The operator feeds the co-ordinates of the corners or boundary of the region of interest into a micro-computer (Fig. 3.29) which is programmed to calculate area.

An analogue image processor is by far the most convenient way to measure areas that have a distinct photographic tone. The image is usually fed into the machine with the aid of a TV camera and a tonal range is set that discriminates a region of interest. The total area is then read directly from the visual display unit which is part of the instrument's console (Fig. 6.3). Unfortunately the variability of aerial photographic tones both within and between similar objects often reduces the accuracy of this technique.

**Fig. 3.28**
Key to the equipment commonly used for the interpretation of aerial photographs.

| Equipment \ Application | Aid to interpretation | Stereoscopic viewing | Measuring area | Measuring height | Measuring radiance | Transfer of detail to base maps | Accurate plotting of maps |
|---|---|---|---|---|---|---|---|
| Pocket stereoscope with parallax bar | ■ | ■ | □ | ■ | □ | □ | □ |
| Polar planimeter | □ | □ | ■ | □ | □ | □ | □ |
| Monoscopic transferscope | ■ | □ | □ | □ | □ | ■ | □ |
| Mirror stereoscope with parallax bar | ■ | ■ | □ | ■ | □ | □ | □ |
| Densitometer | ■ | □ | □ | □ | ■ | □ | □ |
| Multiadditive viewer | ■ | □ | □ | □ | □ | □ | □ |
| Stereosketch | ■ | ■ | □ | □ | □ | ■ | □ |
| Table enlarger | ■ | □ | □ | □ | □ | ■ | □ |
| Radial line plotter | ■ | ■ | ■ | ■ | □ | ■ | ■ |
| Monoscopic zoom transferscope | ■ | □ | □ | □ | □ | ■ | □ |
| Analogue image processor | ■ | □ | ■ | □ | ■ | □ | □ |
| Table digitiser | □ | □ | ■ | □ | □ | □ | □ |
| Scanning stereoscope with parallax bar | ■ | ■ | □ | ■ | □ | □ | □ |
| Stereoplotter | ■ | ■ | ■ | ■ | □ | ■ | ■ |

Key ■ Equipment useful for particular application □ Equipment not useful for particular application

With all four techniques there is an implicit assumption that the ground surface is flat; when this is not the case then a radial line plotter or stereoplotter should be used (sect. 3.4.7)

## 3.4.6 Measuring height from aerial photographs

To measure height the interpreter uses the phenomena of parallax. You will recall from section 3.4.4 that this is where a tall object, for example a tree top, moves more between successive aerial photographs than does a short object, for example a tree base. If the amount of movement between the tree base and the tree top is measured on successive aerial photographs then it is possible to estimate the height of the tree (Fig. 3.30). If the displacement is large, it can simply be measured with a ruler but if the displacement is small, a parallax bar is used to measure it (Fig. 3.27). A parallax bar consists of two glass plates separated by an adjustable metal bar. On each plate there is a small spot or cross and this is termed a 'half-mark'. While looking down the mirror stereoscope the half-marks are positioned on the top of the same object in each photograph. The eyes fuse the two half-marks together and give the impression of one mark which is termed the 'floating mark'. The distance between the two half-marks is measured using a micrometer, which is positioned on the end of the adjustable metal bar. The process is repeated as the half-marks are located

**Fig. 3.29**
A table digitiser, made by Summagraphics, UK is being used to measure and record areas on a remotely sensed image.

exactly on the bottom of the object and appear to fuse together. Again the distance between the two half-marks is read from the micrometer screw. The difference between the distance measured from one half-mark to the other when they were on the top as opposed to the base of the object (Fig. 3.30), is a measurement of parallax displacement and is called ($\Delta p$). The degree of parallax displacement is positively related to the distance between the centre of the two photographs (Pa) and the height of the object of interest ($\Delta h$) and negatively related to the height of the photography above the surface of the ground (H-h). In order to calculate the height of the object of interest ($\Delta h$) the parallax formula is used (formula [3.8]) as this allows for the effect of both the distance between the photographs (Pa) and the height of the photography above the ground (H-h).

$$\Delta h = \frac{\Delta p \times (H - h)}{Pa + \Delta p} \qquad [3.8]$$

where $\Delta h$ = height of object in metres
$\Delta p$ = difference in distance between the top and the bottom of the feature on the two photographs in millimetres

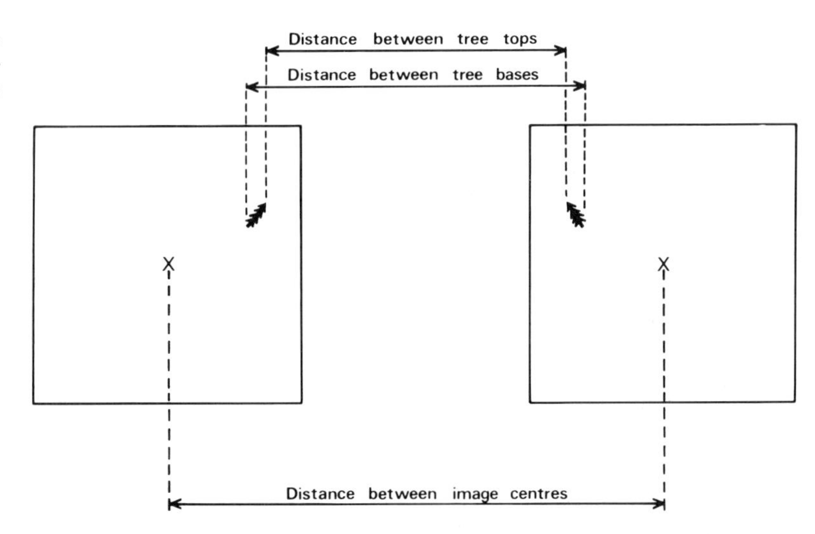


**Fig. 3.30**
The measurements required for the calculation of an object's height on aerial photographs.

Pa = distance between the central (or principal) points of the photographs minus the distance between the feature on the two photographs, in millimetres; this distance is measured by simply using a ruler

(H−h) = aircraft flying height above the surface of the ground in metres.

For example, to measure the height of a tree from 1 : 3,000 scale aerial photographs taken from a height of 500 m (Fig. 3.30).
Distance between top of the tree in the two photographs = 209 mm
Distance between base of the tree in the two photographs = 210 mm
$\Delta p$ 210 − 209 = 1 mm
Distance between the centres of the two photographs = 300 mm
Pa = 300 − 210 = 90 mm
Flying height of aircraft above the surface of the ground (H − h) = 500 m

$$\Delta h = \frac{1 \times 500}{90 + 1} = 5.5 \text{ m}$$

The tree is 5.5 m high.

Height measurements can also be used to derive measurements of angle and volume from aerial photographs. Angular measurements of features like river valleys and coal tips can be calculated once the height of two points has been determined and volume measurements of quarries, buildings and peat bogs can be calculated from measurements of height (or depth) and area.

## 3.4.7 Transfer of planimetric details from aerial photographs to maps

Once details have been interpreted on aerial photographs they can be traced onto a base map using a monoscopic transferscope, a monoscopic zoom transferscope, or a stereoscopic stereosketch or they can be

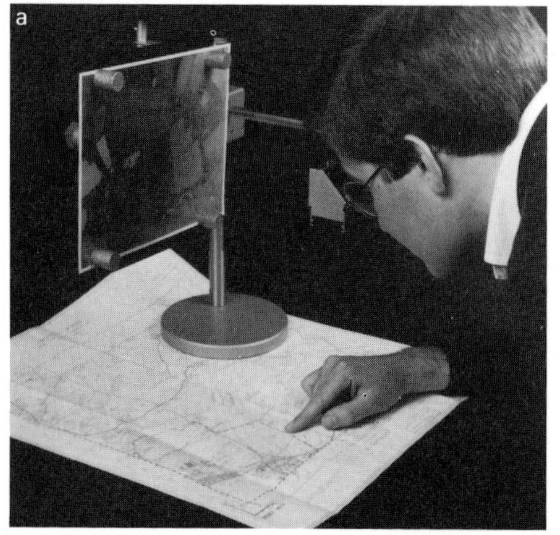

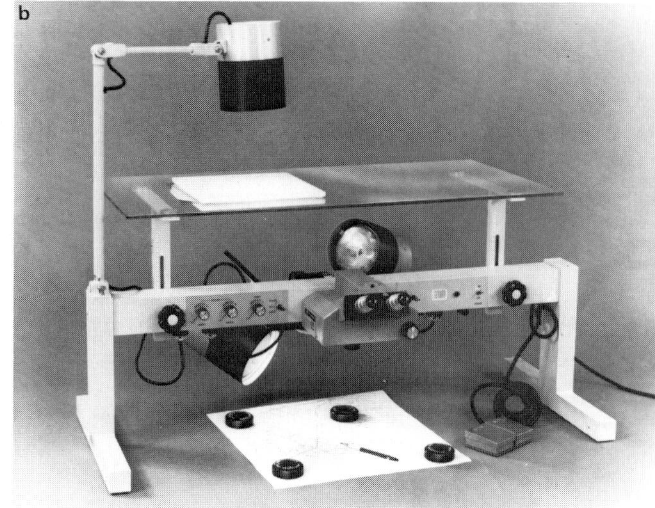

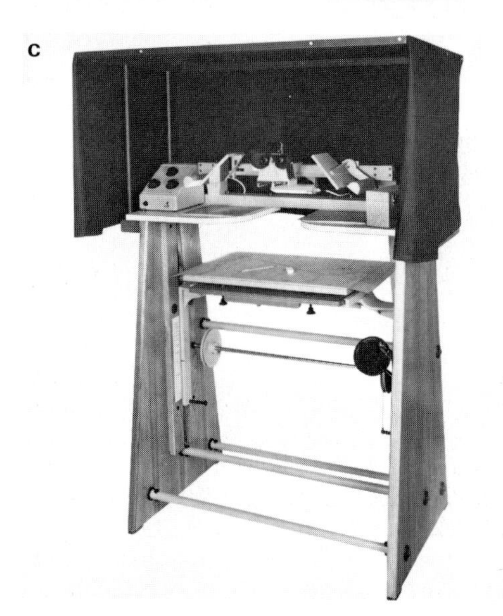

**Fig. 3.31**
Equipment for the transfer of planimetric details to maps: (a) A monoscopic transferscope, made by Carl Zeiss, Oberkochen Ltd, West Germany; (b) A monoscopic zoom transferscope, made by Bausch and Lomb Co. Ltd, USA. (Courtesy, Survey and General Instrument Co. Ltd, UK); (c) Stereosketch, made by Hilger Watts, UK. (Courtesy, Cartographic Engineering Ltd, UK); and (d) a stereoscopic radial line plotter made by Hilger Watts, UK.

plotted onto a base map using a stereoscopic radial line plotter (Fig. 3.31).

The monoscopic transferscope is a very simple instrument that enables an operator to view an aerial photograph and a map simultaneously by looking through a semi-transparent prism in the eyepiece. The aerial photograph and map can be registered at the same scale by careful manipulation of both the eyepiece to aerial photograph and eyepiece to map distance (Kilford 1975). The advantages of this equipment are that it is cheap, easy to use, can readily compensate for tilt and illumination and can be manipulated to emphasise either the aerial photograph or map. The disadvantages of this equipment are that relief displacement cannot be seen or compensated for, thus limiting its application to areas of relatively flat terrain and the use of only one eyepiece can cause eyestrain.

A zoom transferscope is also a non-stereoscopic device, it has a double eyepiece, a large range of continuously variable magnifications and a wide range of tilt adjustments (Fig. 3.31). It is very easy to use and has proved to be most suitable for the transfer of details from single aerial photographs and satellite sensor images to maps (Lo 1976).

A stereoscopic stereosketch enables a pair of stereoscopic aerial photographs and a map to be viewed at the same time. The scale of the map can be matched to the scale of the aerial photographs by altering the height of the map table. The advantages of this equipment are the ease with which tilt and relief displacement on the aerial photograph can be suppressed and illumination can be manipulated to emphasise either the aerial photograph or map (Fig. 3.31).

A stereoscopic radial line plotter enables the radial displacement of objects from the image centre to be corrected as they are plotted onto a base map. This instrument is therefore more accurate than the tracing instruments discussed thus far. The instrument is easy to use as the operator simply maintains the intersection of two cursors over the top of the feature of interest and this point of intersection is recorded by a pencil on a base map. The scale difference between the map and the aerial photograph must be small and the instrument accuracy is modest and varies over the stereomodel, nevertheless the stereoscopic radial line plotter remains the cheapest means of plotting details onto base maps (Spurr 1960).

### 3.4.8 Accurate plotting of topographic and other details from aerial photographs

The measurement of approximate distance, area and height on aerial photographs and the transfer of planimetric detail from aerial photographs to maps are skills readily practised by environmental scientists. However, measuring aerial photographs with sufficient accuracy in two dimensions to enable features to be plotted with respect to a national or international grid co-ordinate system and with sufficient accuracy in three dimensions to enable them to be located in relation to their height above sea level, is the task of the photogrammetist. As this book discusses techniques, rather than the products of techniques there will only be a brief introduction to photogrammetry by means of a review of the main piece of photogrammetric instrumentation: the stereoplotter (ASP 1981).

There are four types of stereoplotter and all are designed to produce topographic contour maps from aerial photographs. They are the optical stereoplotter, the mechanical stereoplotter, the optical-mechanical stereoplotter and the analytical stereoplotter (Fig. 3.32). The most simple stereoplotter is the optical stereoplotter where the aerial photographs are projected onto a small mobile table which is viewed by the operator who traces details from the perceived three dimensional stereomodel onto a base map.

However, the mechanical and optical-mechanical stereoplotters are more accurate instruments. With these the operator views the photograph directly and any details or height measurements are transferred mechanically onto a plotting table (Wolf 1974).

The most accurate instrument is the analytical plotter. The co-ordinate measurements of all points on the stereomodel are fed into it

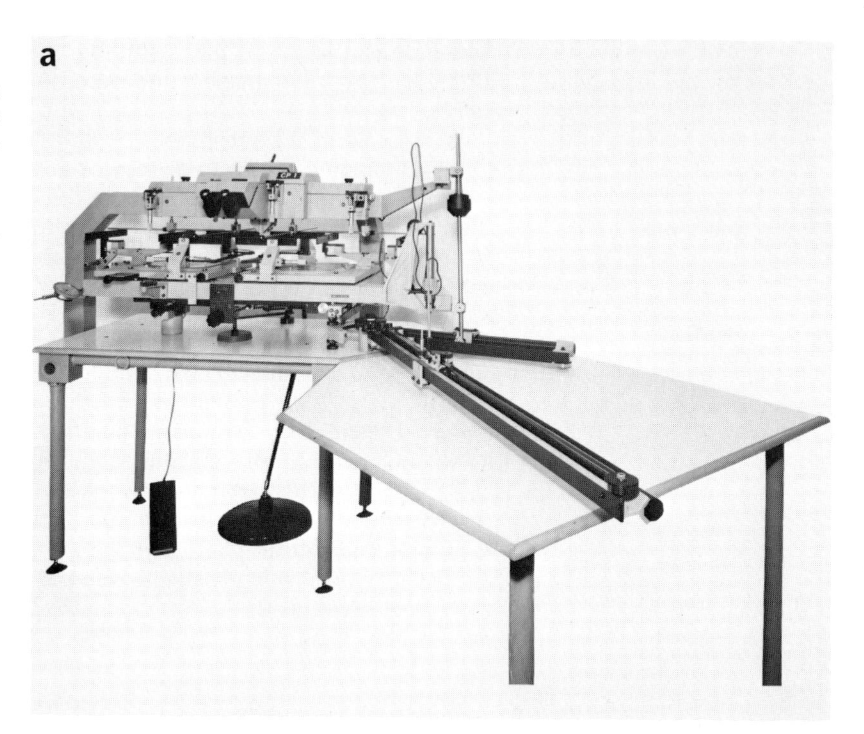

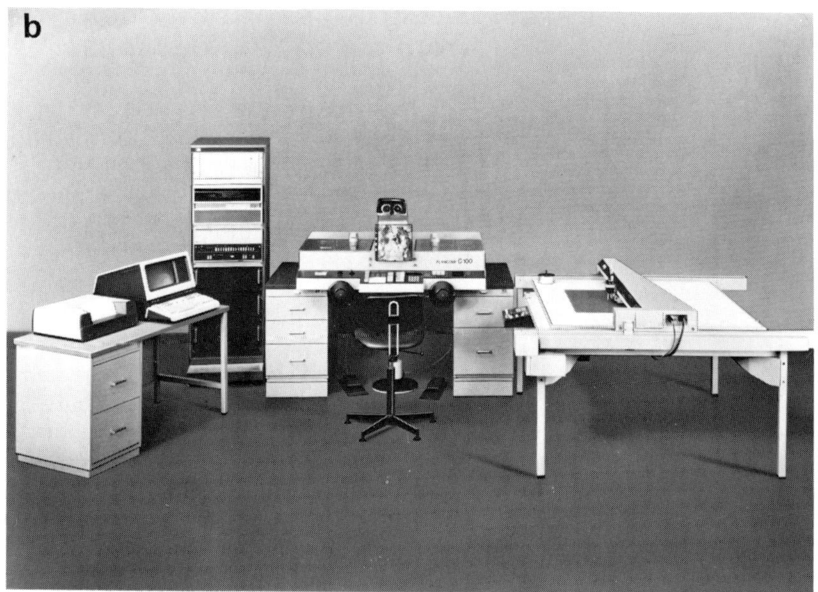

**Fig. 3.32**
Two types of stereoplotter: (a) An optical/mechanical stereoplotter, the CPI cartographic plotter. (Courtesy, Cartographic Engineering Ltd, UK); (b) An analytical stereoplotter, the Carl Zeiss C-100 Planicomp, controlled by a Hewlett Packard HP 1000 minicomputor. (Courtesy, Carl Zeiss, Oberkochen Ltd, West Germany)

to produce a mathematical model of the terrain. This mathematical model is more flexible than the stereomodel as it can be automatically corrected for tilt and other geometric distortions and then interrogated to yield information on any aspect of the stereomodel (Wolf 1974).

There are many makes and designs of the more simple optical stereoplotters but all have three common features. First, a projection system to produce a stereomodel; second, an observation system to allow the operator to view the stereomodel and third, a measurement and transfer system to enable the operator to measure features and then transfer this detail onto a map. The projection system passes light through positive transparencies of the stereoscopic pair of aerial photographs. The transparencies can be moved which gives the operator flexibility to correct the stereomodel for errors brought about by changes in aircraft tilt, height and speed between overlapping pairs of aerial photographs. To see the resultant overlapping aerial photographs as a stereomodel an anaglyph viewing and projection system is used on the simpler optical stereoplotters. In the anaglyph system, green light is projected through a transparency of the left hand aerial photograph and red light is projected through a transparency of the right hand aerial photograph. The resulting green-red overlap is viewed through a rather odd looking pair of spectacles in which the left eye has a red lens and the right eye has a green lens. Through these the operator sees only the left hand image with the left eye and the right hand image with the right eye and these are mentally combined to produce the stereomodel (Fig. 3.33).

In many of the modern optical stereoplotters an image alternator or polarised plate system is used in place of the anaglyph system. In the image alternator system the projector and the viewing optics flicker in synchronisation for each eye so that the left eye only sees the left hand image and the right eye only sees the right hand image (Kilford 1975). In the polarised plate system, polarised filters are used in place of the green and red waveband filters. Polarising filters are similar to the lenses used in polarised sunglasses and these only allow radiation to be transmitted in one vibrational plane and not at right angles to that plane. This produces a clear stereomodel but as with the anaglyph and image alternator systems there is a loss of illumination.

The optical stereoplotter is unnecessary for the majority of interpretive tasks but for the transfer of accurate details from aerial photographs of rugged terrain to maps or for the production of contour maps from aerial photographs it is essential.

## 3.5 Interpreting aerial photographs

Aerial photographic interpretation is formally defined as the act of examining photographic images for the purpose of identifying objects and judging their significance (Colwell 1960). During the process of interpretation, aerial photographic interpreters usually undertake at least some of the seven tasks of detection, recognition and identification, analysis, deduction, classification, idealisation and accuracy determination.

Detection involves selectively picking out objects that are directly visible, for example rock faces, or areas that are indirectly visible, for example areas of wet soil, on the aerial photographs. Recognition and identification involve naming objects or areas and analysis involves trying to detect the spatial order of the objects or areas. Deduction is rather more complex and involves the principle of convergence of evidence to predict the occurrence of certain relationships on the aerial photograph. Classification comes in to arrange the objects and elements identified into an orderly system before the photographic interpretation is idealised using lines which are drawn to summarise the spatial distribution of objects or areas. The final stage is accuracy

**Fig. 3.33**
A diagrammatic view of the main components of the anaglyph viewing system used in an optical stereoplotter. The operator can see the left aerial photograph with the left eye and the right aerial photograph with the right eye and is able to fuse them into one three-dimensional image called the stereo model.

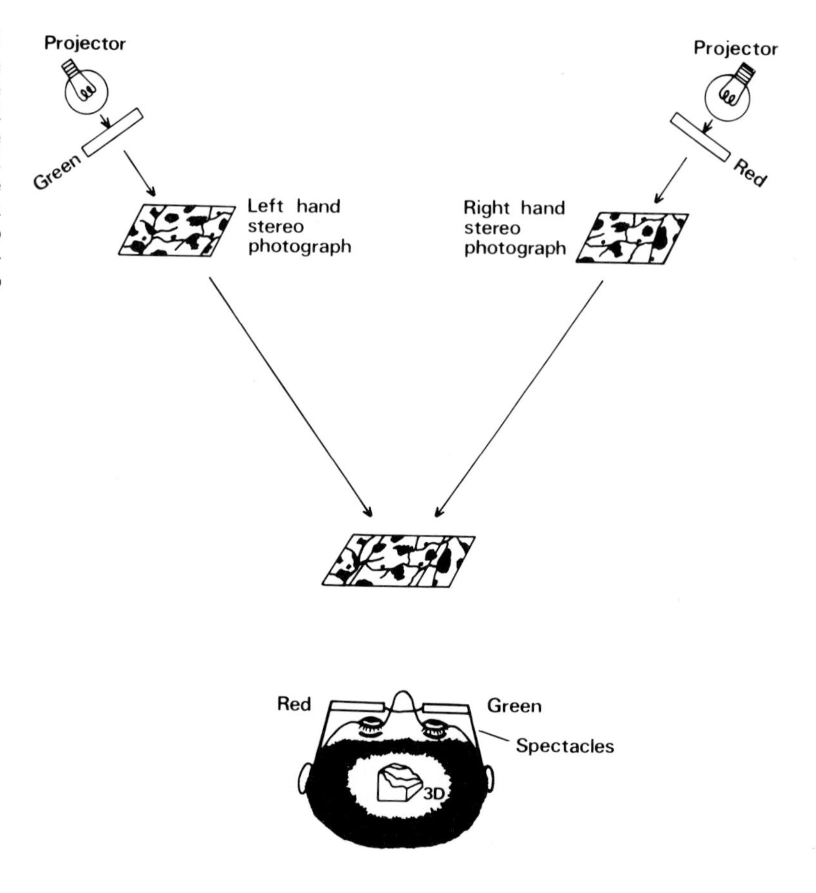


determination where random points are visited in the field to confirm or refute the interpretation (sect. 6.3.7.16).

Recognition and identification of objects or areas is the most important link in this chain of events (Walker 1964; Stanley 1982). An interpreter uses seven characteristics of the aerial photograph to help with this stage and these are tone, texture, pattern, place, shape, shadow and size.

**1 Tone** is the single most important characteristic of the aerial photograph. It represents a record of the radiation that has been reflected from the Earth's surface onto the photographic film. Light tones represent areas with a high radiance and dark tones represent areas with a low radiance. The nature of the materials on the Earth's surface affects the amount of light reflected, for example, in Fig. 3.9 the areas of peaty soil are dark grey in tone and the areas of calcareous soil are light grey in tone.

**2 Texture** is the frequency of tonal changes within an aerial photograph that arise when a number of features are viewed together (Fig. 3.8). This is dependent upon the scale of the photograph. For example, the grass embankment in Fig. 3.24 appears to have a rougher texture when photographed at lower altitudes.

**3 Pattern** is the spatial arrangement of objects. Some of the patterns illustrated by the aerial photographs in this book are road patterns (Fig. 6.2), drainage patterns (Figs 3.8 and 3.21), crop disease patterns (Fig. 3.11) and lithological patterns (Fig. 3.9).

**4 Place** is a statement of an object's position in relation to others in its vicinity and this usually aids in its identification. For example, the stream cannot be seen in Fig. 3.10 but its likely location along a line of trees gives a clue to its course.

**5 Shape** is a qualitative statement of the general form configuration or outline of an object. It is easier to determine the shape of an object if it is viewed stereoscopically. For example, the drainage channels in Fig. 3.26 are deeply incised 'V' shaped valleys when viewed stereoscopically.

**6 Shadows** of objects aid in their identification. For example shadows help emphasise the geological boundaries in Fig. 3.10 and would help in identifying the pylons in Plate 1 and naming the tree types in Fig. 6.19.

**7 Size** of an object is a function of photoscale. The sizes of objects can be estimated by comparing them with objects for which the size is known. Examples of the relationship between object size and photoscale can be seen in Fig. 3.24.

Fortunately there are very few areas of the world for which aerial photographic interpretation has to be undertaken only on the basis of these seven image characteristics as the interpreter will usually visit the field area and use collateral material. For example, in developed countries the aerial photographic interpreter will have access to at least topographic, geological, soil and land use maps and in developing countries topographic maps and reports are sometimes available (Estes and Simonett 1975).

### 3.5.1 Organisational aids to the interpretation of aerial photographs

The interpretation procedure outlined at the start of this section is adequate if you wish to study a relatively small area with which you are familiar. However, one of the important uses of remotely sensed imagery is in the mapping of large areas of the Earth's surface and as this usually requires several interpreters, the procedures need to be well organised to ensure that information is extracted in a consistent manner. Two methods for organising aerial photographic interpretation are the use of a fixed key or classification. Keys are widely used by environmental scientists to identify plants, animals, rocks and minerals and can also be used to help the interpreter identify features or conditions on an aerial photograph. Such a key comprises two parts, first, annotated stereopairs showing the major features to be identified and second, a written description of the photographic characteristics of the features to be identified. This information can be arranged in one of two ways, either as a selection key or as an elimination key

(Estes and Simonett 1975). A selection key consists of numerous examples with a text to which the interpreter tries to associate the feature under study beneath the stereoscope. In an elimination key the interpreter is lead through a series of choices such as light tone or dark tone and rough texture or smooth texture that gradually lead to the object being identified. Their disadvantage is that they can force an interpreter into making a decision on a characteristic about which there is uncertainty, but then again that's life!

The most reliable selection and elimination keys have traditionally been for urban areas within a limited area of the world. However, very good keys are now being produced which will span complete ecosystems (Philipson and Liang 1982).

Classifications are less specific than keys and are used to provide a framework within which interpreters can classify their aerial photographs (Howard 1970). One of the better known classifications is the land use and land cover classification prepared by Anderson *et al.* (1976). It is based upon four levels of classification, three of which are illustrated in Table 3.2. Level I is the most general and could be used with small scale aerial photographs or satellite sensor images, while level 4 is the most detailed and could be used with large scale aerial photographs. This particular classification which is designed around a number of easily measured criteria of accuracy and repeatability has the advantage of applicability in almost any environment.

### 3.5.2 Instrumental aids to the interpretation of aerial photographs

Most interpretation of aerial photographs is undertaken using a stereoscope (sect. 3.4.4) and wax based pencils. However, there are occasions when an interpreter may wish to enlarge, improve, quantify or simplify an aerial photograph and for this other aids are required. For enlargement a hand magnifier or table magnifier can be used on single photographs and the inbuilt devices on stereoscopes can be used on stereopairs (Fig. 3.27). If a permanent record is required then part of

**Table 3.2**
Part of a land use and land cover classification system, designed for use with remotely sensed data.

| *All of level I* | *An example from level II* | *An example from level III* |
|---|---|---|
| 1. Urban or built up land | 11. Residential | 111. Single family units |
| 2. Agricultural land | 12. Commercial and services | 112. Multifamily units |
| 3. Rangeland | 13. Industrial | 113. Group headquarters |
| 4. Forestland | 14. Transportation communication and utilities | 114. Residential hotels |
| 5. Water | 15. Industrial and commercial complexes | 115. Mobile home parks |
| 6. Wetland | 16. Mixed urban or built-up land | 116. Transient lodgings |
| 7. Barren land | 17. Other urban or built-up land | 117. Other |
| 8. Tundra | | |
| 9. Perennial snow or ice | | |

*Source*: Anderson *et al.* 1976.

a single image can be photographically enlarged and despite some loss of detail can give very good results (Fig. 6.2). To improve the appearance of an image an interpreter may wish to alter the contrast photographically, as is discussed in section 6.2. To observe more detail on the aerial photograph than is visible to the naked eye a densitometer is required to measure image tone (sect 3.3.1). A microdensitometer can be used to scan along photographic transects to record the tone for which field data are available or a macrodensitometer can be used to sample the tone of larger areas like fields where the surface conditions are known. Both of these techniques aid in the correlation between field data and image tone (Doverspike *et al.* 1965; Lillesand and Kiefer 1979).

To simplify the aerial photograph the number of tonal levels can be reduced by a process called density slicing (sect. 6.2) and this can be done either photographically using special films, or by a combination of optical and mechanical means using a scanning microdensitometer, or electronically using an analogue image processor (Genderen 1975).

## 3.6 A final thought on aerial photography

It is worth bearing in mind that aerial photography is the last bastion of 'do-it-yourself' remote sensing. For the cost of ten out-of-date, commercially flown panchromatic aerial photographs an aircraft and pilot can be hired for an hour, and you can take as many types of aerial photographs as you wish. The lower spatial resolution and non-vertical view offered by these hand-held aerial photographs is usually more than compensated for by their controllability (Curran 1981c).

## 3.7 Recommended reading

**A.S.P.** (1981) *Manual of Photogrammetry* (4th edn). American Society of Photogrammetry, Falls Church, Virginia.

**Lo, C. P.** (1976) *Geographical Applications of Aerial Photography*. David and Charles, London; Crane, Russak, New York.

**Paine, D. P.** (1981) *Aerial Photography and Image Interpretation for Resource Management*. Wiley, New York.

**Slater, P. N.** (1983) Photographic systems for remote sensing: *In* Colwell, R. N. (ed.) *Manual of Remote Sensing*, (2nd edn). American Society of Photogrammetry, Falls Church, Virginia, pp. 231–91.

**Wenderoth, S.** and **Yost, E.** (1975) *Multispectral Photography for Earth Resources*. Remote Sensing Information Centre, New York.

**Wolf, P. F.** (1974) *Elements of Photogrammetry*. McGraw Hill, New York.

# Chapter 4 Aerial sensor imagery

*'In 1942 . . . the new aid to our bombers . . . was a centimetric radar device which scanned the terrain for several miles around the aircraft and which presented the navigator with what was virtually a map of the ground, showing towns (which gave rise to large radar echoes) rivers, lakes and coastlines. This idea was taken up enthusiastically and developed by our Telecommunications Research Establishment, and noticeably by Philip Dee and Bernard Lovell, under the code name TF (Town Finding).'* (Jones 1978)

## 4.1 Introduction

Four types of aircraft-borne instrumentation are used to record images of the electromagnetic radiation that is reflected or emitted from the Earth's surface. These are multispectral scanners and thermal infrared linescanners, which record radiation at wavelengths shorter than 14 μm and sideways-looking airborne radars and passive microwave scanners, which record radiation at wavelengths longer than 5 mm (Fig. 4.1).

Passive microwave scanners have insufficient advantages to justify their high cost and so unlike the other three instruments, will not be

**Fig. 4.1**
The value of four types of aircraft-borne remote sensors to environmental scientists.

| | Waveband | Value to environmental scientists | | Cost |
|---|---|---|---|---|
| | | Actual | Potential | |
| Multispectral scanner | Visible to thermal infrared (0·30 – 14 μm) | ⧄ | ◪ | ■ |
| Thermal infrared linescanner | Thermal infrared (3 – 14 μm) | ◪ | ◪ | ◪ |
| Sideways-looking airborne radar | Microwave (5 – 500mm) | ◪ | ■ | ■ |
| Passive microwave scanner | Microwave (5 – 500mm) | □ | ⧄ | ■ |

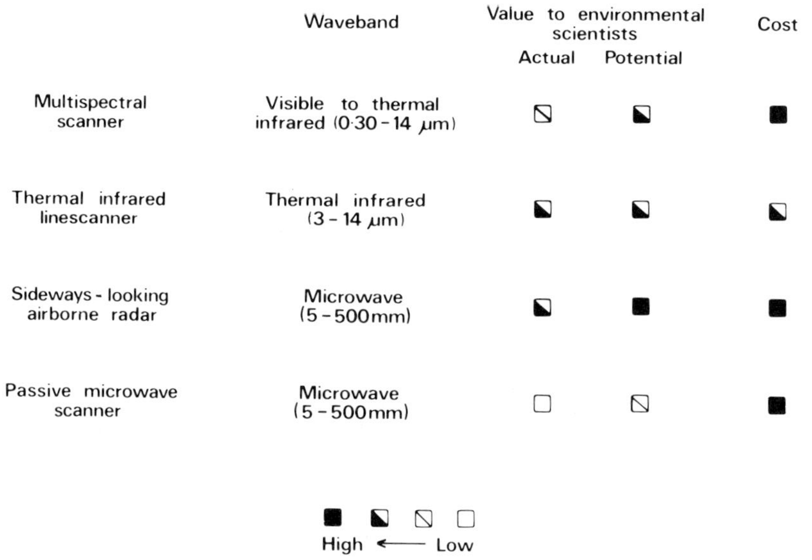

discussed. For those with an interest in this instrument refer to Schanda (1976) and Ulaby *et al.* (1981, 1982).

## 4.2 Multispectral scanner

Airborne multispectral scanners are a relatively unfamiliar method of obtaining images of the Earth's surface (Fitzgerald 1972). This may seem surprising when one considers the three advantages of mulit-spectral scanner data over aerial photography. First, they have a very high radiometric resolution in narrow and simultaneously recorded wavebands; second, these wavebands span a relatively large portion of the electromagnetic spectrum from ultra-violet wavelengths (0.3 $\mu$m) to thermal infrared wavelengths (14 $\mu$m) and third, these data can be stored in digital form for correction and quantitative analysis. However, these advantages are often swept aside by two disadvantages – those of limited sensor availability and high hire costs.

In recent years three factors have encouraged the use of multispectral scanner data. First, the increased availability of these data from government sponsored projects. Second, the industrial use of digital thermal infrared linescanner data for quantitative heat loss surveys and third, a rapid increase in the availability of equipment for processing digital multispectral scanner data (sect. 6.3).

### 4.2.1 How a multispectral scanner works

A multispectral scanner measures the radiance of the Earth's surface along a scan line, perpendicular to the line of aircraft flight. As the aircraft moves forward repeated measurement of radiance enables a two-dimensional image to be built up (Lowe 1976). These scanners have been available since the mid-1960s (Braithwaite and Lowe 1966), and have remained unchanged in their basic design, comprising a collecting section, a detecting section and a recording section (Fig. 4.2).

### 4.2.2 Collectors and detectors

1. A telescope directs the radiation onto the rotating mirror. This is used to control the instantaneous field of view (IFOV) of the sensor, which is the field of view at any one instant. The IFOV has an important effect on the resultant data, for at a given flying height the IFOV is positively related to the size of the ground sampling element but negatively related to the spatial resolution.

The IFOV can be used to calculate the spatial resolution of the image using formula [4.1].

$$D = H\beta \qquad [4.1]$$

$D$ = diameter of ground sampling element (metres)
$H$ = flying height of aircraft above ground (metres)
$\beta$ = IFOV (radians)

For example, for a multispectral scanner flying at a height (H of 1000 m, with an IFOV ($\beta$) of 2.0 milliradians, the diameter of the ground sampling element ($D$) will be 2 m.

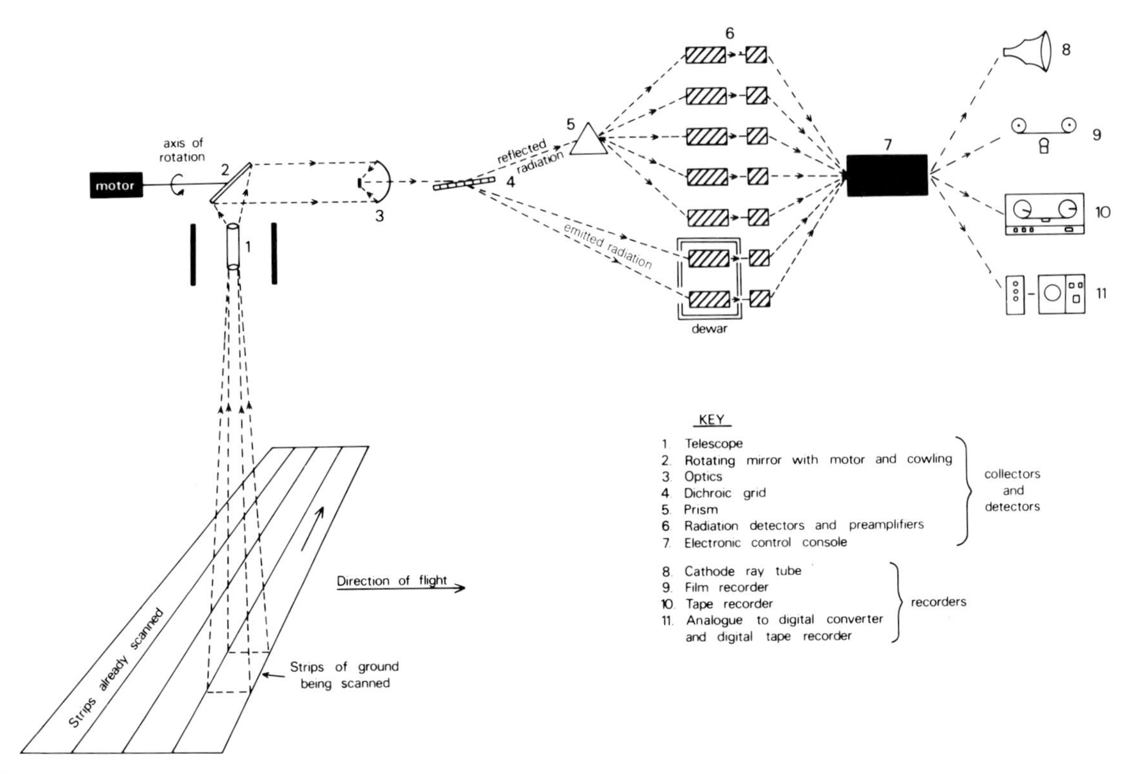


**Fig. 4.2**
A diagrammatic representation of a multispectral scanner. The radiation from the Earth's surface passes through a telescope (1), to be focused onto a rotating mirror (2), which reflects the radiation onto a set of optics (3), where it is passed to a dichroic grid (4), to be split into reflected and emitted radiation. The reflected radiation is divided into its spectral components using a prism (5) and is detected and amplified by the detectors and pre-amplifiers (6), while the emitted thermal radiation goes straight to the thermal recorders and pre-amplifiers (6). All of the information is in electronic form and is controlled by the electronic control console (7), into which can be plugged one or several recording devices.

2. The rotating mirror, which is powered by a small electric motor, reflects the radiation passing through the telescope into the optics. The mirror is set at an angle of 45 °, so that as it revolves it scans the terrain from side to side. The operator has control of the length and timing of these scan lines. The length of the scan line which is the field of view (FOV), is determined by the length of the cowling in front of the mirror, with a long cowling giving a small FOV. As it is usual to have a scan line that is around twice as long the flying height of the aircraft, the FOV is often set at 90 °. The timing of the scan line is positively related to the speed of the aircraft and negatively related to the band widths. The slowest scan is often set at around 1/10 of a second and the fastest scan is often set at around 1/100 of a second.
3. The optics focus the radiation into a narrow beam.
4. The dichroic doubly refracting grid splits the narrow beam of radiation into its reflected and emitted components. The reflected radiation is further divided and the emitted radiation goes straight to the thermal infrared detectors.
5. A prism is placed in the path of the reflected radiation, to divide it into its spectral components.
6. The radiation detectors sense the reflected and emitted radiation. The reflected radiation is usually detected by silicon photodiodes that are placed in their correct geometric position behind the prism. The emitted (thermal infrared) radiation is usually detected by photon detectors, either made from mercury doped germanium (MDG) which is sensitive to wavelengths from 3 to 14 $\mu$m or indium antimonide (IA) which is sensitive to wavelengths from 3 to 5 $\mu$m and mercury cadmium telluride (MCT) which is sensitive to wavelengths from 8 to

14 μm. All of the thermal detectors require cooling, the MDG to −243 °C and the IA and MCT to −196 °C. This is achieved by using liquid helium or nitrogen held in an upmarket coffee flask called a Dewar (Fig. 4.2). There are usually between 7 to 12 reflected and emitted radiation detectors in a multispectral scanner. Although some research instruments have up to 24 detectors (Lowe *et al.* 1975; Norwood and Lansing, 1983)

After detection the signal is amplified by the pre-amplifier and is passed in electronic form, to the electronic control console.

7. The electronic control console has three components: a signal processor, to format the data as required for the recorders, an amplifier, to boost the signal level even further and a power distribution unit, to balance the signal strength in each waveband (Fig. 4.3).

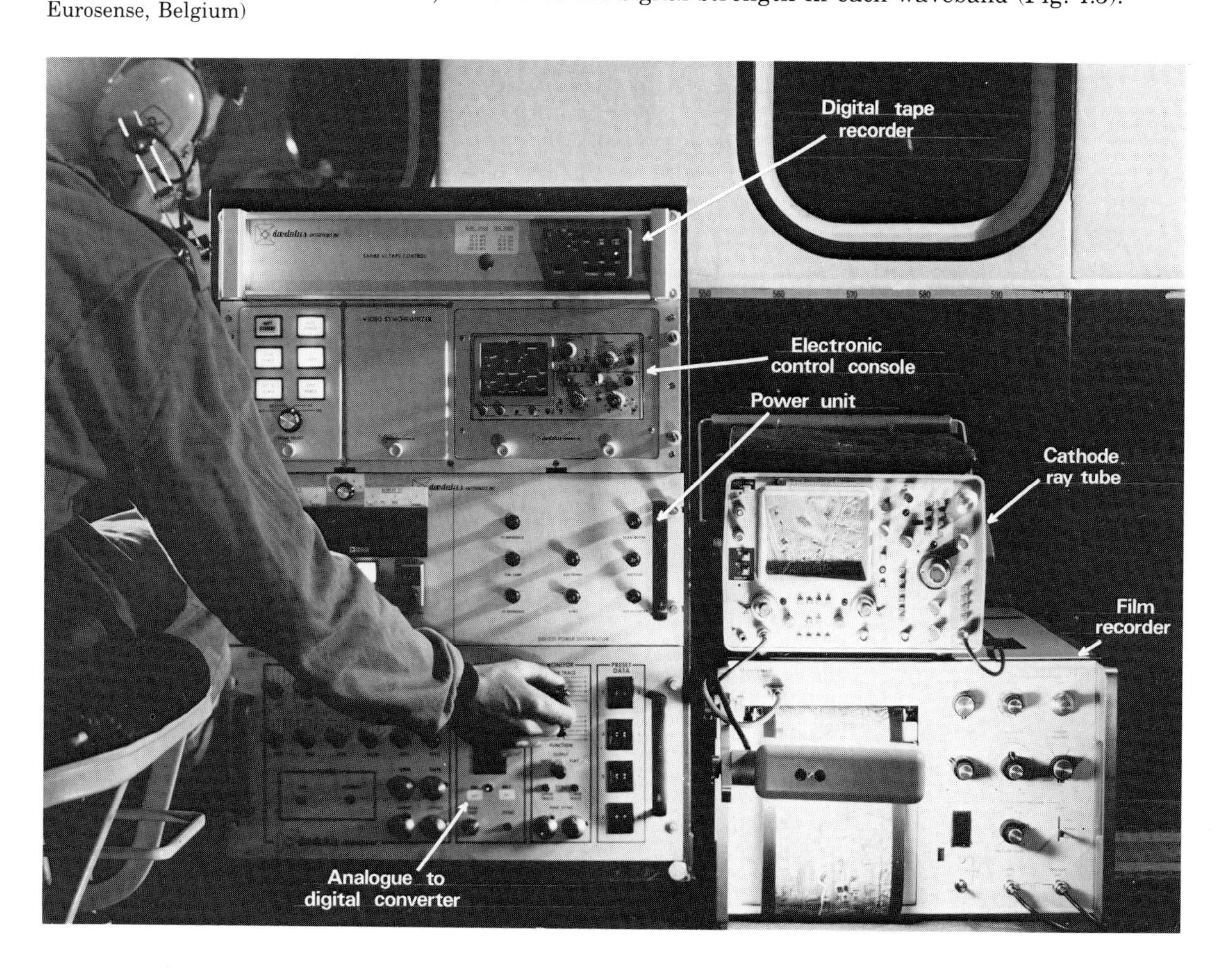


**Fig. 4.3**
A close-up of the operating controls of a multispectral scanner. The aircraft is a Dornier Skyservant DO-28, the multispectral scanner is the Daedalus model DS 1240/1260, the static cathode ray tube is a Hewlett Packard 1741A oscilliscope and the film recorder is made by Honeywell. (Courtesy, Eurosense, Belgium)

## 4.2.3 Recorders

The type of recorder depends upon the make and model of the multispectral scanner. The majority of multispectral scanners have a cathode ray tube (8), to enable the operator to observe the data as they are recorded while providing a source of 'quick look' information prior to digital analysis (Fig. 4.3). Simple multispectral scanners also tend to have analogue recorders; either a film recorder (9), where the electrical impulses are recorded directly onto film or an analogue tape recorder (10), where the electrical impulses are stored on magnetic tape. The film recorder uses the intensity of the electrical signal to modulate the intensity of a light source that moves across an unexposed photographic film at a speed that is proportional to the scanning speed of the rotating mirror. The brightness of the light source is directly related to the strength of the electrical signal so that an area of high radiance gives a large signal, a bright light and a photographic negative with a high optical density. This photographic negative can then be used to produce photographic positive prints.

The analogue tape recorder (10) records the electronic signal in the aircraft. The signals can either be fed into a film recorder (9) to produce images, or can be digitised for later digital analysis.

The newer scanners tend to use analogue to digital converters to produce a digital output which is recorded on high density digital tape (HDDT) in a digital tape recorder (11). On the ground this high density digital tape can be read onto low density computer compatible tapes (CCTs) for digital image processing (sect. 6.3), (Lowe 1980).

### 4.2.4 Characteristics of multispectral scanner images

While not wishing to delve into the technicalities of a multispectral scanner or the aircraft that carries it, there are four characteristics of this platform and sensor which affect multispectral scanner images to such an extent that environmental scientists should be aware of them. These characteristics are the presence of signal noise, the type of signal calibration, the geometry of the scanner and the stability of the aircraft.

### 4.2.5 Signal noise

All multispectral scanners produce extraneous and unwanted signal which is termed noise. This can be attributed to three factors: first, there is modulation in the average signal strength because the radiation arrives at the sensor in bunches known as quanta and not as a continuous beam. This effect is noticeable at longer wavelengths where less energy and fewer quanta are available for detection (sect. 2.4.1). Second, the signal level varies even when the radiation flow is constant because the probability of an electronic signal being caused by radiation hitting a detector is stochastic. The third effect is less predictable and arises from rattling such a sophisticated piece of equipment around in an aircraft. Fortunately, for the majority of shock-stabilised modern scanners the effect of this latter factor is now minimal (Slater 1980).

As all scanners have this noise, it is advisable to maximise the signal strength, in order to increase the signal to noise ratio (S/N). This can be done by either increasing the size of the ground sampling element, increasing the width of the sensed wavebands, or decreasing the scan speed, so that a longer time can be spent over each ground sampling element. While these actions would improve the signal to noise ratio they would also decrease the spatial and spectral resolution an increase geometric distortion. Therefore the problem of a sub-optimal signal to noise ratio is often accepted in preference to the cures.

## 4.2.6 Signal calibration

The majority of early scanners, many of which are still in use, employ an alternating current for the electronic signal that passes from the detector via the control box to the recorders. Unfortunately for the user this means that every scan line is referenced to a different average (Fig. 4.10). This is illustrated in Fig. 4.4 for a thermal infrared waveband, where it can be seen that the radiant temperature ($T_{rad}$) of the water determines the recorded signal level from the power station and grassland. In scan line (a) the water is warm, the $T_{rad}$ is high and the recorded signal is above average, likewise the grassland and power station are cool, the $T_{rad}$ is low and the recorded signal is *below average*. In scan line (*b*) the water is cold, the $T_{rad}$ is very low and the recorded signal is below average, likewise the grassland and power station which are cool have a low $T_{rad}$ but now have a signal that is *above average*. This problem can be overcome by using a direct current and relating each signal to a reference standard (Lowe *et al.* 1975).

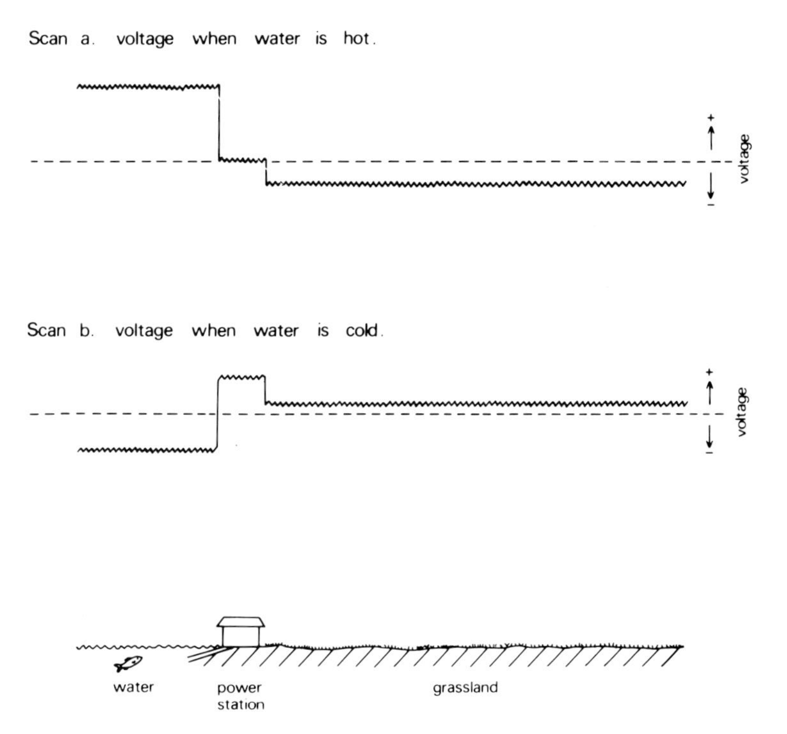


**Fig. 4.4**
Diagrammatic representation of two signals from a thermal infrared detector in a scanner with an alternating current. The $T_{rad}$ of the water but not the power station or grassland changed between the two scan lines (a and b). As the signal s related to an average for each scan line the decrease in the $T_{rad}$ of water results in a decrease in the signal for water but an increase in the signal for the power station and grassland. This also applies to the other wavebands detected by the sensor.

## 4.2.7 Scanning geometry

Multispectral scanner images are distorted both laterally and in the direction of flight.

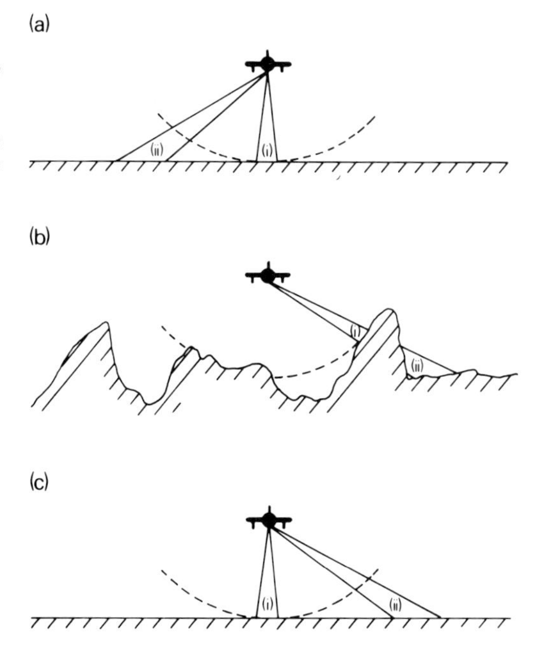


**Fig. 4.5**
Lateral distortion in multispectral scanner images: (a) The area of the ground sampling element increases outwards from the centre line; (b) Objects are displaced laterally; (c) The reflectance properties of non-Lambertian surfaces change with view angle.

## 4.2.7.1 Lateral distortion

Beneath the aircraft, the scanner 'looks' vertically downwards while at the edges of the scan it usually 'looks' at an angle of around 45 °, this has three effects on the resultant data. First, (Fig. 4.5a) the area of the ground sampling element is larger for off-vertical view angles thus squashing the terrain on the image edge. This effect is not all bad, as the changing size of the sampling element compensates for ratiometric fall-off from the aircraft track, so that it can be assumed that radiance is recorded over radiometrically equivalent surfaces, whether they be small and near to the sensor or large and far away from the sensor.

Second, objects are displaced laterally outwards. While this helps in the interpretation of features in urban areas, such displacement reduces overall geometric accuracy. For example, in Fig. 4.5b the sensor is measuring the radiance from the aircraft side of the mountain (i) but on the image the sensed area would be shown as if it were located on the other side of the mountain (ii).

Third, the reflectance and emittance properties of non-Lambertian surfaces are very sensitive to look angle (sect. 2.5 and 2.6). For example, when viewing cereal crops, field (ii) in Fig. 4.5c would have a higher near infrared and lower red reflectance than field (i), even if the canopies were identical, as off-vertical viewing decreases the area of soil seen by the sensor (Fig. 2.16). For this reason, much of the image classification work reported in the literature is based on radiance data recorded near vertically (Plate 3).

## 4.2.7.2. Directional distortion

The early scanners produced images with severe directional distortion as by the time the scan had moved from one side of the swath to the other, the aircraft had moved forward. For example, if an aircraft was travelling at a speed of 200 km/hr and the scanner was scanning 10 lines per second, then by the time the sensor had finished a scan, the aircraft would be approximately 6 m further forward than when it started. This effect is illustrated in Fig. 4.10. Today this distortion is minimised by having very fast scans and automatic correction (Lillesand and Kiefer 1979; Kindelan *et al.* 1981).

### 4.2.8 Aircraft stability

A multispectral scanner will scan the terrain line by line at a constant speed while the pilot tries to keep the aircraft on a straight and level course. This is not always possible (Fig. 4.6) and as a result aircraft pitch either overlaps or parts the scan lines, aircraft yaw either smears or leaves gaps between the scan lines and aircraft roll displaces objects in relation to the ground track (Silva 1978). Fortunately on modern scanners the effect of roll is partially corrected on a line by line basis with the aid of a gyroscope. Yaw is kept to a minimum by avoiding flights during cross-winds and pitch is usually minor enough to be ignored (Sabins 1978).

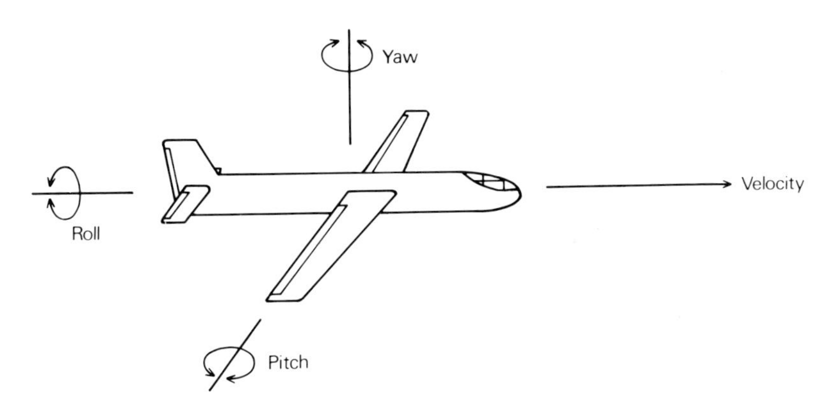


**Fig. 4.6**
The axes of movement for an aircraft.

### 4.2.9 Interpretation and application of multispectral scanner imagery

Early multispectral scanner imagery was visually interpreted band by band (Simonett 1976) using the aerial photographic interpretation techniques discussed in section 3.5. Increased availability of image processors encouraged workers to reduce and simplify the many wavebands of digital data as a prerequisite to interpretation. Such reduction involves either compressing the data into one image (sect. 6.3.6.12) or more commonly combining just three wavebands into a colour or false colour composite (Plate 3). Further processing can involve image enhancement and classification as is discussed in Phillips and Swain (1978) and section 6.3.7.

Two popular applications of digitally processed multispectral scanner imagery have been the mapping of vegetation and the moni-

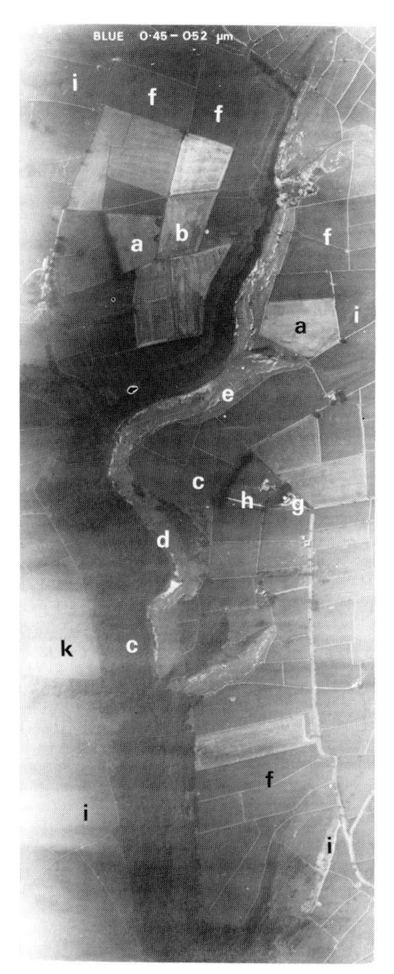


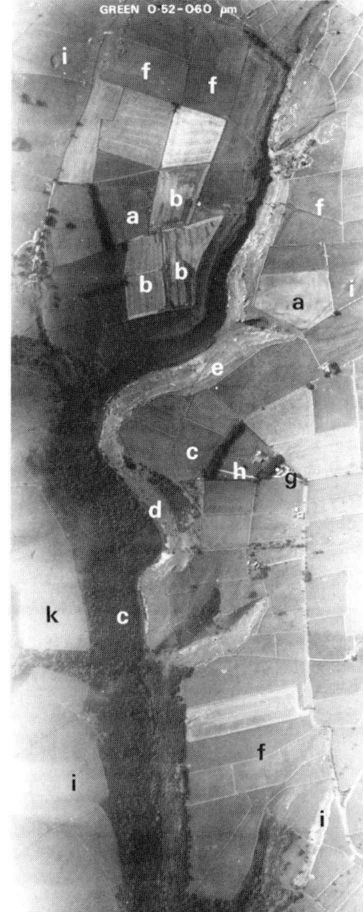


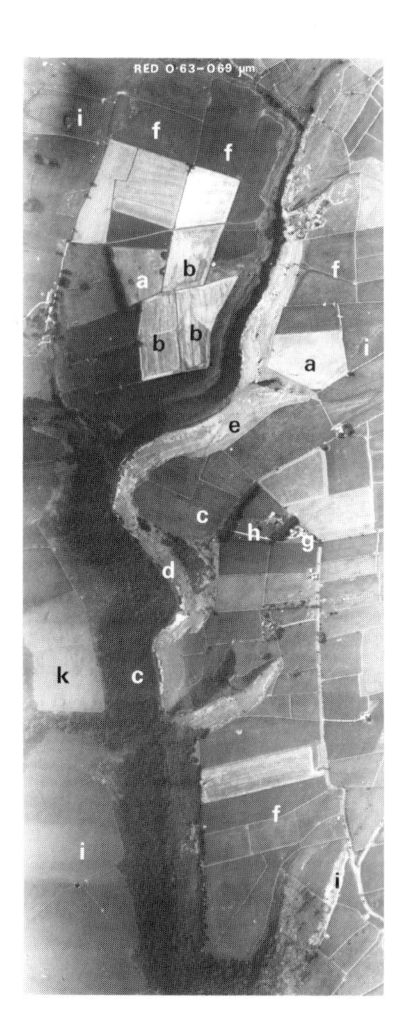


**Fig. 4.7**
Six multispectral scanner images of Lathkill Dale Derbyshire, UK. The images cover an area 7 × 4 km and were recorded on 17.9.1982 (see Plate 2). The major features of the scene are: (a) bare soil, (b) cereal stubble, (c) woodland, (d) water, (e) semi-natural grassland, (f) agricultural grassland, (g) farm, (h) caravan site and (i) mining rake or bolehill, north is to the right. (Data courtesy, Natural Environment Research Council, UK)

toring of waterbodies. Vegetation mapping is frequently used as the first step to many other applications (Williams *et al.* 1983), (Fig. 4.7) for example, the mapping of all land covers (Weber *et al.* 1972; Berg *et al.* 1978) or specific land covers like grassland (Pearson *et al.* 1976; Williams *et al.* 1983) or forests (Weber and Polcyn 1972; Aldrich 1979; Teillet *et al.* 1981). Early studies of waterbodies involved the monitoring of sediment plumes (Miller *et al.* 1977), but today several determinants of water quality can be successfully estimated, including sediment load, water colour and organic matter concentration (Zwick 1979; Khorram 1981).

The use of multispectral scanner images for geological and pedalogical applications is usually limited to occasions when digital data offer distinct advantages over cheaper stereoscopic aerial photography (sect. 3.4.4). This is the case for the studies reported by Gornitz (1979), Slegrist and Schnetcler (1980) and Rowan and Kahle (1982).

For further recent examples, refer to Gustafson (1982), Schweitzer (1982) and recent issues of the major remote sensing journals and symposia (Appendix B).

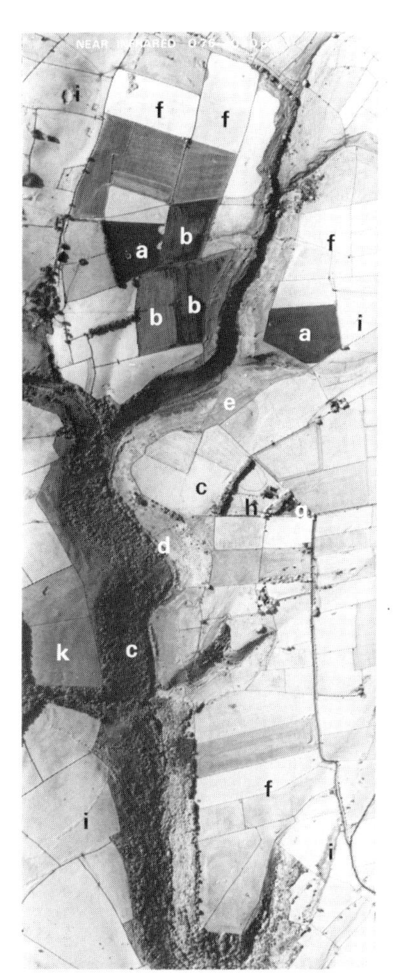


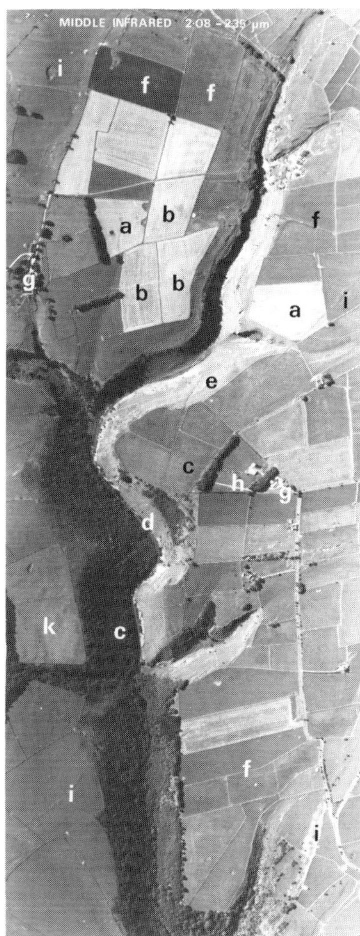


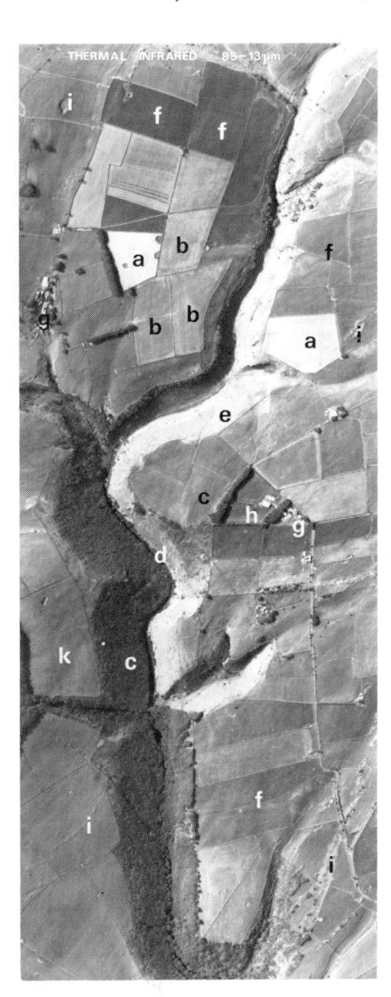


## 4.3 Thermal infrared linescanner

Thermal infrared linescanners were developed for military use in the 1940s and 1950s, initially for nocturnal snooping and later for the location of vehicles and camp fires. They were released for civilian use in the early 1960s and have since been employed for many applications in the environmental sciences (Fischer 1975).

### 4.3.1 How a thermal infrared linescanner works

The early thermal infrared linescanners have two thermal detectors and record an image onto a photographic film (Fig. 4.8), while the more recent thermal infrared linescanners are often part of a multispectral scanner (sect. 4.2) in which data are recorded digitally. With both of these systems the environmental scientist must decide upon the waveband to be used, the time of day when the image is to be recorded and the method of approximate calibration.

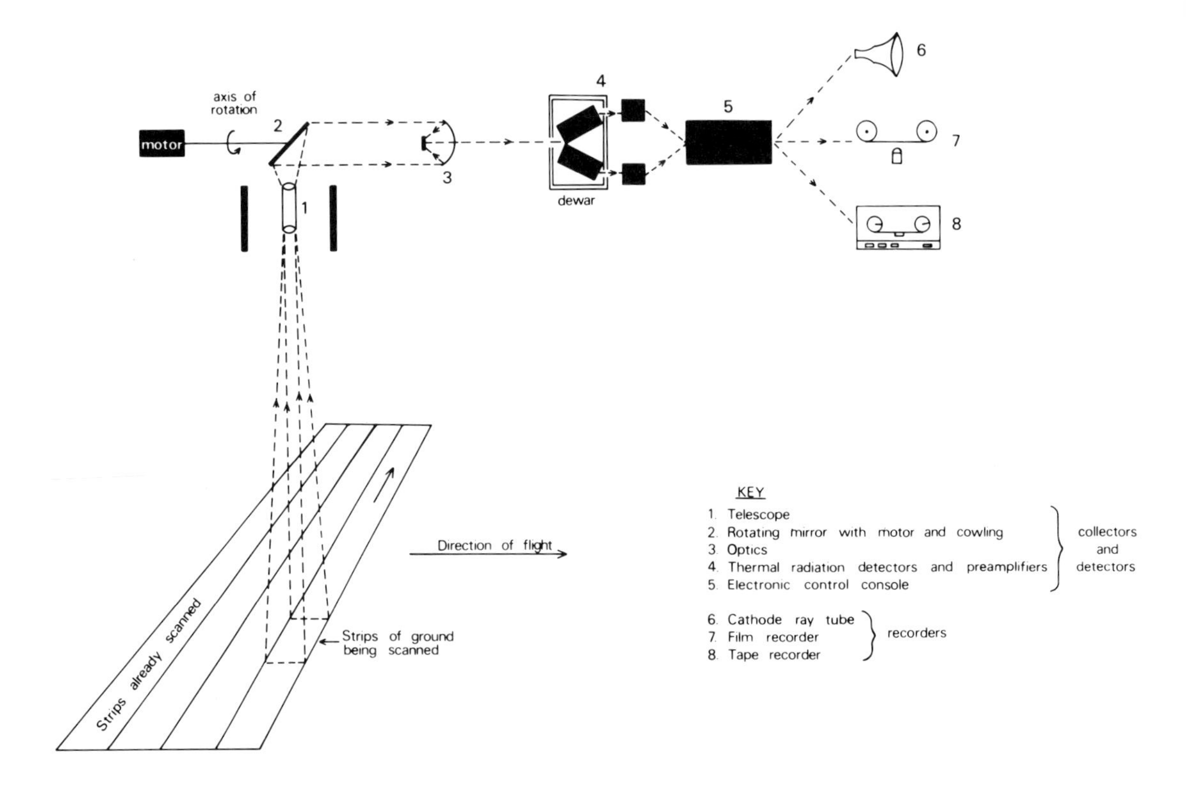


## 4.3.2 Selection of a waveband

The two thermal infrared wavebands used by thermal infrared linescanners are defined by the two transmitting atmospheric windows located between the wavelengths of 3–5 μm and 8–14 μm. The choice of which of these wavebands to employ depends upon the application. The peak of radiant emission from the Earth's surface occurs in the 8–14 μm region and so this waveband is used for the majority of applications. In the shorter wavelength band of 3–5 μm the mapping of terrain and other surface details is less successful, due to the influence of reflected solar radiation. However, this waveband has proved to be useful for sensing very hot objects which have a peak of radiant emission in these shorter wavelengths.

## 4.3.3 Selection of a time of day

Thermal infrared linescanners are used most frequently at night when there is no interference from reflected solar radiation. The usual flying time is just before dawn, when the effects of differential solar heating are at their lowest level. Occasionally, flights are made during daylight hours, either because aerial photography is to be taken or because it is advantageous to have terrain details enhanced by differential solar heating and shadowing (Offield 1975).

### 4.3.4 Selection of an approximate thermal calibration

**Fig. 4.8** *(Opposite)* A diagrammatic representation of a thermal infrared linescanner. The radiation emitted by the Earth's surface passes through a telescope (1), to be focused onto a rotating mirror (2), which reflects the radiation onto a set of optics (3), that focus the radiation onto the two thermal infrared detectors (4). One of these detectors is sensitive to radiation in the 3–5 μm waveband and one is sensitive to radiation in the 8–14 μm waveband. The electronic signal is amplified by the preamplifiers before being passed into the electronic control console (5) into which can be plugged one or several recording devices.

Due to the variation in emissivity across a scene and the presence of a thermally variable atmosphere between the sensor and the ground, it is not possible to calibrate absolutely the $T_{rad}$ signal from a thermal infrared linescanner, (sect. 2.5.5.2). As such a calibration is necessary for the production of $T_{kin}$ images, various approximate methods have been employed (Scarpace *et al.* 1975).

There are two inexpensive methods of approximately calibrating a thermal infrared image, both of which only work for data collected on haze-free days. The first involves flying a well calibrated non-imaging thermal radiometer alongside the thermal infrared linescanner, on the assumption that it measures radiant temperature $T_{rad}$ with greater precision than does the thermal infrared linescanner. The relationship which is determined between the tone of the thermal infrared image and the $T_{rad}$ of the radiometer for numerous points of known emissivity is extrapolated to calibrate the whole thermal infrared image.

The second inexpensive method and the one common to most modern thermal infrared linescanners, involves recording a warm and a cool blackbody at the beginning and end of each scan line, to provide a reference to which all other signals can be calibrated. As this method ignores the effect of variations in surface emissivity and the presence of the atmosphere, the difference between the calibrated $T_{rad}$ and the $T_{rad}$ just above the surface can often be as high as 2 °C (Lillesand and Kiefer 1979).

There are two expensive methods of approximately calibrating a thermal image. The first involves the measurement of emissivity and $T_{rad}$ for a number of sites as the aircraft passes overhead and the use of these data to calibrate the resultant imagery (Bartholic *et al.* 1972). As the logistical problems are enormous, the environment is disturbed during measurement and the sample points are usually too small to be located on the images, this technique has proved to be unpopular (Bowman and Jack 1979; Hatfield *et al.* 1982). The second expensive method involves repeated flights at altitudes ranging from the optimum altitude for the imagery down to as low as the aircraft can physically or legally fly. By using the $T_{rad}$ of thermally stable objects of known emissivity as standards, a graph is constructed of $T_{rad}$ versus height and extrapolated to the ground to give the $T_{kin}$ of enough points to enable calibration of the resultant imagery. This is currently the most accurate method of calibration (Schott and Tourin 1975).

### 4.3.5 Interpretation of thermal infrared linescanner imagery

The approach used in the interpretation of thermal infrared linescanner imagery depends whether the aim is to *detect* temperature differences or to *estimate* temperature.

### 4.3.6 Detection of temperature differences

The majority of thermal infrared linescanner data are presented in image form, to be interpreted using aerial photographic interpretation

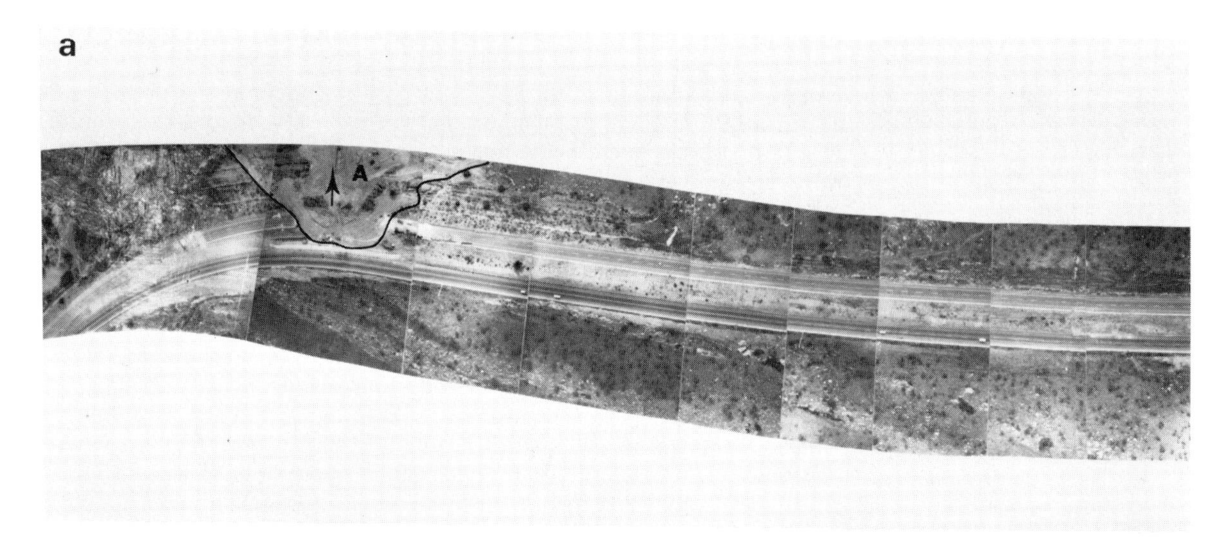


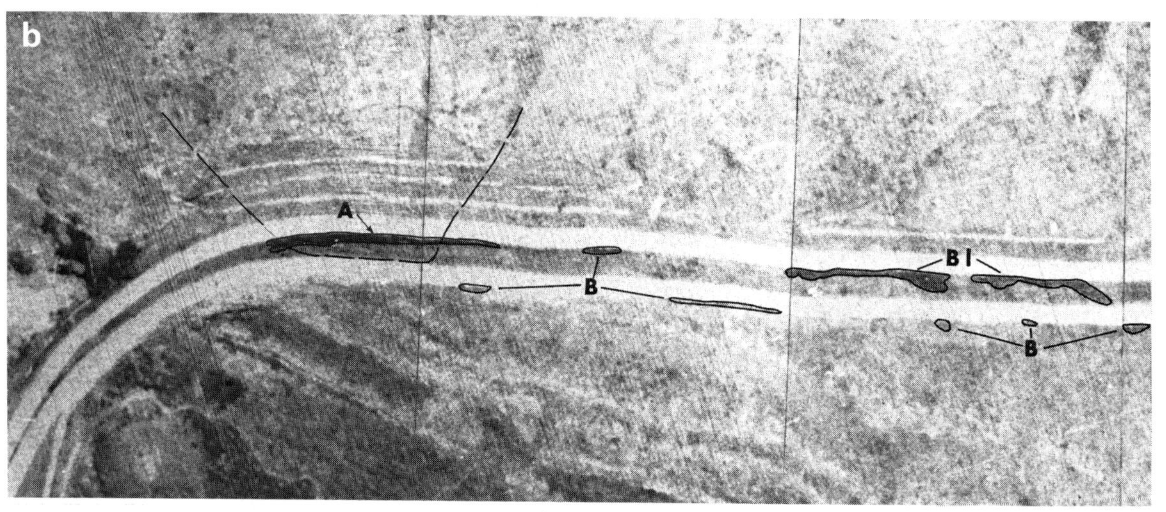


**Fig. 4.9**
Image (a) is an aerial photographic mosaic showing a section of the road between Krugersdorp and Pretoria in South Africa, following a landslip in 1977. Image (b) is a thermal infrared linescanner image, recorded in the 8–14 μm waveband, of the same area just prior to slope failure. A, is a large area of moisture build-up prior to failure and more importantly B and B1 are smaller areas of moisture build-up where the slope has not failed but cracks have been reported. (Courtesy, Spectral Africa (Pty) Ltd, South Africa)

techniques, based on the tone, texture, pattern, place, shape, shadows and size of the image components (sect. 3.5). The interpreter usually uses the images, not to map an area but to search for thermal patterns that give a clue to some past, present or future environmental process. For example, in Fig. 4.9, an interpreter would note that the soil surrounding the roads is cooler. In this case the actual temperature is not important but the inference that a cooler soil probably means a wet and subsidence prone soil is important (Bakker *et al.* 1978). Thermal infrared linescanner data have been used to detect very small temperature differences, for the differentiation of crop type (Fig. 4.10), the location of frost hollows (Stewart *et al.* 1978; Sutherland *et al.* 1981), the evaluation of shelter belts (Barrett and Curtis 1982), the estimation of water stress and irrigation efficiency (Bartholic *et al.* 1972), and the census of range animals (Parker and Driscoll 1972; Best *et al.* 1982). Thermal infrared linescanner data have also been used to detect very large temperature differences for the study of effusive vulcanism (Gawarecki *et al.* 1980; Friedman *et al.* 1981), the location of hot

**Fig. 4.10**
A comparison between a thermal infrared linescanner image (a), recorded in the 8–14 μm waveband, and a near infrared aerial photograph (b) recorded in the 0.7–0.9 μm waveband. Both images are of a 2 km$^2$ agricultural area in Cambridgeshire, UK; the thermal infrared linescanner image was taken on 19.7.1979 and the near infrared aerial photograph was taken on 13.7.1979. Two characteristics of early scanners are particularly well illustrated on the thermal infrared linescanner image. The first is the presence of a scan line distortion; note the roads. The second is the effect of using an alternating current for the signal output. For example, at the top of the thermal infrared linescanner image the warm and light toned field on the right has the effect of decreasing the signal and therefore the resultant image tone of the wheat fields at the centre and left.

The image tone of the two images is dependent upon the degree to which the cool vegetation canopy which is highly reflective in near infrared wavelengths is covering the warm soil which is poorly reflective in near infrared wavelengths. This is summarised below where it can be seen that the almost complete potato canopy has a dark tone on the thermal infrared linescanner image and a light tone on the near infrared photograph, while the reverse is true for the poorly developed carrot canopy.

| Crop type arranged in order of decreasing vegetation cover | Thermal infrared linescanner image (a) | | Near infrared aerial photograph (b) | |
|---|---|---|---|---|
| | $T_{rad}$ | Image tone | Near infrared reflectance | Image tone |
| Potatoes | Low | Dark | High | Light |
| Sugar beet | Low/medium | Dark/ medium | High/medium | Light/ medium |
| Wheat | Medium | Medium | Medium | Medium |
| Carrots | Very high | Light | Low | Dark |

(Images, courtesy, Aerial Photography Unit, Ministry of Agriculture, Fisheries and Food, UK)

springs, geysers and gas seepage (Dean *et al.* 1982) and the location of fires that occur both above ground (Shaw 1981) and below ground (Hirsch *et al.* 1971).

These images have also been used to map the spatial distribution of $T_{rad}$, this has proved useful in four areas of investigation. First, for mapping geological structures and studying geological processes, (Sabins 1969, 1976; Wolfe 1971). Second, for the mapping of water

**Fig. 4.11**
Thermal pollution of water recorded on thermal infrared linescanner images in the 8–14 μm waveband: (a) A large-scale image of an industrial area aside a canal at Botleg in the Netherlands. Warm water (with a light tone) flowing down the canal and the path of an underground heating pipe can be seen. (Courtesy, Clyde Surveys Ltd, UK); (b) A small-scale image of Cork Harbour in Eire. This is one of a series of images that were collected to monitor the movement of hot water from a power station during a tidal cycle. This particular image was recorded at low tide and shows the warm water starting to spread throughout the bay. (Courtesy, Electricity Supply Board, Eire)

pollution (Green and Crouch 1979), either indirectly by locating illicit outfalls into major rivers and canals or directly by monitoring the thermal plumes caused by industry in general (Fig. 4.11a) and power stations in particular (Fig. 4.11b). Third, for the location of excessive heat loss from buildings (Bjorkland *et al.* 1975; Brown *et al.* 1981) and underground heating systems (Fig. 4.11a). Fourth, the location and mapping of soil moisture and groundwater (Gagnon 1975; Skibitzke 1976).

## 4.3.7 Estimation of temperature

Thermal infrared linescanners are designed to detect differences in $T_{rad}$ and although images have been successfully used to estimate $T_{kin}$, the accuracy will be low. This is due to the approximate calibration of the sensor and the variable emissivity of the terrain, which conspire to produce a $T_{kin}$ estimation error of around ±3.5 °C.

To reduce this error, workers have either corrected for atmospheric and emissivity effects (Elkington and Hogg 1981), or concentrated only on areas of uniform emissivity (Heilman *et al.* 1976;

Millard *et al.* 1981; Peacock *et al.* 1981), or worked only in relative differences of $T_{\mathbf{kin}}$, by calculating thermal inertia as discussed in section 2.5.5.3 (Kahle *et al.* 1976; Axelsson 1980).

For further recent examples of the application of thermal infrared linescanner imagery, refer to recent issues of the major remote sensing journals and symposia (Appendix B).

## 4.4 Sideways-looking airborne radar (SLAR)

A sideways-looking airborne radar (SLAR) senses the terrain to the side of an aircraft's track. It does this by pulsing out long, up to *radio*

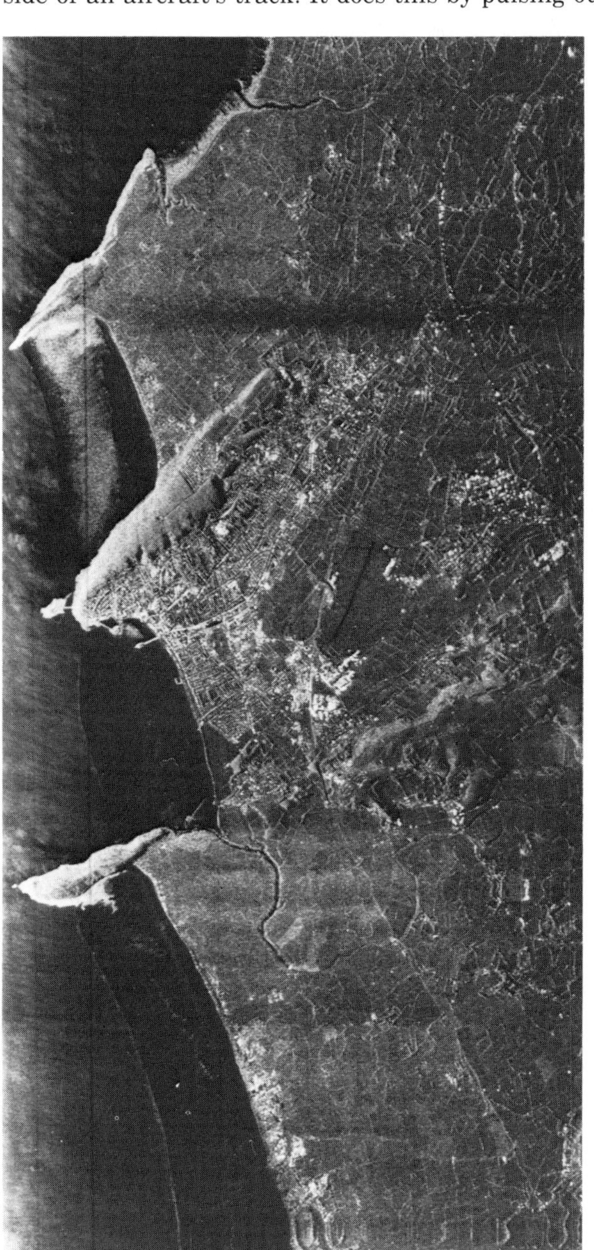

**Fig. 4.12**
SLAR real aperture image of Weston-super-Mare, UK. The image was taken in the Q band from an oblique range of 8 km. (British Crown copyright reserved)

wavelengths of electromagnetic radiation and then recording, first the strength of the pulse return to the aircraft, to *detect* objects and second, the time it takes for the pulse to return, to give the *range* of objects from the aircraft. The name radar is the cunning acronym of these functions of *ra*dio *d*etection *a*nd *r*anging. As these pulses are emitted at right angles to the aircraft track, the movement of the aircraft enables pulse lines to be built up to form an image (Fig. 4.12), (Skolnik 1970; Corless 1977; Henderson and Merchant 1978).

Radar, although relatively new to the environmental sciences, has been around for over a century. As early as the mid-1880s laboratory equipment had been developed which sent out pulses to measure the size and location of distant objects (Waite 1976; Ulaby *et al.* 1981). However, the idea of producing an image of the Earth's surface, albeit with a rotating antenna, was first developed in Britain during Second World War (Jones 1978). These early radars, produced very low resolution images of large areas. To obtain higher resolution images of smaller areas the antenna was locked beneath the aircraft and the sideways-looking airborne radar (SLAR) was born. During the 1950s the military development of SLAR was rapid as its independence of sunlight and weather conditions made it an ideal reconnaissance tool. SLARs were released for civilian use in the 1960s (Fischer 1975) but did not gain widespread use until the 1970s.

## 4.4.1 How a SLAR works

Like a multispectral scanner or thermal infrared linescanner a SLAR possesses collectors, detectors and recorders and in addition, a transmitter and antenna (Fig. 4.13). The transmitter produces pulses of microwave energy which are timed by a synchroniser and standardised to a known power by a modulator. For a fraction of a second the transmit/receive switch is switched to transmit, as the transmitter releases a microwave pulse from the antenna (Fig. 4.14). The transmit/receive switch then returns to its original position and the

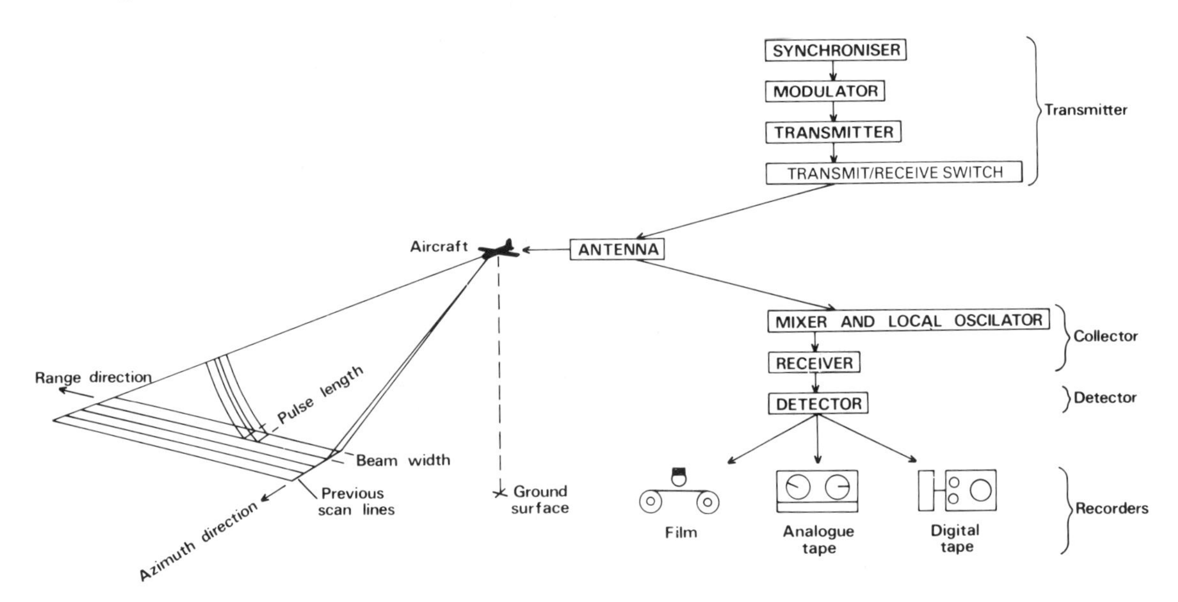


**Fig. 4.13**
A diagrammatic respresentation of a sideways-looking airborne radar (SLAR). Note that the SLAR transmits a microwave pulse out via the antenna and collects, detects and records microwaves backscattered from the Earth's surface using the same antenna.

**Fig. 4.14**
An aircraft and SLAR installation: (a) A turboprop aircraft, the Convair 580, which is used to carry a SAR. The antenna is inside the black pod beneath the aircraft; (b) The interior of (a) showing the instrumentation of an X and L band dual polarisation SAR. (Courtesy, Canada Centre for Remote Sensing, Canada)

antenna continues to receive pulses that have been backscattered from the Earth's surface. These pulses are converted to a form suitable for amplification and further processing by a mixer and local oscillator, before being passed to a receiver. The receiver amplifies the signal before passing it to the detector which produces an electronic signal suitable for recording onto photographic film or, analogue or digital tape (Fig. 4.13), (Moore 1975, 1976, 1983).

## 4.4.2 Spatial resolution

The spatial resolution of SLAR is determined by the pulse length and antenna beam width. The pulse length controls the dimension of the

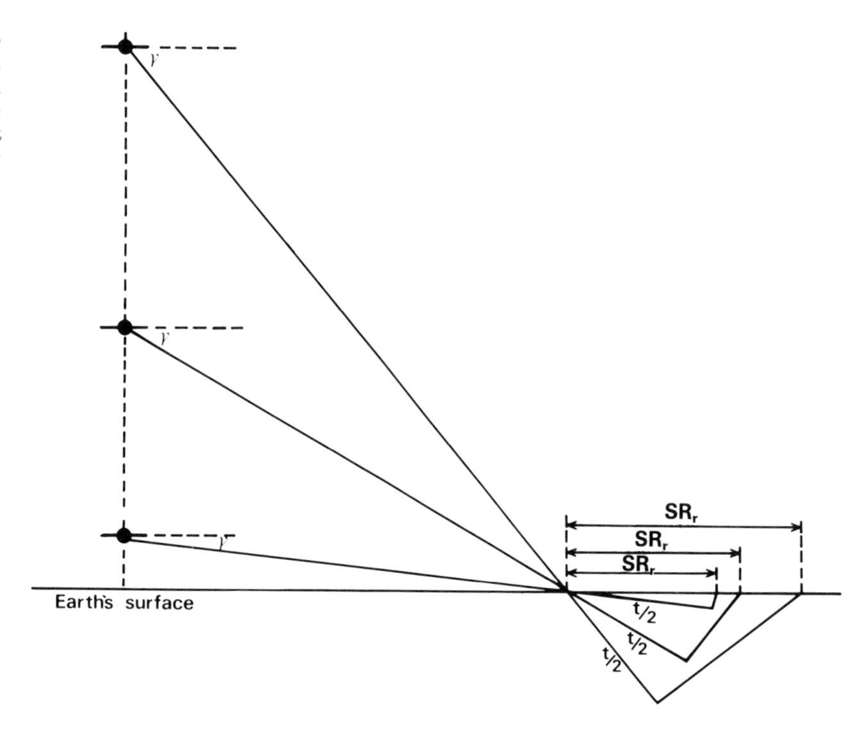


**Fig. 4.15**
The effect of SLAR depression angle ($\gamma$) and pulse length (t) on the spatial resolution in the range direction ($SR_r$). Note that ($\gamma$) depression angle is positively related to ($SR_r$).

ground sampling element away from the aircraft track in the range direction (Fig. 4.15) and the antenna beam width controls the dimension of the ground sampling element along the track of the aircraft in the azimuth direction (Fig. 4.16).

### 4.4.2.1 Spatial resolution in the range direction

The spatial resolution in the range direction is equal to half of the transmitted pulse length. For example, if two fields are under half a pulse length apart in the range direction of a SLAR, the first pulse released from the antenna will have reached the more distant field and will be returning to the antenna just in time to mingle with the second pulse released from the antenna, which is now returning to the antenna from the nearer field. As a result the signals from the two fields merge and they are recorded as one. If the two fields were over half a pulse length apart in the range direction, then the first pulse released from the antenna will have reached the more distant field but will not make it back in time to mingle with the second pulse released from the antenna, which is now returning to the antenna from the nearer field. As a result the signal from the two fields would remain separate.

The spatial resolution in the range direction is also dependent upon the depression angle of the antenna, for as the depression angle decreases, the pulse length represents progressively shorter distances on the ground (Fig. 4.15), (de Loor 1976).

The spatial resolution in the range direction ($SR_r$) can be calculated using formula [4.2].

$$SR_r = \frac{c\ t}{2 \cos \gamma} \qquad [4.2]$$

where the depression angle is ($\gamma$), the pulse length is (t) and the speed of the microwave pulse is (c), which at the speed of light is $3 \times 10^8$ $ms^{-1}$. Many of the early SLARs used a depression angle ($\gamma$) of 30 ° and a pulse length (t) of 0.1 $\mu$s. Therefore the spatial resolution in the range direction ($SR_r$) was:

$$SR_r = \frac{(3 \times 10^8)\ (0.1 \times 10^{-6})}{2 \times 0.866} = 17.3 \text{ m}$$

### 4.4.2.2 Spatial resolution in the azimuth direction

The width of the antenna beam ($\beta$) determines the spatial resolution in the azimuth direction ($SR_a$), (Fig. 4.16). As the beam fans out from the antenna the spatial resolution decreases with ground distance (GD), from a minimum directly below the aircraft. The spatial resolution in the azimuth direction can be calculated using formula [4.3].

$$SR_a = GD\beta \qquad [4.3]$$

Many of the early SLARs used a 1 milliradian beam width ($\beta$) and at a ground distance ($GD_1$ in Fig. 4.16) of 5 km this would produce a spatial resolution in the azimuth direction of:

$$SR_a = (5 \times 10^3)\ (1 \times 10^{-3}) = 5 \text{ m}$$

and at a ground distance ($GD_2$ in Fig. 4.16) of 8 km this would produce a spatial resolution in the azimuth direction of:

$$SR_a = (8 \times 10^3)\ (1 \times 10^{-3}) = 8 \text{ m}$$

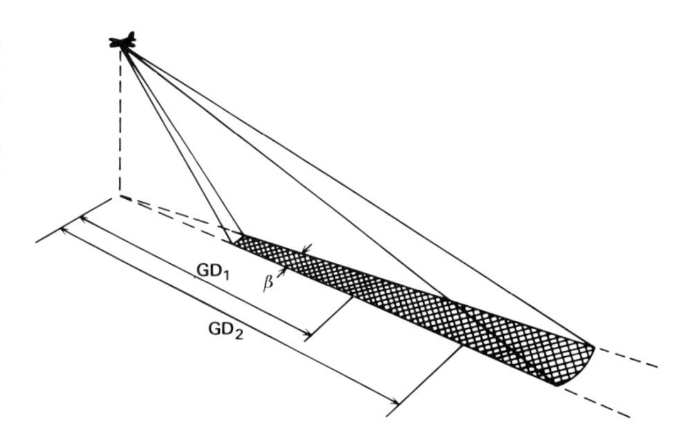


**Fig. 4.16**
Dependence of spatial resolution in the azimuth direction, on the beam width ($\beta$) and the ground distance (GD) from the aircraft track. (Modified from Lillesand and Kiefer 1979)

### 4.4.2.3 Improving spatial resolution

A SLAR is usually operated within a limited range of altitudes and depression angles and at the fastest possible pulse rate. Therefore the only way left to improve spatial resolution is to decrease the beam width.

However, the beam width (β) is determined by the wavelength of the microwaves (λ) and the antenna length (AL), (formula [4.4]).

$$\beta = \frac{\lambda}{AL} \qquad [4.4]$$

Therefore to decrease the beam width (β) there must be a large decrease in the wavelength (λ) used, or a large increase in the antenna length (AL). A large decrease in the wavelength is impractical as it would make the microwaves sensitive to atmospheric effects like rain clouds (Deane 1973) and a large increase in the antenna length would constitute an aviation hazard. To overcome this problem the *effective* length of the antenna can be increased by taking advantage of the motion of the aircraft (Harger 1970). As an aircraft flies past an object on the ground, the object will enter the antenna's beam, move through and then leave it (Fig. 4.17). During the time the object is within the beam it will have backscattered a number of microwave pulses, each at a slightly different frequency. The information on backscatter and frequency can be used to synthesise electronically, an antenna that is as long as the flight line of the aircraft (measured from where the sensor first and finally detects the object) and has a beam that is as wide at the source as it is at the sensor (Fig. 4.18). Such a SLAR is called a synthetic aperture radar or SAR (Lodge 1981).

Although SARs are currently few in number, they are destined to be the prime SLAR sensor in the late 1980s and 1990s. For this reason

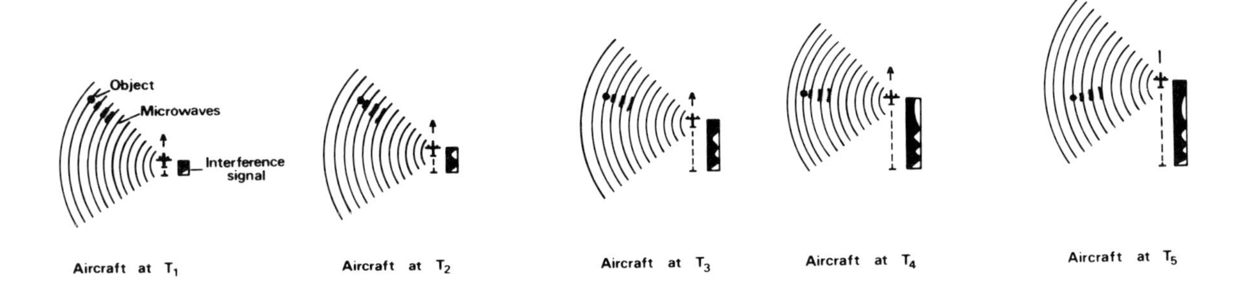


**Fig. 4.17**
Synthetic aperture radar (SAR). As the aircraft passes an object it receives backscattered microwaves that are successively in and out of phase with the transmitted microwaves. This is used to develop an interference signal for later image construction. (Modified from Jensen *et al.* 1977)

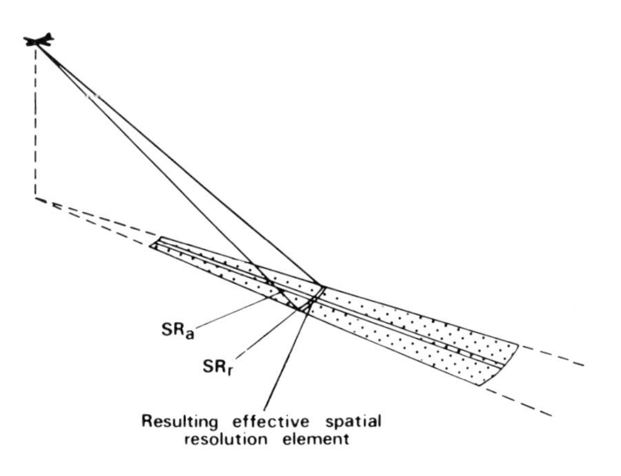


**Fig. 4.18**
The spatial resolution in the azimuth direction ($SR_a$) and the spatial resolution in the range direction ($SR_r$) for a synthetic aperture radar (SAR). (Modified from Lillesand and Kiefer 1979)

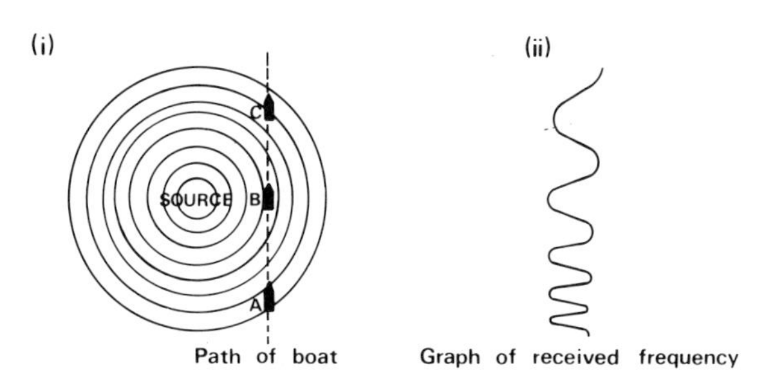


**Fig. 4.19**
A diagrammatic representation of the doppler frequency effect: (i) The path of waves transmitted at a frequency ($f_t$) of 10 cycles per second; (ii) The frequency of the received wave ($f_r$), this is less than 10 cycles per second at A and more than 10 cycles per second at C. (Modified from MacDonald 1980)

the rather complex methods by which frequency data are used by the SAR as a means of improving spatial resolution will be discussed, but first, an analogy. Imagine an object bobbing up and down in a pond, producing waves at a frequency of 10 cycles per second (Fig. 4.19i). A passenger on a boat moving *towards* the source at position A would count *more than 10* waves per second, as the boat *passed* the source at position B the passenger would *count 10* waves per second and as the boat moved *away* from the source at position C the passenger would *count less than 10* waves per second. If the passenger were to record these observations on a graph, then the curves produced would resemble those in Fig. 4.19ii. This difference between the transmitted frequency of 10 cycles per second and the received frequency is called the doppler frequency.

Now imagine the wave source is an aircraft transmitting a microwave pulse from an antenna and the boat is a reflecting object moving through the antenna beam. The microwaves backscattered to the antenna would have a shift in frequency due to the apparent motion of the target relative to the aircraft (Fig. 4.17) and a record of wave frequency received by the antenna would be similar to that obtained by the boat's passenger (Fig. 4.19ii).

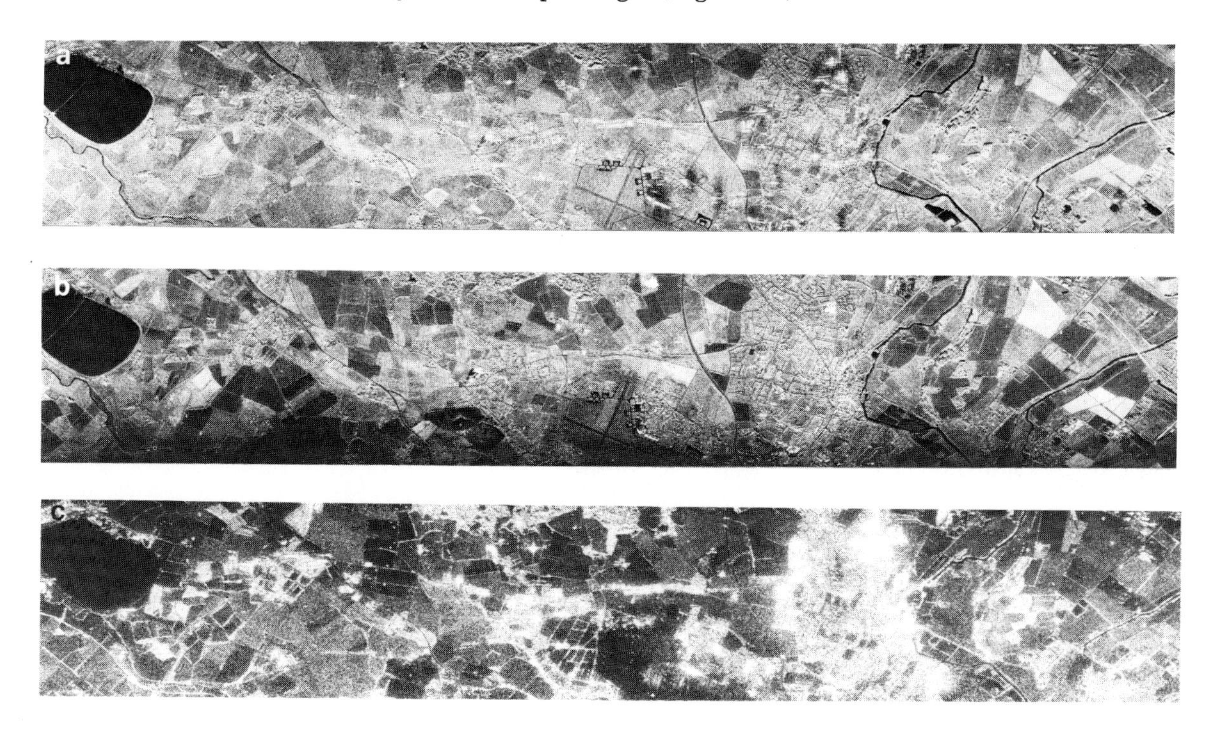


**Fig. 4.20**
A comparison between three SLAR images taken using a synthetic aperture radar (SAR) of part of the Thames Valley, UK on 27.6.1981. Images (a) and (b) are taken in the X band: (a) with HH polarisation and (b) with HV polarisation; (c) is taken in the L band with HH polarisation. (Courtesy, ESA)

These waves are combined in the sensor with waves at the transmitted frequency to create interference patterns. When the backscattered waves are in phase with the transmitted waves they mutually reinforce each other and produce a strong signal; when the backscattered waves are out of phase with each other then they cancel each other out and produce no signal. This variation in signal strength is recorded onto photographic film (Jensen *et al.* 1977). Back on the ground this photographic film is illuminated by a beam of coherent light, from a laser (Kozma *et al.* 1972). The interference pattern from each area on the film turns a small portion of the light transmitted through the film into a diverging wave that is brought into a point focus by a lens. Many overlapping areas on the film are similarly reformed producing the SAR image (Fig. 4.20), (Ulaby *et al.* 1982).

The main disadvantage with the SAR system is the long lag time between recording backscatter and producing an image, as can well be imagined from the previous discussion! Despite the many advantages of SAR, conventional SLAR with its lower spatial resolution (Fig. 4.12), remains the preferred sensor where data are required as they are recorded, for example in monitoring sea ice, floods or cropping patterns.

### 4.4.3 Scale distortion

The early SLAR systems recorded the image onto film during flight. The geometric scale of these images in the range direction was related to the time it took for the microwave pulse to complete a round trip to an object. Because this measure of time is related to the distance between the antenna and an object and not the planimetric distance between the aircraft and an object, the image scale decreases with range. This is illustrated in Fig. 4.21, where it can be seen that while objects A, B and C are equally spaced on the ground, a SLAR image would represent B as further away from C, than it is from A (Sabins 1978). Fortunately modern SLARs correct for this scale distortion during data recording (MacDonald 1980).

### 4.4.4 Topographic effects

Topography has four effects on SLAR images: corner reflection, long shadows, terrain layover and terrain foreshortening. Corner reflection

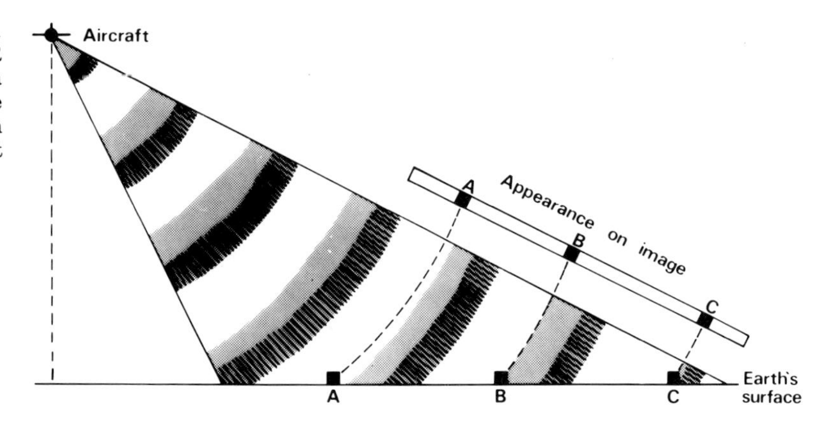


**Fig. 4.21**
Scale distortion on a SLAR image. By using time as a measure of distance, image scale decreases with range. On the ground A to B = B to C but on the image A to B < B to C.

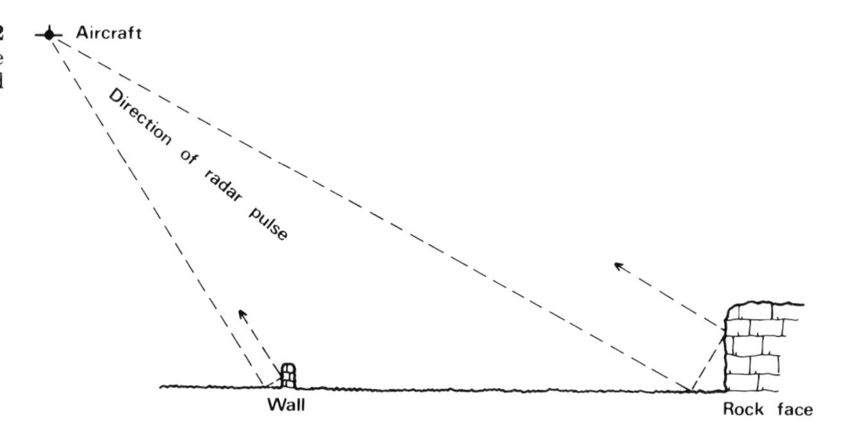


**Fig. 4.22**
Corner reflection, where the microwave radiation is reflected straight back to the antenna.

occurs when a microwave pulse is reflected straight back to the antenna, as a result of double reflection from a horizontal and then a vertical surface (Fig. 4.22). The returning pulse has so much energy that it saturates the receiver resulting in a white spot on the radar image. Walls and most buildings act as corner reflectors (Fig. 4.20), as do the metal pyramids used by researchers as ground control points (Fig. 4.23).

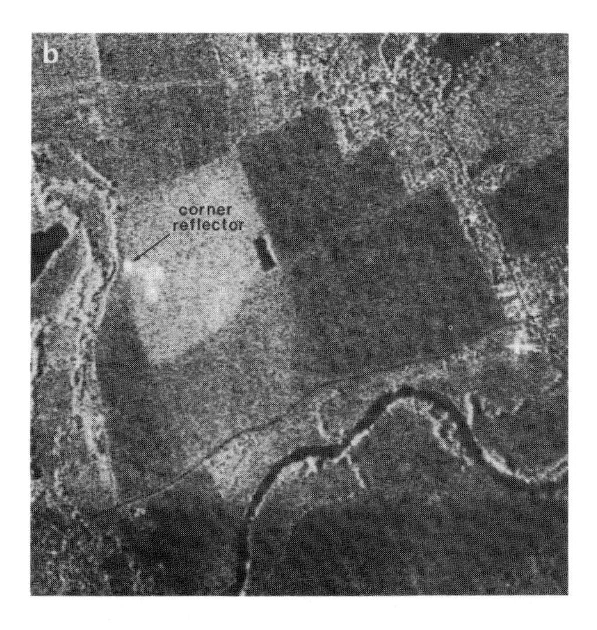


**Fig 4.23**
The use of a corner reflector in a cereal crop to provide geometric control during mapping and radiometric control during data manipulation. (a) Corner reflector and (b) a SLAR image of the same area taken using an X band, HH polarisation synthetic aperture radar (SAR) on 13.7.1981. This image covers the same area as Fig. 6.19. (Image (b) courtesy, ESA)

Long shadows are the result of an object shielding an area from the oblique microwave pulse (Fig. 4.24). The large area of shadowing on SLAR images of undulating terrain is a mixed blessing. Geologists and geomorphologists praise this effect for giving an impression of relief and enhancing microtopography (Lewis and Waite 1973) while others lament the loss of data (MacDonald 1980).

Terrain layover is the result of a microwave pulse hitting a high point like a mountain top before it hits a low point like a river valley, which is further from the aircraft ground track but nearer to the

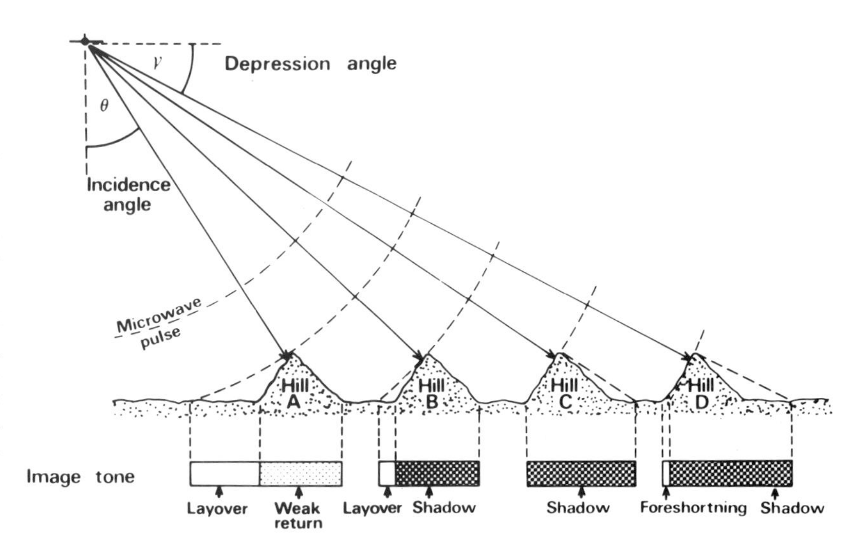


**Fig. 4.24**
The effect of topography on a radar image. Once upon a time there were four identical hills lying in a line in the range direction of a SLAR carrying aircraft. Hill A, which was nearest to the aircraft had a far-side slope that was steeper than the SLAR depression angle and as a result this slope backscattered little to the antenna. Hill B's far-side slope was the same angle as the SLAR depression angle and so produced no backscatter to the antenna. The far-side of hills C and D were even further from the aircraft track, the shadows they generated were so big that they hid not only themselves but other terrain as well. Hills A and B were hit by a radar pulse first on the crest and then on the base causing them to 'layover' or lean towards the aircraft track on the SLAR image. Hill C was hit all down its nearside producing a simultaneous backscatter and just one line on the SLAR image. Hill D was hit by a radar pulse first on the base and then on the crest resulting in the nearside being squashed or 'foreshortened' on the radar image. (Modified from Lillesand and Kiefer 1979)

**Fig. 4.25**
Portion of a SLAR swath collected by an X band SAR of the Evergreen Forest in northern Arizona, USA: (a) Original scale; (b) enlarged × 2; (c) enlarged × 4; and (d) enlarged × 8. (Courtesy, Goodyear Aerospace Corporation, Texas, USA)

antenna (Fig. 4.24). As a result tall objects lean towards the aircraft ground track on SLAR images.

Terrain foreshortening is a result of a microwave pulse hitting a point like a valley before, or at the same time, that it hits a higher point like a mountain (Fig. 4.24). As a result a slope facing the antenna will be compressed on a SLAR image.

## 4.4.5 Speckle

When microwave images are enlarged homogeneous areas appear speckled (Fig. 4.25). This is a result of the antenna transmitting many minor pulses along with each major pulse. Individually these pulses are insignificant but when they reinforce or suppress the backscatter of the major pulse, then speckle is the result (Deane 1973). There are three ways to suppress speckle. The first is to transmit the microwave pulse over a broad waveband so that the minor pulses are less likely to interfere with the major pulses (Moore and Thomann 1971). The second method is to smooth the image (sect. 6.3.6.8) and the third method applies to SAR only and involves the generation of three or more independent images of the same scene each centred at a different wavelength which can then be averaged together (Lodge 1981). As the first method is prohibitively expensive and the second and third methods result in a decrease in spatial resolution, speckle is tolerated, with reluctance, by many users.

## 4.4.6 Interpretation of SLAR imagery

SLAR images are usually interpreted visually, using the aerial photographic interpretation techniques discussed in section 3.5 and the continuous image processing techniques discussed in section 6.2. Of particular value to environmental scientists have been the two interpretation aids of mosaics and stereo SLAR.

Mosaics are employed for the mapping of large areas of what is often cloud ridden terrain. For example, between 1967 and 1969 SLAR mosaics were produced for eastern Panama (MacDonald 1969) as an aid to the mapping of structural geology (Wing 1971). Since then radar mosaics have been produced for many cloud-covered countries like Bolivia, Brazil, Columbia, Indonesia, Guatemala, New Guinea, Nicaragua, Nigeria, Philippines, Togo and Venezuela (Fig. 4.26), (Roessel and de Godoy 1974; Jensen *et al.* 1977; MacDonald 1980).

Stereo pairs of SLAR images can be produced by viewing the same area from two parallel flight lines (A and C, in Fig. 4.27). Unfortunately this results in images which appear to be very different, as they are illuminated from opposite directions. To overcome this problem stereoscopic pairs of SLAR images can be produced by viewing the same area from the same direction at two altitudes (A and B, in Fig. 4.27). With both methods, images are produced that can be used under a sterescope to produce a stereomodel of the terrain (sect. 3.4.4). For example, Koopmans (1973) used stereo SLAR images to map drainage patterns, and a number of organisations now use stereo SLAR images for topographic mapping (Leberl 1978, 1979).

An increasing proportion of SLAR data are being collected in digital form by calibrated multiwavelength and polarisation SARs (Fig. 4.20). (Inkster *et al.* 1979; Brisco and Protz 1980, 1982; Ulaby *et al.* 1980; Lodge 1981). Recent developments in our understanding of how microwaves interact with the Earth's surface (sect. 2.5.9) and an increase in the availability of digital image processors (sect. 6.3.3) are likely to lead to a boom in the use of digital SLAR data in the 1980s.

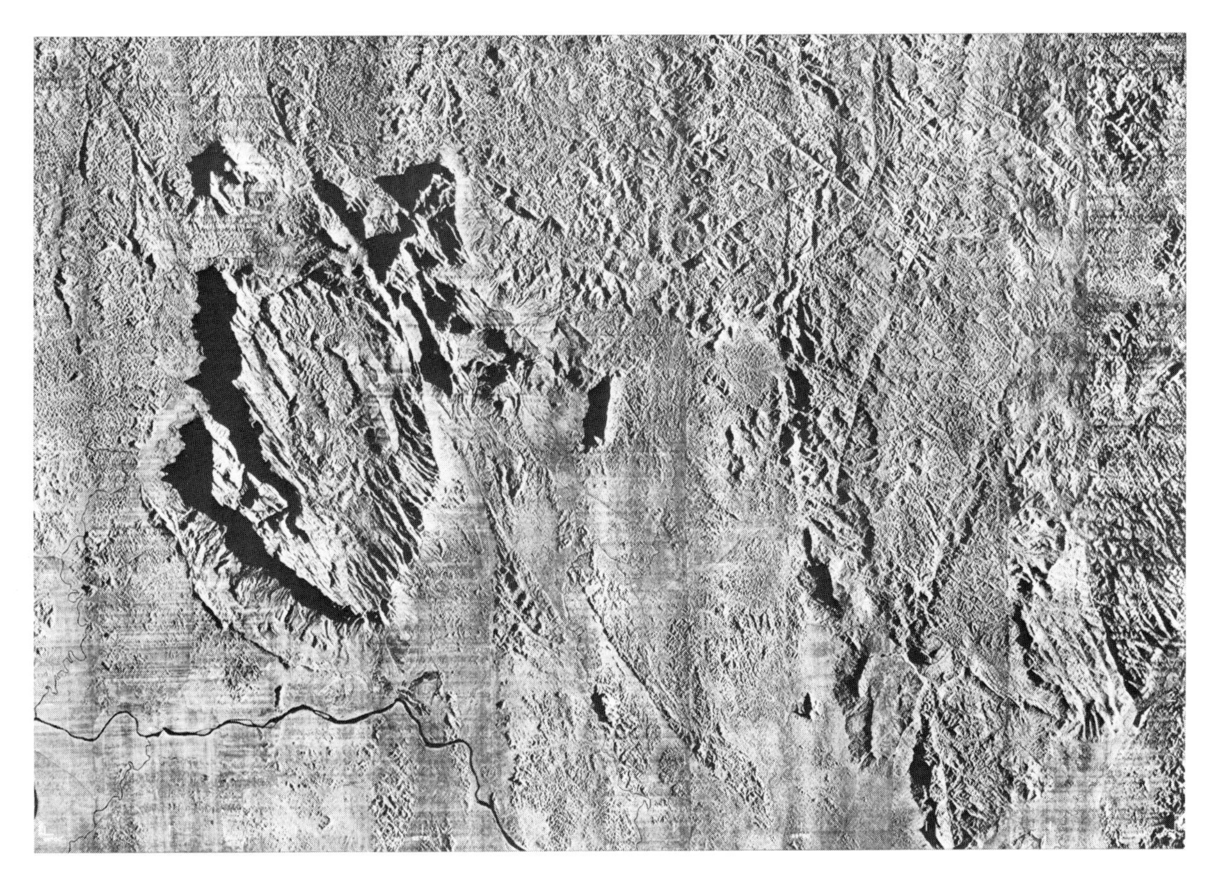

**Fig. 4.26**
A SLAR X band mosaic of Venezuela. This mosaic which was compiled from synthetic aperture radar (SAR) images was published commercially at a scale of 1:250,000. It is one of a series of 'radar maps' that provided the first overview of this perenially cloud-covered area. (Courtesy, Goodyear Aerospace Corporation and Aero Service Corporation, Texas, USA)

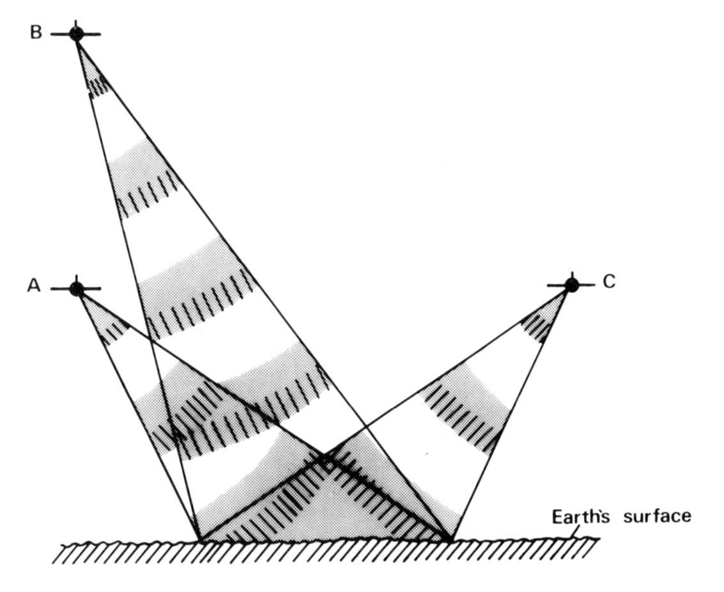


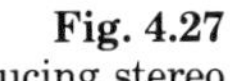

**Fig. 4.27**
Two methods of producing stereo SLAR images. To obtain images with opposite directions of illumination but similar depression angles the aircraft flies along flight lines A and C. To obtain images with similar directions of illumination but dissimilar depression angles the aircraft flies along flight lines A and B.

### 4.4.7 Applications in the environmental sciences

There are four characteristics of SLAR imagery that determine its fields of application, they are: its relatively high cost (Fig. 4.1), its rapid rate of data acquisition and its sensitivity to both surface roughness and surface moisture content. SLAR was initially used for geological exploration as the likely financial returns were high, the areas to be covered were large and surface roughness and moisture content often varied between the areas of interest (Dellwig *et al.* 1968; Crandall 1969; Allum 1981).

Closely linked to the acceptance of SLAR by geologists has been an increase in the use of SLAR in geomorphology (Nunnally 1969; Barr and Miles 1970; McCoy and Lewis 1976); especially in geomorphological mapping and terrain analysis (Lewis 1974; Blom and Elachi 1981; Koopmans, 1982) and the study of drainage pattern and catchment hydrology (Feder and Barks 1972; Parry *et al.* 1980; Bernard *et al.* 1981).

The sensitivity of microwaves to surface moisture content has been used in studies of water and vegetation. In studies of water, the main applications of SLAR have been in the mapping of surface soil moisture by visual interpretation (MacDonald and Waite 1971); the estimation

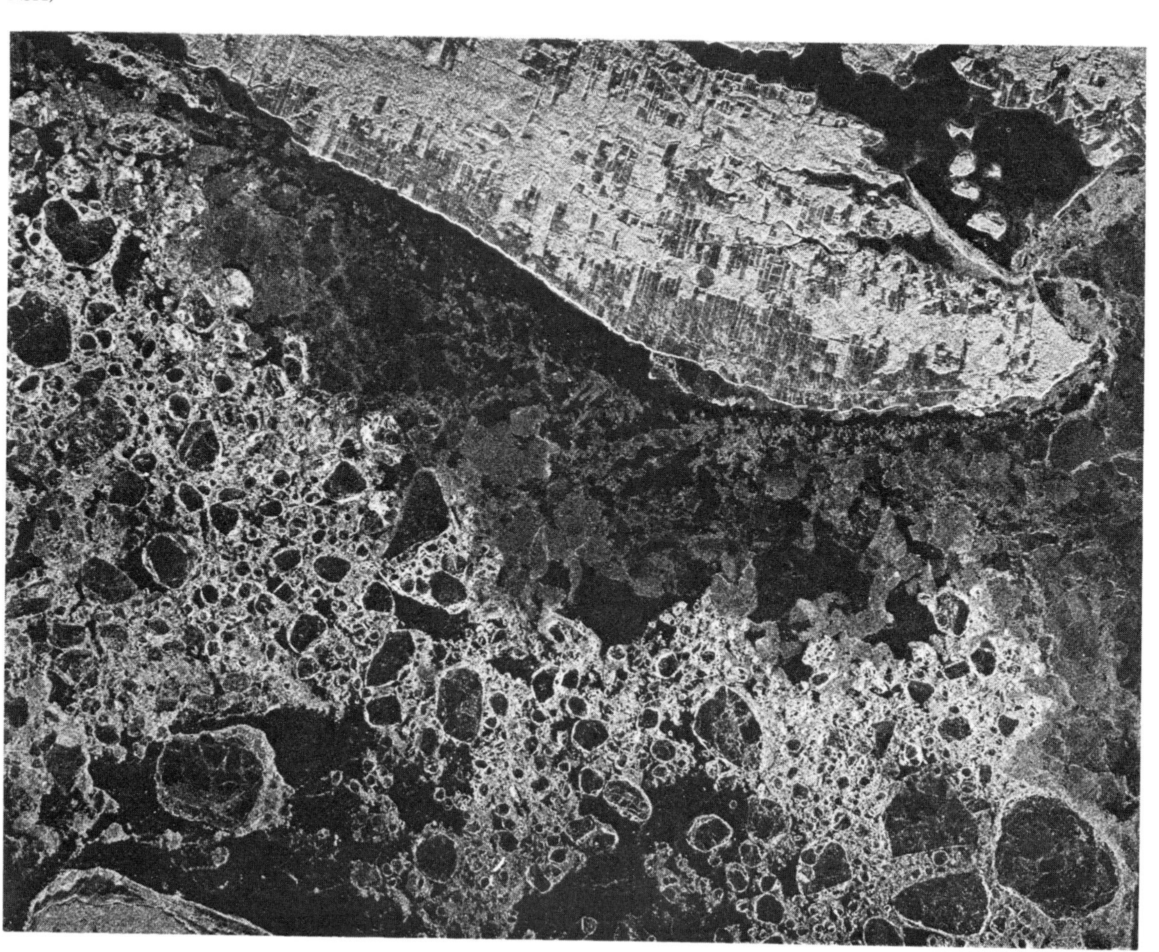

**Fig. 4.28**
A SLAR image taken using a synthetic aperture radar (SAR) of the channel to the south of Prince Edward Island (top central) in Canada. A variety of sea ice conditions are illustrated including rapidly melting sea ice near to the island with a mixture of brash ice and ice flows further away from the island. (Courtesy, Goodyear Aerospace Corporation and Aero Service Corporation, Texas, USA)

of surface soil moisture using calibrated SAR (Choudhury *et al.* 1979); Dobson and Ulaby 1981; Bernard *et al.* 1982); the monitoring of sea ice (Eckhart and Geerders, 1975; Chizhov *et al.* 1978), (Fig. 4.28) and the location of oil spills, where the pouring of oil on troubled water decreases the surface roughness (Kuilenberg 1975; Fischer *et al.* 1976; Klaus *et al.* 1977).

In studies of vegetation, emphasis has tended to be on visual interpretation for mapping, either semi-natural vegetation (Morain and Simonett 1967; Morain 1976; Parry and Trevett 1979); forest (Aldrich 1979; Hardy 1981), or land cover (Schwarz and Caspall 1969; Henderson, 1975). In recent years increased availability of calibrated SAR has also resulted in its application to the identification of agricultural vegetation (Fig. 4.20), (Bush and Ulaby 1978; Ulaby *et al.* 1980).

There are many fields in which the use of SLAR has yet to be fully developed. These include cartography (Dowman and Morris 1982), archaeology (Adams *et al.* 1981) and urban planning (Bryan 1975, 1983).

For further recent examples of the application of SLAR imagery, refer to Gustafson (1982) and to recent issues of the major remote sensing journals and symposia (Appendix B).

## 4.5 Recommended Reading

**Jensen, H., Graham, L. C., Porcello, L. J.** and **Leith, E. N.** (1977) Side-looking airborne radar. *Scientific American,* **237**: 84–95.

**Lillesand, T. M.** and **Kiefer, R. W.** (1979) *Remote Sensing and Image Interpretation.* Wiley, New York, pp. 382–527.

**Lowe, D. S.** (1976) Nonphotographic optical sensors: *In* Lintz, J. and Simonett, D. S. (eds) *Remote Sensing of Environment.* Addison-Wesley, Reading, Massachusetts; London, pp. 155–93.

**MacDonald, H. C.** (1980) Techniques and applications of imaging radars: *In* Siegal, B. S. and Gillespie, A. R. (eds) *Remote Sensing in Geology.* Wiley, New York, pp. 297–336.

**Moore, R. K.** (1983) Imaging radar systems: *In* Colwell, R. N. (ed.) *Manual of Remote Sensing* (2nd edn). American Society of Photogrammetry, Falls Church, Virginia, pp. 429–74.

**Norwood, V.T.** and **Lancing, J.C.** (1983) Electro-optical imaging sensors: *In* Colwell, R. N. (ed.) *Manual of Remote Sensing* (2nd edn). American Society of Photogrammetry, Falls Church, Virginia, pp. 335–67.

**Slater, P. N.** (1980) *Remote sensing: Optics and Optical Systems.* Addison-Wesley, Reading, Massachusetts; London, pp. 401–38.

**Ulaby, F. T., Moore, R. K.** and **Fung, A. K.** (1981) *Microwave Remote Sensing, Active and Passive, Volume 1, Fundamentals and Radiometry.* Addison-Wesley, Reading, Massachusetts; London.

# Chapter 5 Satellite sensor imagery

*'I wouldn't want to be quoted on this, but we spent 35–40 billion (dollars) on the space program and, if nothing else has come of it except knowledge that we've gained from space photography, it would be worth ten times what the whole program has cost.'*
Lyndon B. Johnson, former President of the USA.
(Baker 1981)

## 5.1 Introduction

To those new to the field of remote sensing, the large number of remote sensing satellites must be confusing. In this chapter a satellite taxonomy is provided in which development of remote sensing from space can be seen and future developments studied (Fig. 5.1). To evaluate the utility of the satellites in this taxonomy it is necessary to start

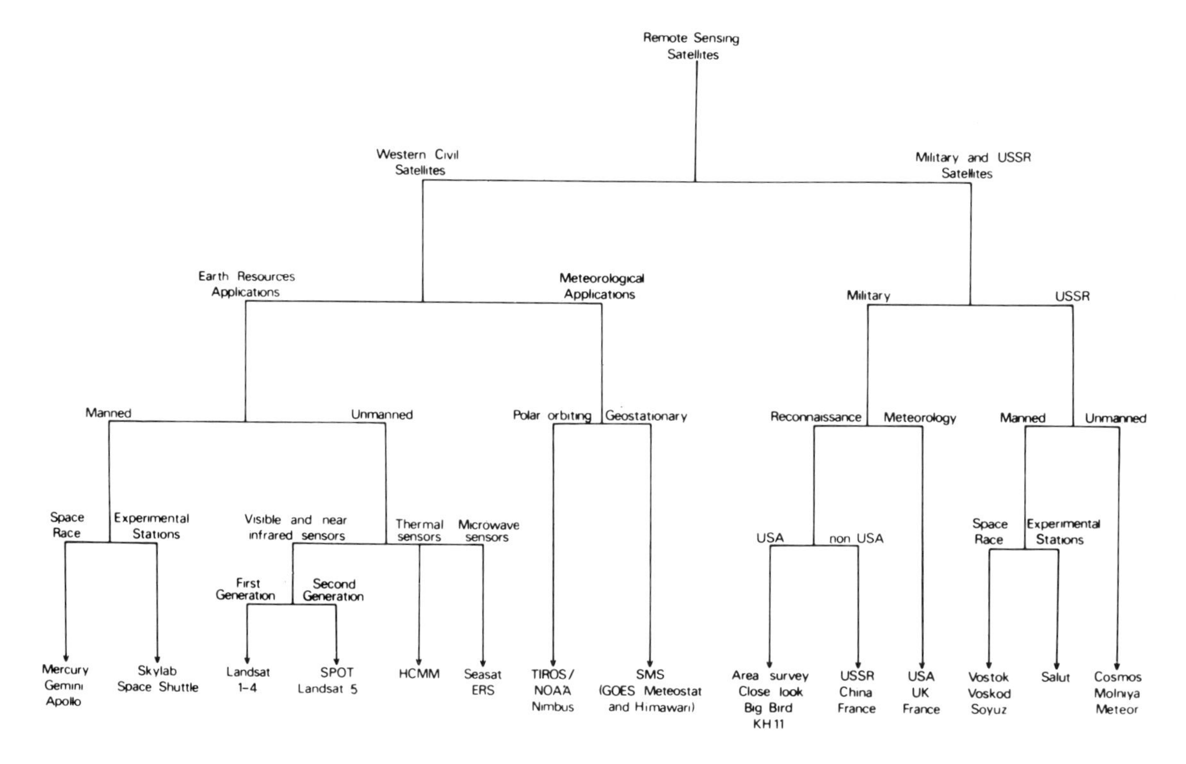


**Fig. 5.1**
A taxonomy of remote sensing satellites.

by describing a satellite that has never been launched; this is the mythical 'ideal remote sensing satellite'. This satellite should have a high altitude orbit that does not need adjusting, with an azimuth sufficient to give global coverage and a repeat cycle of a few hours. The many high quality sensors on board should produce stereoscopic images that cover vast areas of the Earth's surface in one sweep at a high spatial resolution in many narrow wavebands, from visible to microwave wavelengths.

However, such virtues cannot, for technical reasons, be put into one satellite and so remote sensing satellites have been designed according to their application, for use in either the study of the Earth's resources or for meteorology. The Earth resources satellites, which can be manned or unmanned, tend to carry sensors with a medium spatial resolution, of better than 0.25 km and a slow repeat cycle of less than two weeks. Satellites designed for meteorological applications are all unmanned and tend to carry sensors with a low spatial resolution of 0.75 km or less and a fast repeat cycle of better than a day.

Within these broad divisions remote sensing satellites can be grouped by the orbits they follow and the sensors they carry, as is indicated in Fig. 5.1. The characteristics of the USA's and European remote sensing satellites noted in Fig. 5.2 can be used as an indication of the similarity of any remote sensing satellite to the ideal remote sensing satellite.

For obvious political reasons there is a further division in the hierarchy of remote sensing satellites between available and unavailable imagery. For practical reasons this book concentrates on available data from Earth resources satellites and to a lesser extent meteorological satellites. However, as those interested in remote sensing need to be aware of the vast quantity of unavailable data, if only as an indication of the potential of remote sensing technology then the data collected by military and USSR remote sensing satellites will be discussed towards the end of this chapter.

## 5.2 Earth resources satellites

There are two groups of Earth resources satellites; first, the manned satellites which carry photographic and other sensors for the production of images of the Earth's surface. These images can be interpreted with the aid of aerial photographic interpretation techniques, discussed in section 3.5. Second, the unmanned satellites that carry a wide range of non-photographic sensors for the production of images of the Earth's surface. These images can be interpreted with the aid of both aerial photographic interpretation techniques discussed in section 3.5 and digital image processing techniques discussed in section 6.3.

## 5.3 Manned Earth resources satellites

The first manned satellites (Fig. 5.2) were designed, and are known, for their participation in the 'space race' while the later satellites,

| | Satellite series | | | Orbit details | | | Sensor details | | | | | | | | | |
|---|---|---|---|---|---|---|---|---|---|---|---|---|---|---|---|---|
| | Satellite series names | Actual and Potential relative importance to Environmental scientists | Era | Altitude | World coverage | Repeat cycle | Prime sensor(s) | Image width | Spatial resolution | Wavebands number | Wavebands type | Secondary sensor(s) | Image width | Spatial resolution | Wavebands number | Wavebands type |
| Manned Earth Resources Satellites | MERCURY and GEMINI | □ | 1960 s | □ | □ | □ | C | ⧄ | ◪ | □ | V | | | | | |
| | APOLLO | □ | 1960s | □ | □ | □ | C | ⧄ | ◪ | □ | V | | | | | |
| | SKYLAB | ⧄ | 1970 s | □ | □ | ◪ | MbC | ⧄ | ◪ | ◪ | V | C | □ | ■ | □ | V |
| | SPACE SHUTTLE | ⧄ | 1980 s | □ | □ | ⧄ | C | ⧄ | ■ | □ | V | SAR | □ | ⧄ | □ | M |
| Unmanned Earth Resources Satellites | LANDSAT 1 and 2 | ■ | 1970 s | ⧄ | ■ | ⧄ | MSS | ⧄ | ◪ | ⧄ | V | RBV | ⧄ | ◪ | ⧄ | V |
| | LANDSAT 3 | ■ | 1970 s | ⧄ | ■ | ⧄ | MSS | ⧄ | ◪ | ◪ | V | RBV | □ | ■ | □ | V |
| | LANDSAT 4 and 5 | ■ | 1980 s | ⧄ | ■ | ⧄ | MSS | ⧄ | ◪ | ⧄ | V+T | TM | ⧄ | ■ | ■ | V+T |
| | SPOT | ■ | 1980 s | ⧄ | ■ | ◪ | HRV | □ | ■ | ⧄ | V | | | | | |
| | LANDSAT D'+ (?) | ■ | 1990 s | ⧄ | ■ | ⧄ | MRS | □ | ■ | ⧄ | V | | | | | |
| | HCMM | ◪ | 1970 s | ⧄ | ■ | ■ | Rad | ⧄ | □ | □ | V+T | | | | | |
| | SEASAT | ◪ | 1970 s | ⧄ | ◪ | □ | SAR | □ | ■ | □ | M | Rad | ◪ | □ | □ | V+T |
| | ERS | ⧄ | 1980 s | ⧄ | ■ | ◪ | SAR | □ | ◪ | □ | M | | | | | |
| Meteorological Satellites | 2nd generation TIROS/NOAA | □ | 1970 s | ◪ | ■ | ■ | VHRR | ◪ | ⧄ | □ | V+T | | | | | |
| | 3rd generation TIROS/NOAA | ◪ | 1980 s | ⧄ | ■ | ■ | AVHRR | ◪ | ⧄ | ◪ | V+T | | | | | |
| | NIMBUS | □ | 1980 s | ⧄ | ■ | □ | CZCS | ◪ | ⧄ | ◪ | V+T | | | | | |
| | SMS | ⧄ | 1970 s | ■ | ◪ | ■ | Rad | ■ | □ | ⧄ | V+T | | | | | |

**Fig. 5.2**
The characteristics of the satellites that carry environmentally useful sensors.

| Symbol key | □ | ⧄ | ◪ | ■ |
|---|---|---|---|---|
| Actual and potential relative importance for environmental scientists | Low | Medium | High | Very High |
| Altitude (km) | <500 | 500–1000 | 1000–30 000 | >30 000 |
| World coverage | Variable between 50° N and S | 60° N and S | 70° N and S | Near Global |
| Repeat cycle (image time⁻¹) | 1 per month to rarely repeated | 1 per week to 1 per month | 1 per day to 1 per week | <1 per day |
| Image width (km) | <100 | 100–1000 | 1000–5000 | 5000 – hemisphere |
| Spatial resolution (km) | >1 2 | 0 25–1 2 | 0 035–0 25 | <0 035 |
| Number of wavebands | 1 or 2 | 3 or 4 | 5 or 6 | 7 or more |

Code key

| Sensor | Sensor | Waveband type |
|---|---|---|
| C = Camera | Rad = Radiometer | V = Visible and near visible |
| MbC = Multiband Camera | HRV = High Resolution Visible Scanner | T = Thermal infrared |
| MSS = Multispectral Scanning System | VHRR = Very High Resolution Radiometer | M = Microwave |
| MRS = Multispectral Resource Sampler (?) | AVHRR = Advanced Very High Resolution Radiometer | |
| TM = Thematic Mapper | CZCS = Coastal Zone Colour Scanner | |
| SAR = Synthetic Aperture Radar | RBV = Return Beam Vidicon | |

Skylab and the Space Shuttle, were space stations designed specifically for experimentation.

## 5.3.1 An introduction to manned 'space race' satellites

The first photographs of the Earth from space were taken from the USA's National Aeronautics and Space Administration (NASA) satellites of Mercury, Gemini and Apollo. The overt objective of these satellites was to fulfill John F. Kennedy's declaration that an American would be standing on the Moon by the end of the 1960s. As a spin-off to this 'space-race', photographs were taken, initially for reasons of curiosity and publicity and later for the study of the Earth's resources.

## 5.3.2 Mercury series

The first successful photographs of the Earth's surface from space were taken in 1961 from a Mercury satellite on its fourth mission. The interest provoked by these photographs lead to 70 mm format colour photography being taken of geologically interesting areas, during the satellite's later missions. The ninth Mercury mission was photographically the most successful of this series, returning 29 good quality photographs, several of which were of unmapped areas of the world in South West Asia and Tibet (Lowman 1965).

**Fig. 5.3**
Early space photograph taken using a hand-held 70 mm format camera from Gemini IV on 4.6.1965. The photograph was taken looking to the southeast over the sedimentary Hadramawt Plateau in southern Saudi Arabia. The Wadi Hadramawt, a dry sand filled river valley can be seen running across the image. Original in colour. (Courtesy, NASA)

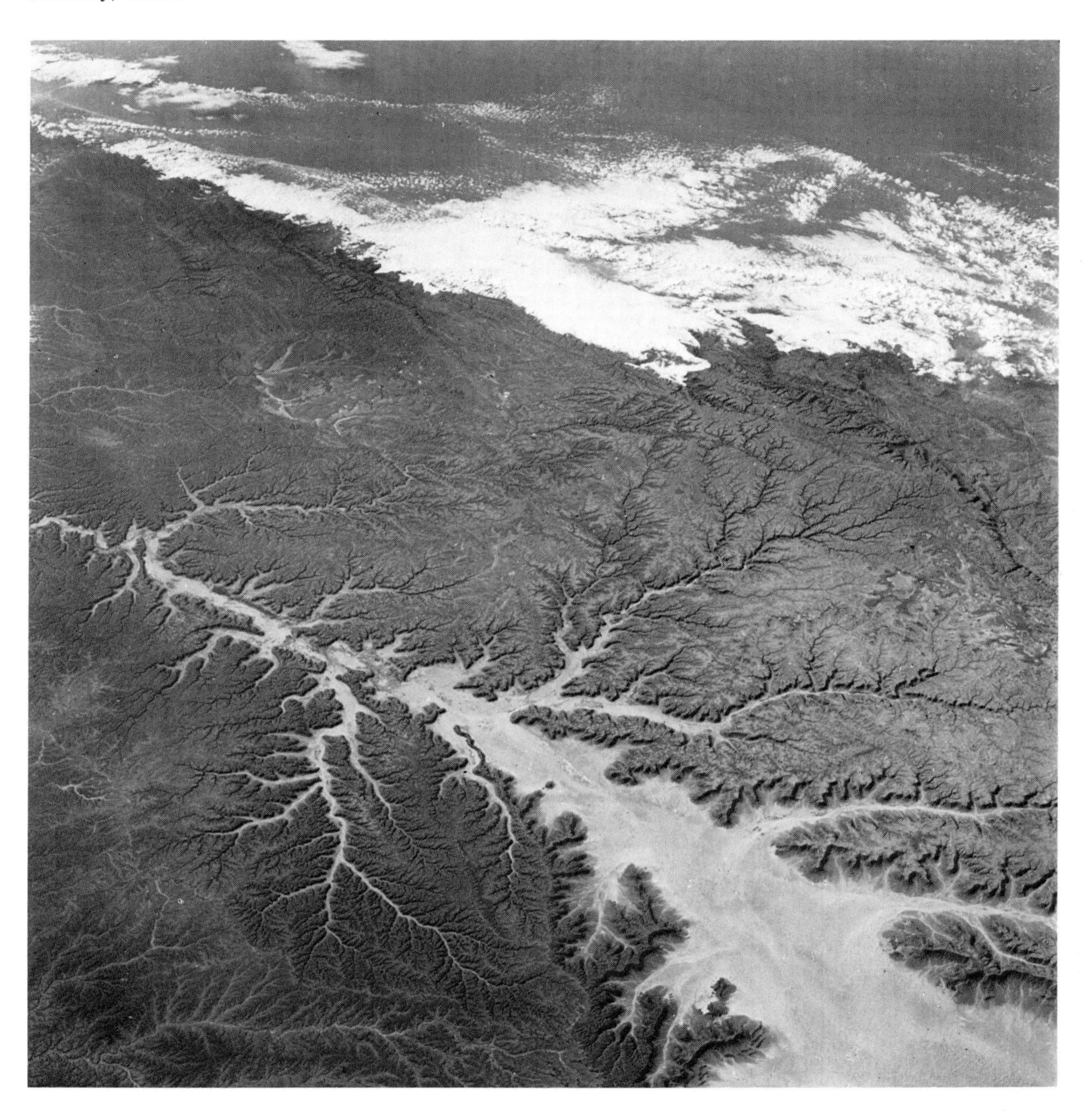

## 5.3.3 Gemini series

Following the photographic success of the Mercury missions the two-man crews of the Gemini series of satellites were asked to obtain colour photographs of 'interesting' areas on the Earth's surface using 70 mm and in some cases 16 mm format cameras (Fig. 5.3). The third Gemini mission in 1965, although short, was the first flight on which photographs were taken and these were of the South West USA. The fourth Gemini mission was much longer and enabled over 100 photographs to be taken (Fig. 5.3) many of which had a spatial resolution of around 70 m and were very nearly vertical (Rowan 1975). The fifth Gemini mission suffered from a lack of power that prevented optimal alignment for ground photography, nevertheless 175 photographs were taken of all over the world. The sixth and seventh Gemini missions are well known for their space rendezvous but they also returned 310 photographs between them, some of which were taken using false colour near infrared film (sect. 3.3.6). During 1966 many more colour photographs were taken as part of a further five Gemini missions, increasing the total number of photographs taken from Gemini satellites to over 2,400 (NASA 1967a). Full catalogues of these photographs, along with some of the more beautiful and interesting examples, are to be found in NASA (1967b) and NASA (1968) with a discussion of their applications in Lowman (1969) and Lowman (1980).

## 5.3.4 Apollo series

By the time the Apollo series of satellites were launched, a remote sensing experiment was formalised. The aims were first, to obtain automatic colour photography from Apollo 6 in 1967. The second was to obtain hand-held colour photography from Apollos 7 and 9 for areas of the Earth for which aerial photography and ground data were also available and the third was to obtain multiband photography using a multicamera array (sect. 3.2.6) from Apollo 9. The first two objectives were fulfilled and over 600 colour photographs were taken (Ordway 1975). The third objective was also successful and provided the stimulus for NASA to push quickly ahead with the development of Landsat, the most important Earth resources satellite series to date (Lowman 1980). A considerable amount of research has been undertaken using the photographs taken from the Mercury, Gemini and Apollo satellites (Bodechtel and Gierloff-Emden 1974) but due to their oblique view, age and limited coverage they are now infrequently used for environmental research.

## 5.3.5 An introduction to manned experimental satellite stations

Manned experimental satellite stations provide scientists with an environment for making observations that could not be made on Earth. For example, observations of other planets or the effect of zero gravity can be readily made from manned experimental satellites. In addition they provide a useful platform for remote sensing instrumentation.

### 5.3.6 Skylab

The space station Skylab was launched by NASA during 1973 and 1974 into a near circular orbit, 435 km above the Earth's surface. It circled the Earth every 93 minutes and passed over each point on the Earth's surface between 50 °N. and 50 °S. once every five days. It gained public notoriety by its return to Earth rather than by its launch, as in 1979 it wobbled, re-entered the Earth's atmosphere and scattered itself across the Australian outback. Skylab was designed for

**Fig. 5.4**
Two cameras used for photography from spacecraft altitudes: (a) The Earth resources terrain camera. (Courtesy, McDonnell Douglas Astronautics Company, USA); (b) A 70 mm format Hasselblad camera. (Courtesy, Victor Hasselblad Aktiebolag, Sweden)

a number of space experiments, one of which was concerned with remote sensing and was named the Earth Resources Experiment Package (EREP). This comprised three types of camera, an infrared spectrometer, a 13 channel multispectral scanner, a microwave radiometer-scatterometer and an L band radiometer (NASA 1977, 1978).

Of greatest interest to environmental scientists was the S190A multicamera array that provided good quality images with a medium spatial resolution of 30 to 80 m in six wavebands and the S190B Earth terrain camera that provided good quality images with a high spatial resolution of 20 m in one waveband (Figs 5.4 and 5.5a). These two cameras along with hand-held 35 mm and 70 mm format cameras took over 35,000 photographs, only a small number of which have yet been interpreted.

The S190A multiband images did prove useful to environmental scientists (Anderson *et al.* 1975) but have now been superseded by multiband images taken by sensors on board the Landsat satellite (Corbett 1974). The S190B and hand-held images are still widely used by environmental scientists because they are taken at a large range of solar angles and have a high spatial resolution (Fig. 5.5), (Lee and Weimer 1975; Klemas *et al.* 1975).

To date these images have been used for a variety of environmental applications which have included the monitoring of ocean currents, volcanoes, snow cover, sea ice and dust storms; the mapping of sand deposits, tectonic features and urban areas; the location of meteorite impact zones and the identification and classification of land cover types.

The S192, 13 channel multispectral scanner has received less attention from environmental scientists. Severe noise and calibration problems prevented the use of these data in the thermal infrared wavebands; however, the visible and near infrared wavebands have been used for some applications including snow mapping and land cover studies (NASA 1978).

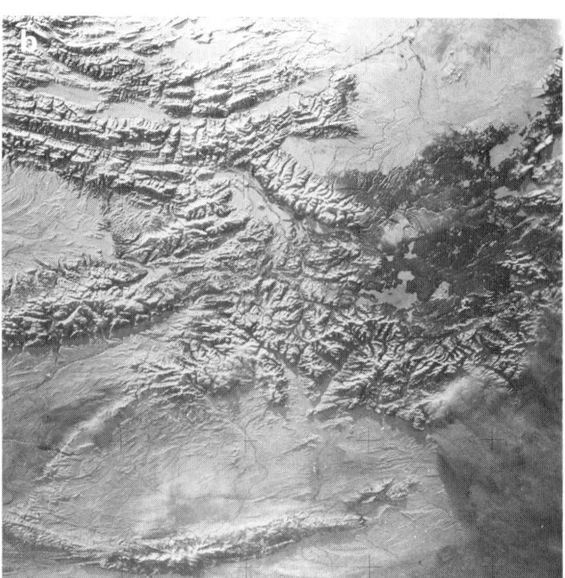


**Fig. 5.5**
Two photographs taken from the satellite Skylab: (a) A high resolution vertical image taken with the Earth resources terrain camera in the summer of 1973. It covers part of the coastline of France from the rivers Durance and Verdon in the north and northeast, the lake Étang de Berre in the northwest, the Golfe du Lion of the Mediterranean in the southwest and the A8 motorway in the southeast; (b) A medium resolution oblique image of northern Wyoming taken with a hand-held, 70 mm format camera in the winter of 1974. The heavy snow cover and low solar angle emphasise the relief. Originals in colour. (Courtesy NASA)

## 5.3.7 Space Shuttle

The Space Shuttle is a series of four NASA spacecraft that are designed to shuttle backwards and forwards between Earth and space. They started flying in April 1981 and are likely to continue well into the 1990s. The Space Shuttle contains three separate components. These are the two rocket boosters, the liquid propellant tank and the orbiter vehicle. The orbiter vehicle, which is almost as large as a jumbo jet, carries three NASA astronauts and up to four scientists. After launch the two rocket boosters are returned to Earth by parachute and the orbiter vehicle burns the liquid fuel of hydrogen and oxygen from the liquid propellant tank until the required orbit at a height of 277 km is reached. The tank is then jettisoned. On return to Earth the orbiter converts from a spacecraft to a glider to land on a runway like an aircraft (NASA 1981a).

The orbiter vehicle will carry and launch many satellites into space. These will include military satellites, possible future Landsat satellites and a new Multimission Modular Spacecraft satellite (MMS) that will probably carry a large format camera and a stereoscopic scanner (Doyle 1978, 1979). The orbiter vehicle will also carry two remote sensing packages named OSTA and Spacelab.

The OSTA package is a collection of remote sensing experiments and is named after its NASA designers, the former Office of Space and Terrestrial Applications. The package contains three environmentally useful sensors, a synthetic aperture radar, a pair of television cameras and an optical imager. The synthetic aperture radar, named the Shuttle Imaging Radar (SIR) operates in the L band and has similar specifications to the synthetic aperture radar carried by the satellite

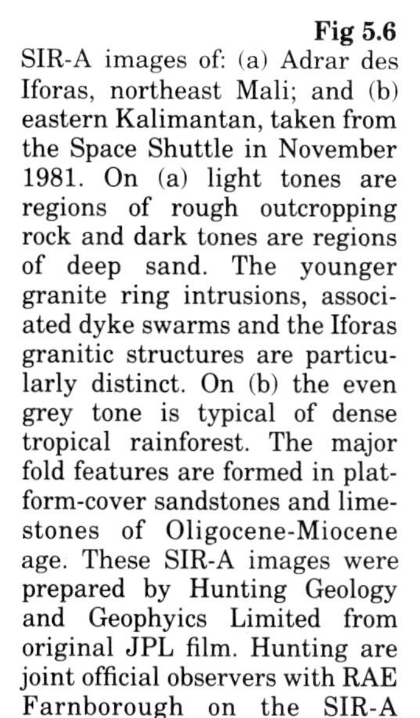

**Fig 5.6**
SIR-A images of: (a) Adrar des Iforas, northeast Mali; and (b) eastern Kalimantan, taken from the Space Shuttle in November 1981. On (a) light tones are regions of rough outcropping rock and dark tones are regions of deep sand. The younger granite ring intrusions, associated dyke swarms and the Iforas granitic structures are particularly distinct. On (b) the even grey tone is typical of dense tropical rainforest. The major fold features are formed in platform-cover sandstones and limestones of Oligocene-Miocene age. These SIR-A images were prepared by Hunting Geology and Geophyics Limited from original JPL film. Hunting are joint official observers with RAE Farnborough on the SIR-A investigation team. (Courtesy, NASA)

Seasat (sect. 5.4.12), from whose spare parts it was constructed. In 1981 SIR-A, the first SIR, recorded for eight hours returning images with a 40 m spatial resolution, of 10 million square km of terrain in 56 km wide swaths (Fig. 5.6), (Holmes 1983). In 1984 SIR-B, the second SIR, will return images in 35 to 50 km wide swaths and due to improved system specifications SIR-B will be the first space borne radar mission that will provide quantitative, calibrated, multiple incidence angle imagery (sect. 4.4).

The pair of television cameras carried in the OSTA package are known as the Feature Identification and Location Experiment (FILE). The red (0.65 $\mu$m) and near infrared (0.85 $\mu$m) wavelengths they record are ratioed together for the classification of the Earth's surface into vegetation, rocks, soils, water, clouds, snow and ice. It is anticipated that FILE will be incorporated into the Earth resource satellites of the 1990s as a means of limiting data acquisition over both oceans and clouds. In 1981 the optical imager carried by the OSTA package was the ocean colour experiment (OCE), an eight channel multispectral scanner which recorded in very narrow wavebands from blue to near infrared wavelengths and produced images with a 3 km spatial resolution and a swath width of 550 km. In 1984 the optical imager will be a large format camera (LFC) designed specifically for mapping applications (Taranik and Settle 1981; NASA 1982a; Edelson 1982; Doyle 1982).

The second remote sensing package to be carried by the Space Shuttle is Spacelab, which is built and operated by the European Space Agency (ESA). As remote sensing is one of the subjects under study by the scientists on board Spacelab, a large exterior pallet has been installed and this will carry a number of sensors, including a mapping camera (sect. 3.2.2) with a 10 m to 20 m spatial resolution and an X band synthetic aperture radar (SAR), (sect. 4.4) with a spatial resolution of around 50 m (Hannover 1981). Both of these will be operated when the Space Shuttle is flying upside down with its doors open (Sobotta 1979). As Spacelab will have a mission time of seven days and a turn around time of only two weeks it should be possible to undertake detailed seasonal monitoring from space (NASA 1982d).

### 5.3.8 Availability of data from manned satellites

Photographs taken during the early manned space missions of Mercury, Gemini, Apollo and Skylab can be purchased from the EROS data centre in the USA (Appendix A). The data recorded during Space Shuttle flights can be obtained from NASA in the USA and ESA in France (Appendix A).

## 5.4 Unmanned Earth resources satellites

There are four distinct groups of unmanned Earth resources satellites. Group one and two both carry sensors that record visible and near visible wavelengths; group one comprise the Landsat series which were the first generation of Earth resources satellites and group two comprise the second generation of Earth resources satellites. These include the French satellite, Systeme Probatoire de l'Observation de la Terre (SPOT) and the Landsat-borne Multispectral Resource Sampler (MRS). Group three carry sensors that record thermal infrared wavelengths and include the Heat Capacity Mapping Mission satellite (HCMM) and group four carry sensors that record microwave wavelengths including Seasat, the Earth Resources Satellite (ERS), (Fig. 5.1) and possibly Radarsat.

## 5.4.1 First generation, unmanned Earth resources satellites: an introduction to Landsat

Following the success of the manned space missions, NASA and the US Department of Interior developed an experimental Earth resources satellite series to evaluate the utility of images collected from an unmanned satellite. The first satellite in this series was a modified Nimbus weather satellite that carried two types of sensor, a four waveband multispectral scanning system (MSS) and three return beam vidicon (RBV) television cameras. When launched in July 1972 it was called the Earth Resources Technology Satellite-1 (ERTS-1), a name that it held until January 1975 when it was renamed Landsat 1 (Freden and Gordon 1983).

This satellite proved to be of great importance as it gave remote sensing world-wide recognition and was the harbinger of the unmanned Earth resources satellites of today. The main advantages

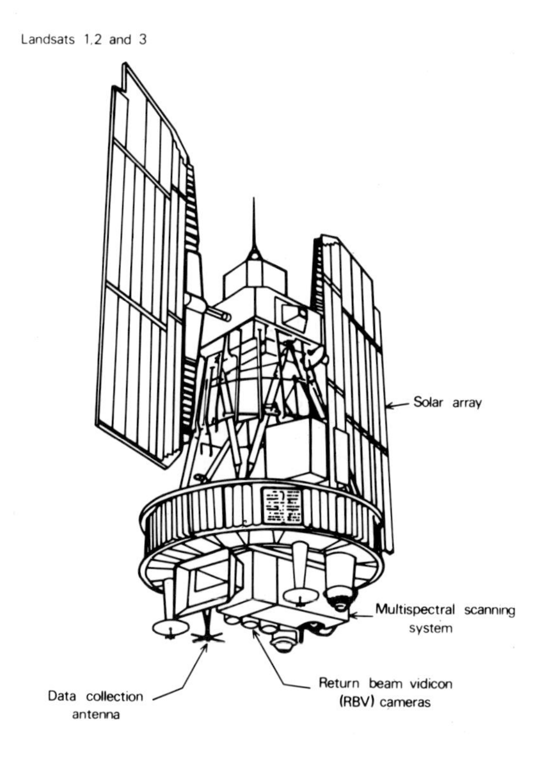


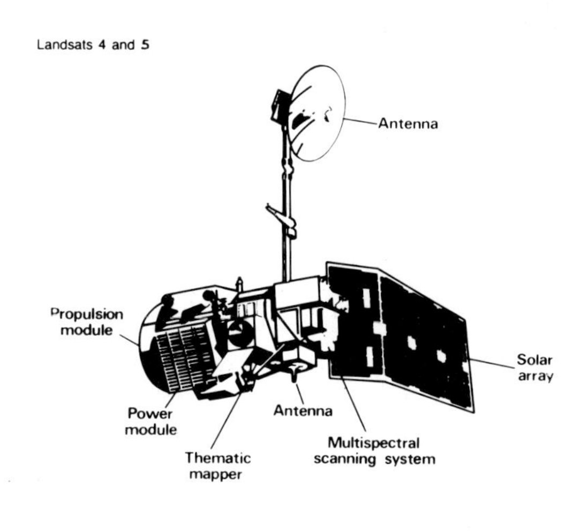


**Fig. 5.7** The two Landsat satellites. The satellite on the left is Landsat 1, 2 and 3, the satellite on the right is Landsat 4 and 5 (Modified from NASA 1976, 1982c)

offered by the imagery collected from this satellite were: ready availability for most of the world, lack of political, security or copyright restrictions, low cost, repetitive multispectral coverage and minimal image distortion.

Landsat 1 was a small satellite. Its body was the size of a British Leyland Mini Metro car and its wing-like solar panels stretched little more than 4 m (Fig. 5.7). The satellite's orbit was high and fast at an altitude of 900 km (± 30 km) and a speed of 6.5 km $sec^{-1}$. The orbit was circular, flying within 9 ° of the North and South Poles and was Sunsynchronous — synchronised with the Earth's motion in relation to the Sun so that it kept pace with the Sun's westward progress as the Earth rotated. To obtain repeat coverage of an area the satellite's orbits were moved westwards each day and this enabled an image to be taken of each area on the Earth's surface once every 18 days (NASA 1976). This orbit proved to be successful and was adopted virtually unchanged by Landsat satellites 2 and 3.

Landsat 1 lasted for almost six years until January 1978 and for part of its life it shared the heavens with Landsat 2 (named ERTS-B while under construction) which was launched in 1975 (Table 5.1). In 1978 Landsat 3 was launched (named Landsat C while under construction) and in 1982 Landsat 4 was launched (named Landsat D while under construction). The current Landsat satellite was named Landsat D′ and became Landsat 5 on launch. For Landsats 4 and 5 the satellite body was changed to improve stability and pay load capability (Fig. 5.7), the orbital altitude was lowered by 215 km to 705 km, thus giving a faster repeat cycle of 16 days and a changed orbit spacing (NASA 1982c).

**Table 5.1**
The operational status of the Landsat series of satellites in February 1984.

| | *Launch* | *Deactivation or reduction in service* | *Activation or increase in service* | *Retired* |
|---|---|---|---|---|
| Landsat 1 | July 1972 | — | — | January 1978 |
| Landsat 2 | January 1975 | November 1979<br>February 1982 | June 1980 | July 1983 |
| Landsat 3 | March 1978 | December 1980<br>March 1983 | April 1981 | September 1983 |
| Landsat 4 | July 1982 | February 1983 (TM only) | December 1983 (TM only) | |
| Landsat 5 | March 1984 | | | |

NASA ran Landsats 1 to 3 for their whole life and Landsat 4 for a few months as these were experimental satellites. The commercial National Oceanic and Atmospherics Administration (NOAA) now runs Landsats 4 and 5 unless a private buyer can be found to relieve the USA government of what has so far been a $1 billion enterprise.

### 5.4.2 Landsat sensors

The Landsat satellites all carry or have carried two sensors; a multispectral scanning system (MSS) and either a Thematic Mapper (TM) scanner or return beam vidicon (RBV) television cameras (Fig. 5.8).

**Plate 1**

False colour near infrared aerial photograph of Kirkby Industrial Estate, East Liverpool, UK. The photograph covers an area of 3.5 x 3.5 km and was taken in 1975. The effect of industrial pollution can be seen by the poor condition of the vegetation in the industrial estate. For example, the two mixed woodlands either side of the image centre are under stress as a result of steam, phosphoric acid and hydrochloric acid emissions from the phosphorus pentoxide factory located between them. Since 1945, when the factory opened, surveys of these two woods (Vick and Handley 1977), have revealed a gradual decrease in the area of the tree canopy within 100 m of the factory as this is the region in which the majority of pollutants descend to the ground. (Courtesy, Merseyside County Council)

**Plate 2** (right)
Digitised colour photograph used in Figs. 6.11, 6.22, 6.23 and 6.26.

**Plate 3**
Multispectral scanner images of Lathkill Dale, Derbyshire, UK. These images cover an area 7 x 4 km and were recorded on 17.9.1982. At bottom left is a colour composite, at bottom centre is a false colour composite and at bottom right is a vegetation index image. In the bottom right image, the woodland and image edge have been removed and the tonal range has been density sliced into four levels of green leaf area index (GLAI), where brown represents 0-1 GLAI, yellow represents 1-2 GLAI, light green represents 2-3 GLAI and dark green represents over 3 GLAI. For further details refer to Fig. 4.7 and Wardley and Curran (1983). (Data courtesy, Natural Environment Research Council, UK)

**Plate 4**
Mosaic of 52 Landsat MSS simulated colour composite images of the UK(Merson 1983). (Courtesy, Space and New Concepts Department, RAE Farnborough, UK)

**Plate 5**
Optical products from an analogue image processor. (a) Oblique colour aerial photograph of a wood near Booker in Buckinghamshire, UK, taken in May 1980. This photograph was converted to a black and white image and density sliced using a colour video processor. The resultant images are two of many possibilities. They both demonstrate the ability of density slicing to enhance and suppress information. For example, (b) and (c) both suppress fine detail, (b) suppresses the grass/wood boundary and both (b) and (c) enhance differences in crop cover

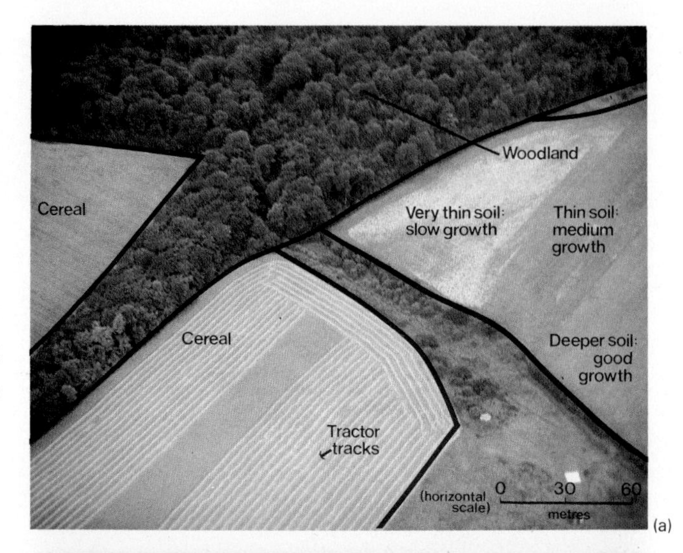


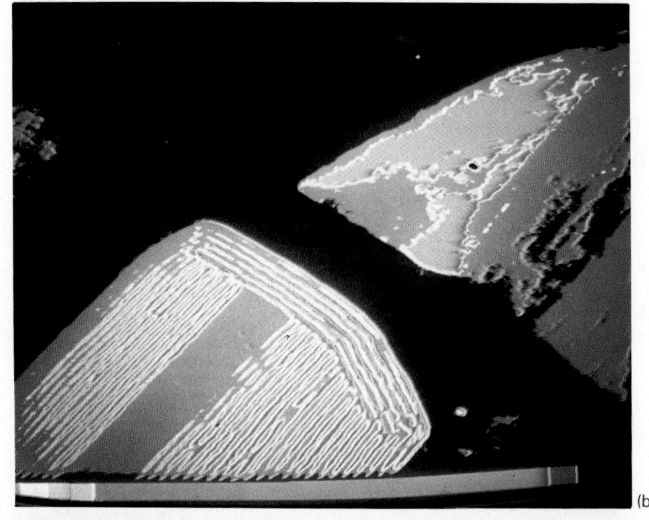

**Fig. 5.8**
The sensors carried, or to be carried, by the Landsat series of satellites. All Landsat satellites carried or carry the multispectral scanning system (MSS); in addition, early satellites carried return beam vidicon (RBV) television cameras and later satellites carry the Thematic Mapper (TM) scanner.

| | Multispectral Scanning System | | Television Cameras | |
|---|---|---|---|---|
| | Multispectral Scanning System (MSS)<br>4 wavebands<br>80m resolution | Thematic Mapper (TM)<br>7 wavebands<br>30 and 120m resolution | Three Return Beam Vidicon (RBV) cameras<br>3 wavebands<br>80m resolution | Two Return Beam Vidicon (RBV) cameras<br>1 waveband<br>40m resolution |
| Landsat 1 | ■ | □ | ◩ | □ |
| Landsat 2 | ■ | □ | ◩ | □ |
| Landsat 3 | ■ | □ | □ | ■ |
| Landsat 4 | ■ | ■ | □ | □ |
| Landsat 5 | ■ | ■ | □ | □ |

KEY
■ Carried
◩ Carried but little used
□ Not carried

## 5.4.2.1 Multispectral scanning system (MSS)

The MSS records four images of a scene, each covering a ground area of 185 km by 185 km at a nominal ground resolution of 79 m. These four images cover green, red, near infrared and near infrared wavebands (Table 5.2) and were identified by the channels they occupied in the satellites telemetry system which were 4, 5, 6 and 7 respectively (Fig. 5.9).

**Table 5.2**
The four wavebands recorded by the multispectral scanning systems (MSS) on board the Landsat series of satellites.

| *Landsat MSS band number* | | | |
|---|---|---|---|
| *Landsat 1, 2 & 3* | *Landsat 4 & 5* | *Band width (μm)* | *Waveband name* |
| 4 | 1 | 0.5–0.6 | green |
| 5 | 2 | 0.6–0.7 | red |
| 6 | 3 | 0.7–0.8 | near infrared |
| 7 | 4 | 0.8–1.0 | near infrared |

For the MSS carried by Landsats 1, 2 and 3 the images are produced by reflecting the radiance, recorded from 79 m wide scan lines on the Earth's surface, to detectors onboard the satellite (Fig. 5.10). To measure radiance for a particular area, the scan line is divided into units by the instantaneous field of view (IFOV) of the sensor. As the satellite moves forward so quickly it is necessary to record 6 scan lines at once, necessitating the use of 24 detectors, 6 in each of the 4 wavebands. Each one of these detectors converts the recorded radiance into a continuous electrical signal which is then sampled at fixed time intervals and converted to a 6 bit number (0–64) and either recorded onto a magnetic tape or transmitted down to Earth, where it is rescaled to a 7 bit number (0–128) for bands 4, 5 and 6 (Lansing and Cline 1975; Slater 1980).

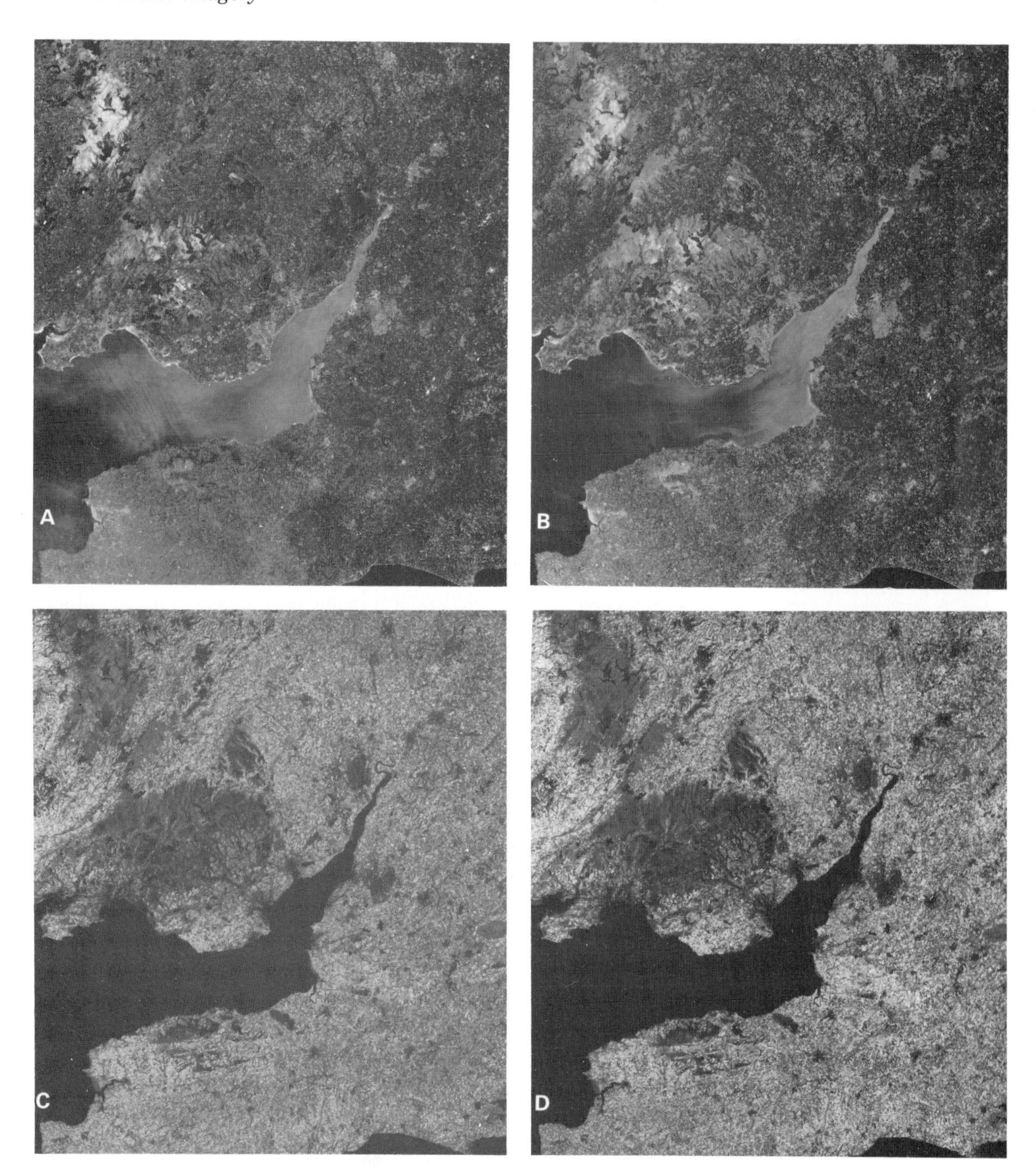


**Fig. 5.9**
Landsat MSS images of south western England. Image (A) is in band 4 (green), image (B) is in band 5 (red), image (C) is in band 6 (near infrared) and image (D) is in band 7 (near infrared). (Courtesy, Space and New Concepts Department, RAE Farnborough, UK)

The sampling procedure, whereby the continuous electrical signal is converted to a discrete digital number, is not the same as the scanning rate. As a result the picture elements (pixels) correspond to a ground area of 56 m by 79 m but the spatial resolution of the image is the 79 m by 79 m area sampled by the IFOV of the sensor. Therefore the pixels which comprise a Landsat MSS image are not square but are rectangular (Slater 1979, 1980). This confusing state of affairs is summarised in Fig. 5.11.

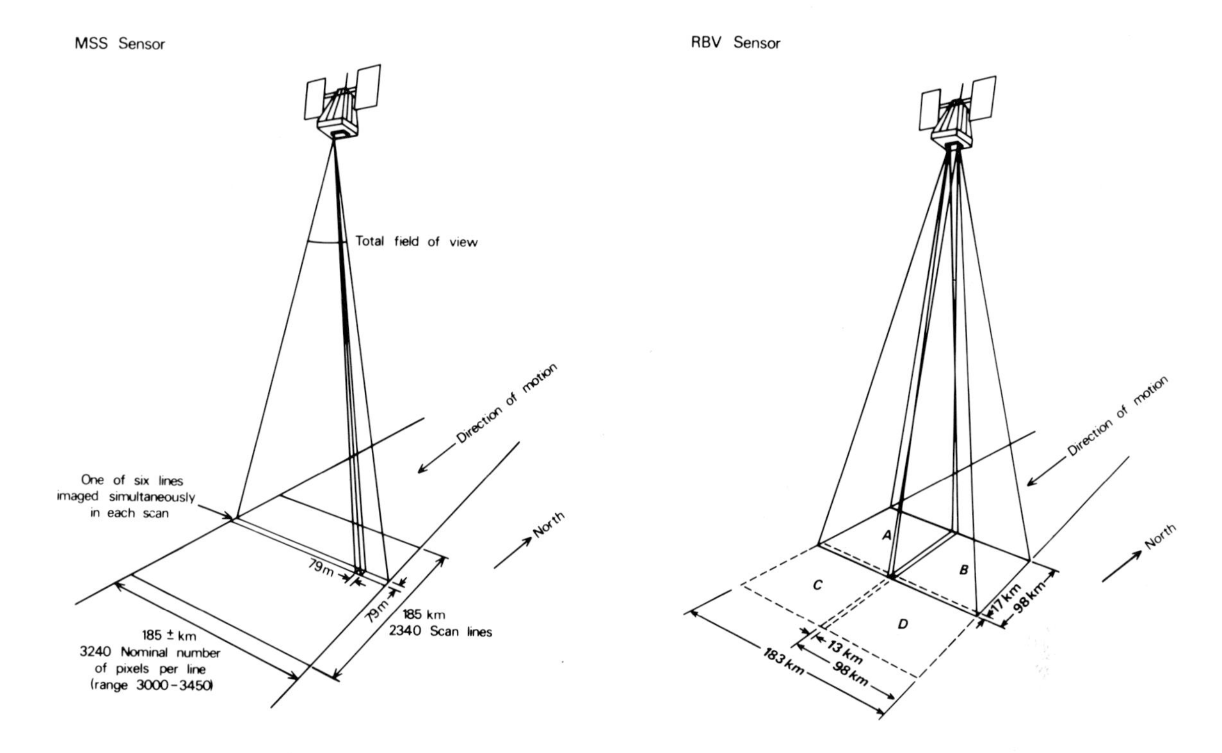


**Fig. 5.10**
Dimensions and configuration of Landsat MSS and RBV images from Landsats 1, 2 and 3. (Modified from NASA 1976)

A complete channel of Landsat MSS imagery comprises 2,340 scan lines and 3,240 pixels per line, a total of around 7.5 million pixels per channel and 30 million pixels per scene of 4 channels. As this awesome amount of data, which fills just one computer compatible tape (CCT), can be collected in just 25 seconds it should come as no surprise to learn that not all of the Landsat MSS scenes are recorded, as to do so would generate 30 million of these data packed CCTs each and every year.

**Fig. 5.11**
The sampling pattern of the Landsat multispectral scanning system. (Modified from NASA 1976)

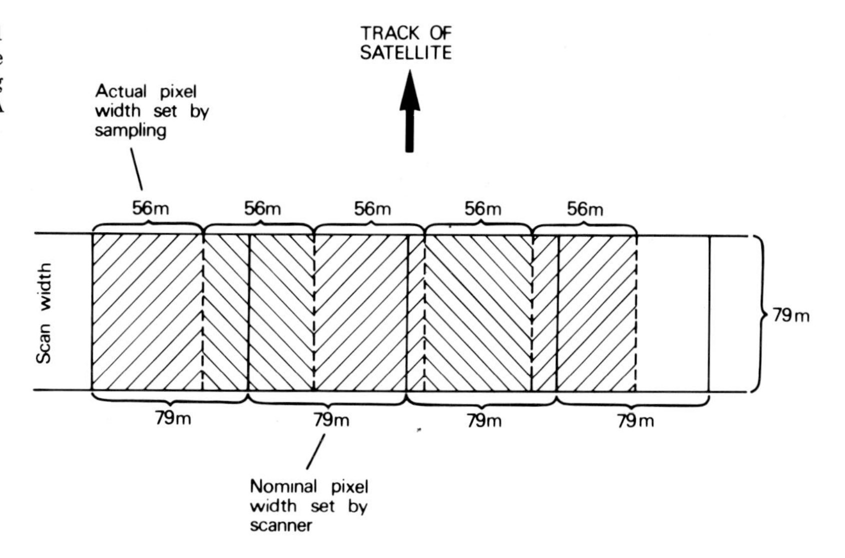

The MSS sensor has undergone very little change since the launch of Landsat 1. The only three changes worthy of note are first, the addition of an extra waveband, known as band 8 to the MSS of Landsat 3. This recorded thermal infrared (10.4–12.6 μm) images but as it failed shortly after launch few of these images have been used for environmental applications (Ormsby 1982; NASA 1982b). Second, to compensate for the lower orbit altitude of Landsat 4 the spatial resolution (IFOV) of the MSS images was decreased by 3 m to 82 m and the field of view (FOV) was increased by 3.41 ° to 14.93 °. Third, the numbering of the MSS wavebands was changed for Landsat 4 and D′, as is indicated in Table 5.2.

Not all of the available Landsat MSS images are of high quality. Of the many possible image defects (Kalush 1979), two have proved to be particularly annoying, these are a line start anomaly and sixth line banding (sect. 6.3.5.2). A line start anomaly occurred when the pulse that activated the MSS detector intermittently failed, so that the scanner occasionally scanned the first 30 per cent of the scene before a second pulse activated the detector. To make matters worse the 70 per cent of the recorded scan line was justified to the left resulting in a virtually unusable image. From 1980 onwards the scan lines were justified to the right so that in such circumstances only 30 per cent of the scene was lost (NASA 1980a).

A more common, but less serious, problem is that of sixth line banding, where due to drifting in the response levels of the six detectors per waveband, every sixth scan line is regularly either lighter or darker than average. It is, however, possible to suppress this effect as is detailed in Table 6.1.

Landsat MSS data were initially used to obtain a synoptic view of a large area of the Earth's surface for interpretation with aid of aerial photographic interpretation techniques discussed in section 3.5. Today due to the increased availability of digital image processors, digital Landsat MSS data are also used for classifying land cover, estimating

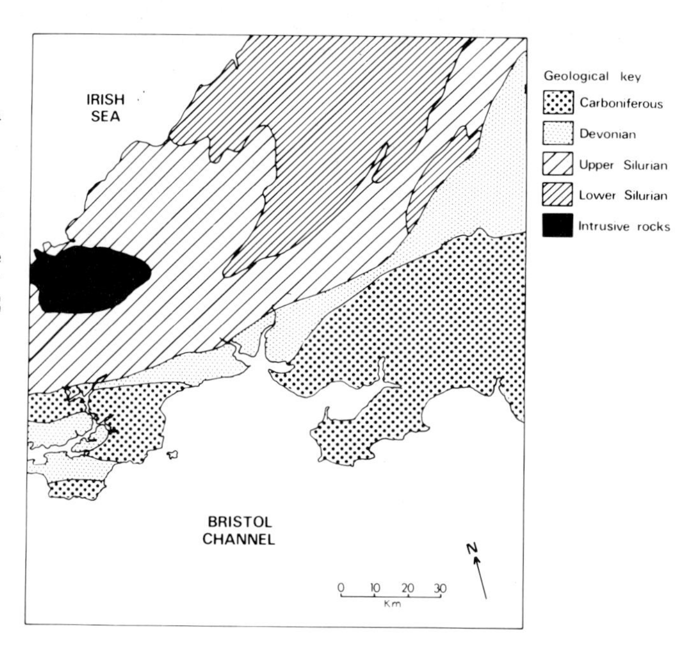


**Fig. 5.12**
Part of a near infrared (MSS 6) Landsat MSS scene of South West Wales, UK. Taken on 13.9.1977. The image is an enlargement of the whole Landsat MSS scene and covers an area with a very complex geology as can be seen in the accompanying map. (Image Courtesy, Space and New Concepts Department, RAE Farnborough, UK)

characteristics of the Earth's surface and for monitoring change (Armstrong and Brimblecombe 1979; NASA 1980b, 1980c). Some of the earliest applications of Landsat MSS data were in geological investigations (Welby 1976; Missallati *et al.* 1979; Iranpanah and Esfaniari 1980), (Fig. 5.12 and 5.13) and geomorphological mapping (Kayan and Kelmas 1978; Sauchyn and Trench 1978; Robinove *et al.* 1981), Fig. 5.14). Other applications have included hydrological studies (Heller and Johnson 1979; Shih 1980; Killpack and McCoy 1981; Green *et al.* 1983), soil surveys (Westin and Frazee 1976; Mitchell 1981; Imhuff *et al.* 1982), forest inventory (You Ching 1980; Hall *et al.* 1980; Strahler

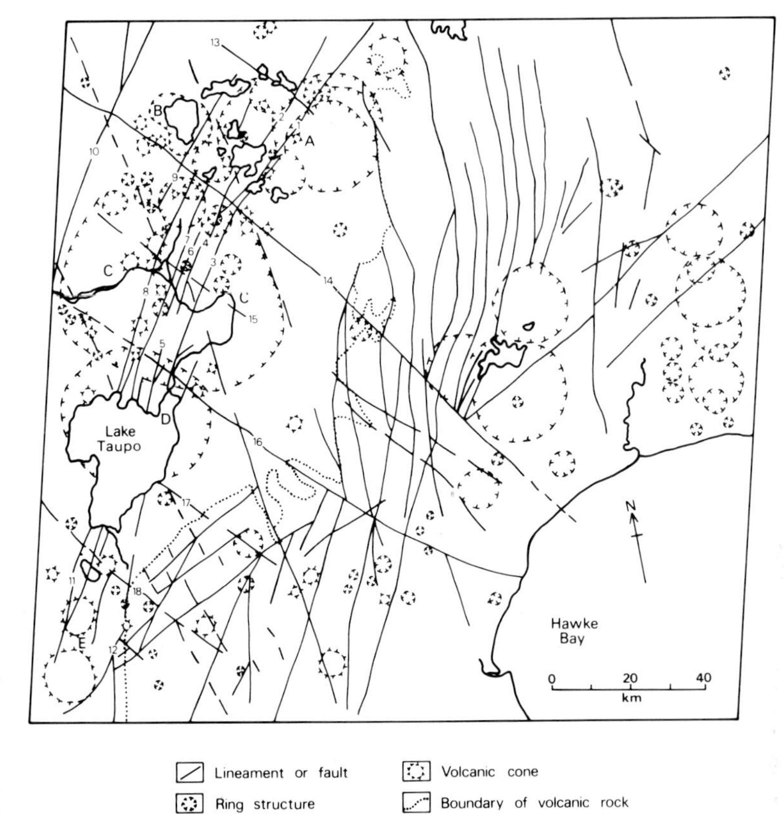


**Fig. 5.13**
A geological interpretation of a Landsat MSS 7 image of part of North Island, New Zealand. (Cochrane and Tianfeng 1983)

**Fig. 5.14**
Two Landsat MSS 7 images of arid terrain. Image (a) is of the folded Toba Kaka Hills in West Pakistan and is produced by the EROS data centre in the USA. (Courtesy, NASA). Image (b) is of the heavily dissected Gebel el Tîh desert in the Sinai peninsula and is produced by the Space and New Concepts Department, RAE Farnborough, UK.

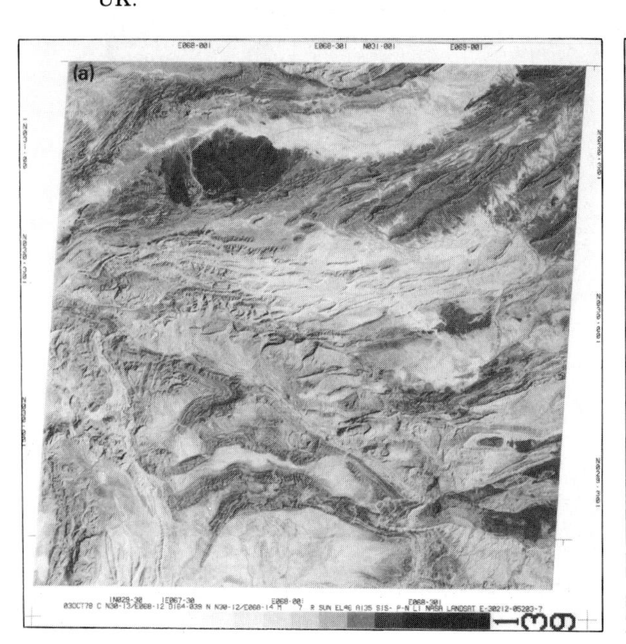


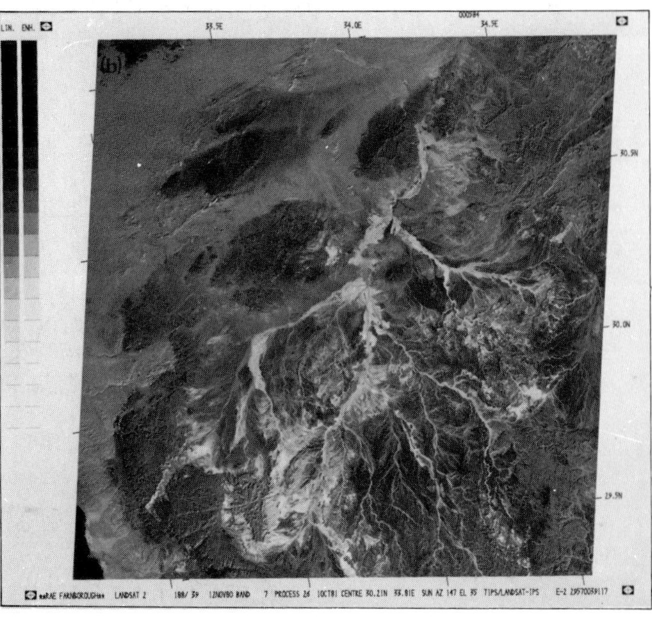

1981; Mayer and Fox 1981) and oceanography (Klemas 1980; Mather 1981; Sturm 1981; Cracknell *et al.* 1982), (Fig. 5.15).

Two large applications of Landsat MSS data are mapping land cover (Fig. 5.15 and 6.34) and monitoring change. Land cover mapping has involved assessment of the total land cover (Stove and Hulme 1980; Sweet *et al.* 1980) the agricultural land cover (Welch *et al.* 1979; Bauer *et al.* 1979; Hixson *et al.* 1981) and the urban land cover (Welch 1980; Jackson *et al.* 1980; Lo 1981; Ilsaka and Hegedus 1982). Monitoring change has involved both aquatic applications (Rango and Martinec 1979; Ohlhorst 1981; Hong and Iisaka 1982; Ito and Muller 1982) and terrestrial applications (Robinove 1975; Coiner 1980; Hamlin 1980; Howarth and Wickware 1981). Landsat MSS data are also being successfully applied to zoological studies (Craighead 1976; Hielkema 1977; Marmelstein 1978), the estimation of crop yield

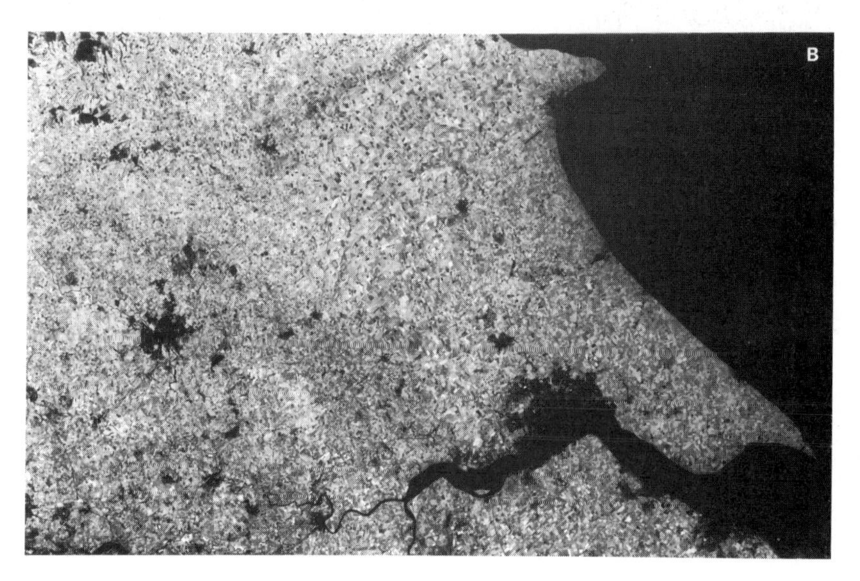


**Fig. 5.15**
Landsat MSS images of the Humber estuary in northeast England. These images are a fifth of a total Landsat MSS scene. They were taken on 27.6.1976, a time when there were high sediment loads in the estuary, as canbe seen in image (A), MSS 5, (red) and when the fields were ready for harvest, as can be seen in image (B), MSS 7, (near infrared). Refer to the cover for a false colour composite. (Courtesy, Space and New Concepts Department, RAE Farnborough, UK)

(Barnett and Thompson 1982; AgRISTARS 1981, 1982, 1983), the monitoring of water quality (Deutsch and Estes 1980; Klemas and Philpot 1981; Aranuvachapun and LeBlond 1981), the identification of clouds (Barrett and Grant 1978; Bagchi 1979), the measurement of atmospheric aerosols (Darnell and Harriss 1983), and also as background data to a number of development projects (Sobur *et al.* 1978; Cyganiak 1980).

For further examples of the application of Landsat MSS refer to Deutsch *et al.* (1981), Lulla (1983) and recent issues of the major remote sensing journals and symposia (Appendix B).

### 5.4.2.2 Thematic Mapper (TM)

The Thematic Mapper (TM) carried by Landsats 4 and 5 is the ultimate in scanner design, for it is unlikely that a sensor with finer tolerances can ever be manufactured. It records 256 radiance levels in 7 wavebands (Table 5.3) with a spatial resolution (IFOV) of around 30 m in 6 of them (Fig. 5.7). This is in contrast to the MSS that records 64 radiance levels in 4 wavebands with a spatial resolution of 79 m (Blanchard and Weinstein 1980), (Table 5.4). There have been severe problems with the collection of TM data (NOAA 1983), however, the data that are available look most exciting (Fig. 5.16 and 5.17). Full evaluation of TM data in terms of its geometric and radiometric performance will not be available until well after the data are released for sale in January 1985 (NASA 1982b; NOAA 1982a). However, current indications are that TM data will be one of the most important types of Earth resources satellite data in the 1980s (Short 1982). For further details refer to Salomonson (1978), Salomonson *et al.* (1980), Colvocoresses (1979a) and NASA (1982c).

**Table 5.3**
The wavebands recorded by the Thematic Mapper sensor carried by Landsat 4 and 5.

| *Band number* | *Band name* | *Band width (μm)* | *Points* |
|---|---|---|---|
| 1 | Blue/Green | 0.45–0.52 | Good water penetration, strong vegetation absorbance |
| 2 | Green | 0.52–0.60 | Strong vegetation reflectance |
| 3 | Red | 0.63–0.69 | Very strong vegetation absorbance |
| 4 | Near infrared | 0.76–0.90 | High land/water contrasts, very strong vegetation reflectance |
| 5 | Near-middle infrared | 1.55–1.75 | Very moisture sensitive |
| 6 | Thermal infrared | 10.4–12.5 | Very sensitive to soil moisture and vegetation |
| 7 | Middle infrared | 2.08–2.35 | Good geological discrimination |

*Sources*: NASA 1982b, 1982c.

**Table 5.4**
A comparison between the multispectral scanning system (MSS) carried by all Landsat satellites and the Thematic Mapper carried by Landsat 4 and 5.

| | *Multispectral Scanning System (MSS)* | *Thematic Mapper (TM)* | *Comment on (TM) details* |
|---|---|---|---|
| Wavebands | 4 | 7 | See table 5.3 |
| Spatial resolution (IFOV) metres | 79 | 30 metres in bands 1–5 and 7 120 metres in band 6 | High spatial resolution provides greater detail but lower classification accuracy |
| Pixel size on standard computer compatible tape (metres) | 56 × 79 | 30 × 30 (all bands) | Data are resampled (sect. 6.3.5.5) |
| Field of view (FOV) in degrees | 11 | 17 | Off-vertical view angle effects are evident (sect. 2.5.2.6) |
| Grey levels | 64 | 256 | Possible due to improved radiometric accuracy |
| Number of detectors | 6 | 16 | Evidence of 16 line banding |
| Mirror recording mode | Forwards | Forwards and backwards | Reduces mirror speed and increases dwell time but causes processing problems |
| Computer compatible tapes per scene (1,600 bpi) | 1 | 7 | Awkward to handle |
| Relative cost per scene | 1 | 5 | In 1984 |

*Sources*: Colvocoresses 1979a; Lillesand and Kiefer 1979; Townshend 1981a (personal communication); and Tucker 1980.

## 5.4.2.3 Return beam vidicon (RBV) camera

Return beam vidicon (RBV) television cameras were carried on Landsats 1, 2 and 3. On Landsats 1 and 2 three cameras were used, each filtered into a different waveband, camera 1 into green (0.48–0.58 μm), camera 2 into red (0.58–0.68 μm) and camera 3 into near infrared (0.69–0.83 μm). The images they produced were similar in appearance to MSS images in that they covered an area of 185 km by 185 km and had a ground resolution of 80 m. The only major difference as far as the user was concerned was their greater geometric fidelity, which can be attributed to the recording method of the camera.

Unfortunately, the RBV on board Landsat 1 returned only 1,690 images before a power problem caused it to be turned off in August 1972. The RBV on Landsat 2 returned even fewer images and so for Landsats 1 and 2 the MSS images were their primary product.

Landsat 3 carried two RBVs and these were both filtered to a broad green to near infrared waveband (0.51–0.75 μm). Their design was similar to the RBVs carried by Landsats 1 and 2 except for their focal

**Fig. 5.16**
Landsat TM 3 (red) image taken on 20.8.1982, of the northeast USA only four days after the launch of the satellite Landsat 4. It shows the northeast quadrant of a whole 185 × 185 km scene and covers Detroit and Lake St Clair in the northeast, Toledo and Lake Eire in the southeast and the agricultural land of Michigan to the west. As the spatial resolution of the image is 30 m, the roads and individual fields can be seen. Note enlargement in Fig. 5.17. (Courtesy, NOAA)

length, which was increased to give a nominal ground resolution of around 40 m (Fig. 5.8) and an image area of 98 km by 98 km. To keep the RBV data spatially comparable with MSS data two RBVs operated in tandem producing four images for each one MSS image and these were labelled A, B, C and D (Fig. 5.10).

Unfortunately, less than half of the RBV images collected are of good quality, due to low radiometric resolution (Dosiére and Justice 1983) and shading (Clark 1981). In 1980 NASA implemented a priority system for RBV data acquisition, with countries like USA, USSR and China at the top of the list (NASA 1980a). As the RBV failed in October 1981 coverage of other areas of the world is relatively poor (NASA 1982a). Where RBV data are available, their high spatial resolution in one waveband (Fig. 5.17 and 5.26) have made them ideal for mapping geomorphology (Cochrane and Browne 1981), land cover (King 1981) and urban form, especially the urban form of Soviet cities (Synder 1982)! It can be used on its own (Welch and Pannell 1982) or in composite with Landsat MSS data (NASA 1981b; Short 1982; Barrett and Curtis 1982).

### 5.4.3 Transmission of Landsat sensor data to Earth

All of the Landsat sensors must transmit their images to Earth through a ground receiving station. If there is a ground receiving station nearby, then the data can be transmitted virtually as they are recorded. If not, then Landsat satellites 1, 2 and 3 could store data on

**Fig. 5.17**
A comparison between the products of the three sensors carried by the Landsat satellites. Image (a) is from the Landsat 3, RBV and is in the visible to near infrared waveband. Image (b) is from the Landsat 3, MSS in band 5, (red). Image (c) is from the Landsat 4, TM in band 3, (red). The area is near Detroit, Michigan, USA. All of the images were taken during the summer months and are subscenes of much larger images. The scale, which is the same for all three images, can be determined by reference to the 3.2 km long main runway at Wayne County Metropolitan airport. (Courtesy, NOAA)

a magnetic tape for playback to a receiving station, later in the satellite's orbit. Landsat 4 does not carry tape recorders; but when out of range of a receiving station it is possible to transmit the data to a receiving station in the USA at White Sands, New Mexico via a USA communications satellite called the Tracking and Data Relay Satellite (TDRS), (NASA 1982c).

Initially all Landsat MSS and RBV data were received by the NASA receiving stations at Fairbanks in Alaska and Goldstone in California but since 1972 many new receiving stations have been set up, each responsible for receiving data over a limited area of the globe (Fig. 5.18). For countries that are not within reach of a receiving station, data acquisition will be limited. For example, the majority of

Saudi Arabia is out of range of its nearest receiving stations in Italy and India and as a result only 4% of all potential Saudi Arabian Landsat MSS images have been recorded (Curran and Adawy 1983). Even for countries that are in reach of a receiving station, the number of available images is usually far less than one every over-pass; for example, in the UK there are few areas of the country that can boast an average of more than one or two good quality cloud-free Landsat MSS scenes each year.

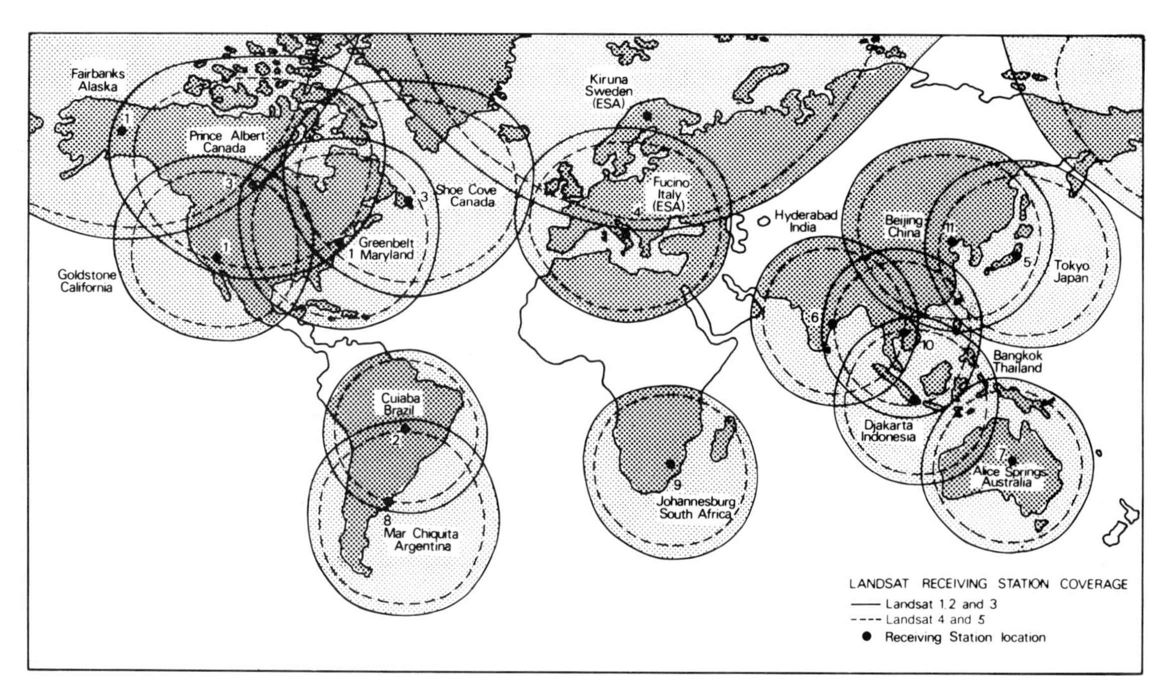


**Fig. 5.18** The location and area coverage of the nine receiving stations that record data derived from Landsat sensors. (Modified from NASA 1982a)

## 5.4.4 Availability of Landsat sensor data

To purchase Landsat sensor data you can either contact the EROS data centre in USA (Appendix A), a receiving station in the area of your interest, a data dissemination service in your own country or a remote sensing company. The choice depends upon your country, the country for which data are required and your specific data requirements. For example, the majority of the Landsat sensor data purchased from the EROS data centre (e.g. Fig. 5.19) is of either the USA or parts of the world that did not have a receiving station when the required imagery was obtained.

A receiving station provides Landsat sensor data for a 4,000 km radius of the station since the station was opened. So for example, to obtain recent cover of Japan you could contact the receiving station in Tokyo (Appendix A). An increasing majority of Landsat sensor data are now purchased from a data dissemination service operated for individual countries. In Europe a group within the European Space Agency called 'Earthnet' obtains Landsat sensor data from the two European receiving stations at Fucino in Italy and Kiruna in Sweden and sells the data through 11 National Points Of Contact (NPOC), (RAE 1983).

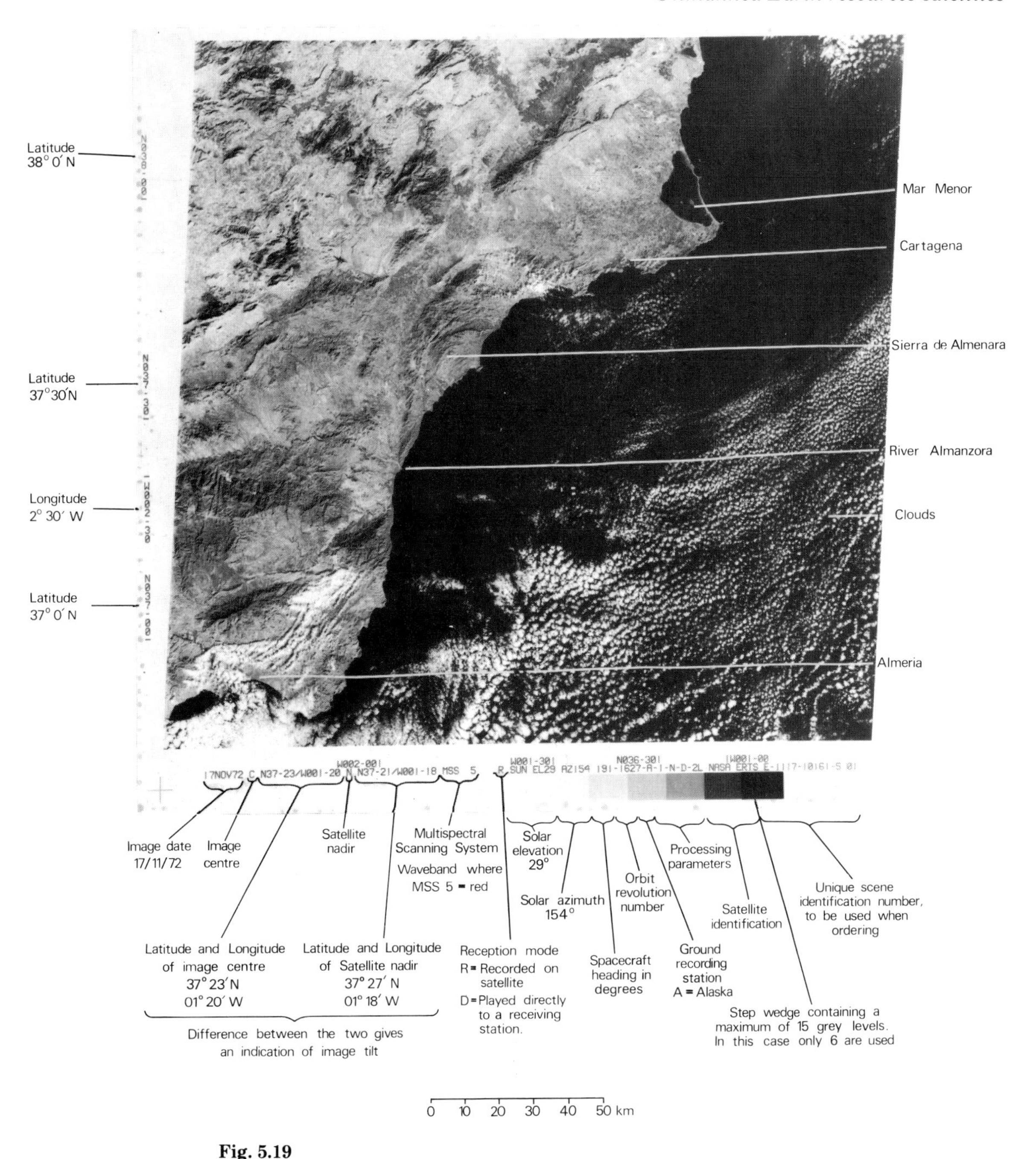


**Fig. 5.19**
An explanation of the annotation used on Landsat MSS imagery purchased from the EROS data centre, prior to 1978. After this date there were minor alterations to include a path and row code, details of processing procedures and the satellite's new name. (Courtesy, NASA)

Commercial firms often sell Landsat sensor data for areas of the world in which they have worked and for this reason the choice of cover is poor but this is often more than compensated by the lower cost of the data.

Wherever the data are purchased from, there is a need to locate the image of interest. This can be done by reference to the flight lines (paths) and the distance down one of the flight lines (rows). The paths and rows of MSS and RBV images from Landsats 1, 2 and 3 for Europe and the UK are given in Fig. 5.20.

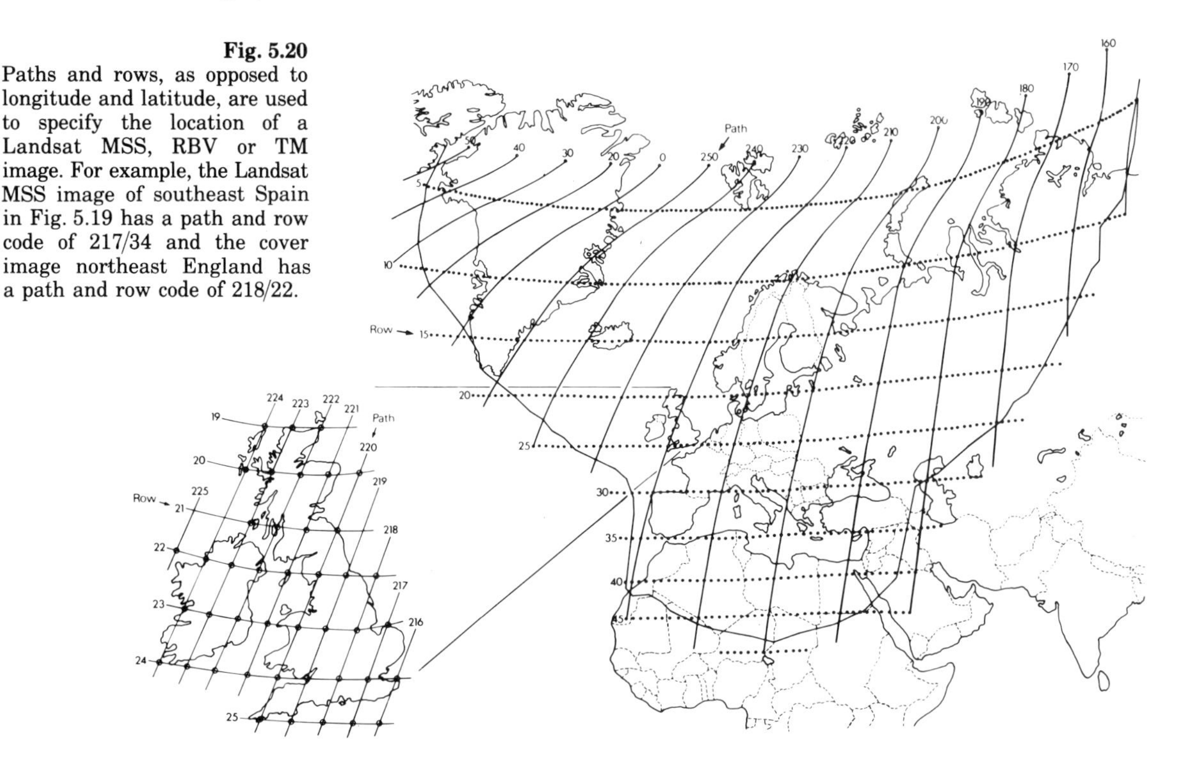


**Fig. 5.20**
Paths and rows, as opposed to longitude and latitude, are used to specify the location of a Landsat MSS, RBV or TM image. For example, the Landsat MSS image of southeast Spain in Fig. 5.19 has a path and row code of 217/34 and the cover image northeast England has a path and row code of 218/22.

Further details are given in NASA (1976) and excellent examples of Landsat MSS and RBV imagery are contained within Short *et al.* (1976), Williams and Carter (1976) and Sheffield (1981).

## 5.4.5 Second generation, unmanned Earth resources satellites

The 1980s and 1990s will see the launch of the second generation of Earth resources satellites. Like the Landsat satellites they will carry optical sensors but these sensors will be linear arrays controlled by microchips that can be pointed in any direction to obtain stereoscopic images or near daily over-passes.

The satellites so far named are the French satellite SPOT, a future Landsat that may carry a sensor called the multispectral resource sampler, and the two Japanese satellites, the Marine Observation Satellite (MOS-1) and the Earth Resources Satellite (ERS-1).

## 5.4.6 SPOT series

The French satellite Systeme Probatoire de l'Observation de la Terre (SPOT) which translated literally means the Earth observation test system, is the first Earth resources satellite to be launched from Europe. It is to be operated by the French Centre National d'Études Spatials (CNES) with participation from both Belgium and Sweden. The SPOT satellites will carry a range of sensors at a number of orbits and altitudes well into the late 1990s (Doyle 1978). To date, two SPOT satellites have been built for launch in 1986 and 1987 (Fig. 5.21). The first of these, SPOT-1, will have a near-polar, Sun-synchronous, 832 km

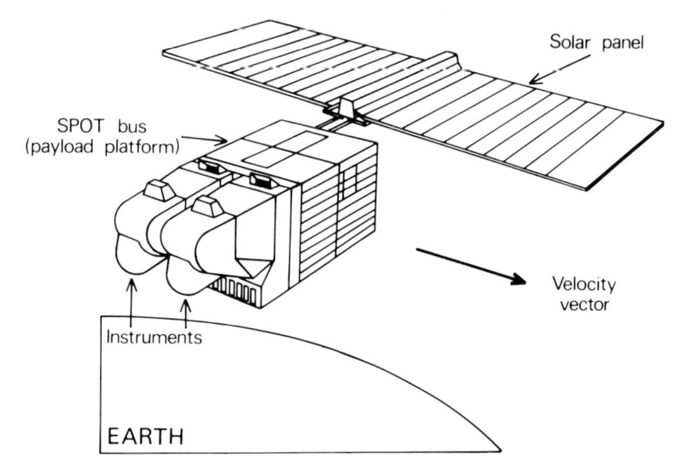


**Fig. 5.21**
An artist's impression of the satellite SPOT showing its three components: a solar panel, a 'bus' or payload platform and the sensor instrumentation. (Modified from CNES 1980)

high orbit, which will repeat every 26 days. It will carry two pushbroom scanners, two tape recorders and telemetry equipment to transmit data to Earth.

The pushbroom scanner is a new generation of multispectral sensor (Wharton *et al.* 1981), the conventional multispectral scanner reaching its peak in the Thematic Mapper carried by Landsats 4 and 5 (Thompson 1979). You will recall from section 5.4.2. that the multispectral scanner scans the scene from side to side and reflects radiation from the ground surface onto a detector (Fig. 5.10). This process is limited by the accuracy of the rotating mirror. To overcome this problem the pushbroom scanner, or 'multispectral solid state linear array' to give its full title, does not have any moving parts and records each scan line at one go by means of a line of detectors – one detector for each area sampled on the ground. The detectors are each controlled by a microchip and are so small that each of the 8 cm long linear arrays carried by the SPOT sensor contains 1,728 detectors. Further details are given in Table 5.5.

**Table 5.5**
A comparison between a pushbroom scanner and a conventional multispectral scanner.

| *Advantages of a pushbroom scanner over a conventional multispectral scanner* | *Disadvantages of a pushbroom scanner over a conventional multispectral scanner* |
|---|---|
| i Lighter weight | i There are many more detectors to calibrate |
| ii Smaller size | ii Cannot sense wavelengths longer than near infrared |
| iii Lower power requirement | |
| iv No moving parts | |
| v Longer life expectancy | |
| vi Greater reliability | |
| vii Higher geometric accuracy | |
| viii Higher radiometric accuracy | |
| ix Higher signal to noise ratio | |
| x Higher spatial resolution | |
| xi Lower cost | |

*Sources*: Colvocoresses 1979b; Thompson 1979; and Slater 1980.

**Table 5.6**
The two operational modes of the High Resolution Visible (HRV) scanner carried by a SPOT satellite.

| *Panchromatic mode* | | *Multispectral mode* | |
|---|---|---|---|
| *Waveband name* | *Waveband width* | *Waveband name* | *Waveband width* |
| visible | 0.51–0.73 μm | green | 0.50–0.59 μm |
| | | red | 0.61–0.69 μm |
| | | near infrared | 0.79–0.89 μm |

*Source*: CNES 1980.

SPOT-1 will carry two identical pushbroom scanners, these are called High Resolution Visible (HRV) scanners and can be used to record in either panchromatic or multispectral mode (Table 5.6), (Begni 1982). When in panchromatic mode all of the detectors are sampled, giving a spatial resolution of around 10 m and when in multispectral mode only half of the detectors are sampled, giving a spatial resolution of around 20 m. The panchromatic mode will be used for planimetric cartography at scales of 1 : 250,000 and cartographic updating at scales of 1 : 50,000 to 1 : 100,000. The multispectral mode will be used for environmental studies ranging from pollution monitoring to vegetation mapping (Chevrel *et al.* 1981; Short 1982).

Of interest to environmental scientists is the inclusion of a mirror in the optical path of the HRV sensor as this enables it to be pointed to ± 27° either side of the ground track (Fig. 5.22) giving three advantages. First, stereoscopic coverage can be obtained by viewing one area in successive orbits; second, the sensor can be pointed to cloud-free areas and third, the sensor can be set to obtain repeated coverage of one area, at a number of look angles, thus decreasing the repeat time to one or two days. The actual repeat time will be dependent upon the latitude and will vary from 7 passes every 26 days at the Equator to 13 passes every 26 days in northern England (CNES 1980; Chevrel *et al.* 1981).

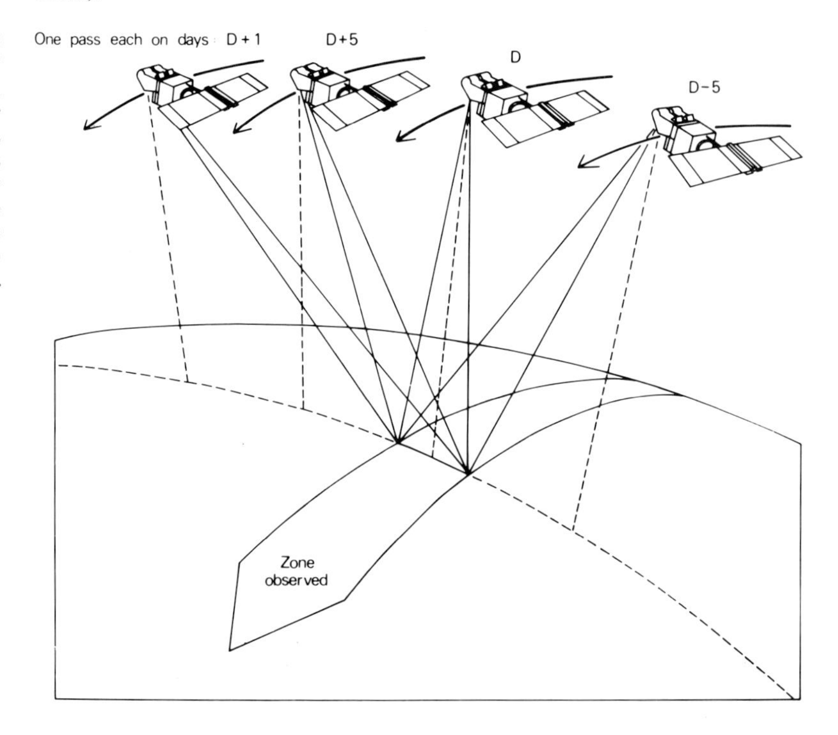


**Fig. 5.22**
An illustration of how the SPOT satellite can be used to obtain images of an area with greater frequency than its 26-day repeat cycle. In this case, the area that is viewed vertically on one day (D), was viewed obliquely five days earlier (D−5) and will be viewed obliquely in five (D+5) and ten (D+10) days time. (Modified from CNES 1980)

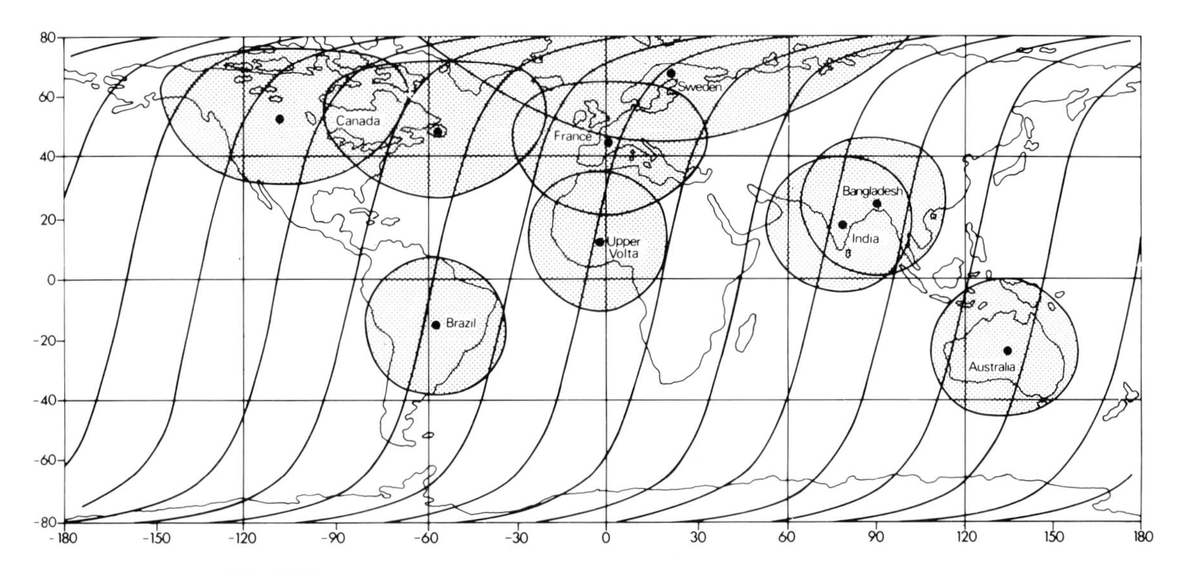


**Fig. 5.23**
The location and area coverage of the 15 receiving stations that can record data from the SPOT sensors. (Modified from CNES/SPOT Image 1981)

Images of Europe from Sweden in the north to the Sahara in the south, the Atlantic in the west and Leningrad in the east will be telemetered in real time to the receiving station at Toulouse in France or occasionally Kiruna in Sweden. Images recorded outside this area will be recorded and then played back when the SPOT satellite passes over Toulouse or transmitted directly to one of the eight SPOT receiving stations (Fig. 5.23), or one of the specially converted Landsat 4 and 5, MSS and TM receiving stations (Fig. 5.18). These data can be purchased from the SPOT marketing service (Appendix A).

## 5.4.7 Landsat with the multispectral resource sampler

The multispectral resource sampler (MRS) sensor may be launched on a Landsat satellite in the late 1980s. If so it will be NASA's answer to the HRV's on board SPOT, but unlike the HRV it will be used to obtain local as opposed to regional coverage. The MRS is a two mode pushbroom scanner with a ground resolution of around 15 m. In mode one it has a swath width of 15 km in four wavebands and in mode two it has a swath width of 30 km in two wavebands. The sensor consists of four arrays each of which can have one of five visible to near infrared filters placed in front of them (Schnetzler and Thompson 1979; Curran 1982b).

## 5.4.8 Japanese Earth resource satellites

The Japanese propose to launch two Earth resource satellites. The first in 1986 is to be the Marine Observation Satellite (MOS-1) and the second in 1990 is to be the Earth Resources Satellite (ERS-1). MOS-1 will have an orbit and altitude similar to that currently used by the Landsat series of satellites and will carry three sensors. The most important of these will be a pushbroom scanner named the Multispectral, Electronic, Self-Scanning Radiometer (MESSR). This will sense in four wavebands from green to near infrared, have a spatial resolution of 50 m and an image area of 200 km by 200 km. MOS-1 will also carry a low resolution multispectral scanner and a microwave scanning radiometer. ERS-1, which is not to be confused with the European satellite with the same name, will carry a stereoscopic linear array camera and a synthetic aperture radar (NASA 1982d).

### 5.4.9 Unmanned Earth resources satellites carrying thermal infrared sensors

Thermal infrared images are required for a number of applications in fields as diverse as pollution monitoring and geological mapping. Unfortunately for remote sensing purposes the intensity of energy emitted from the Earth's surface decreases with wavelength (sect. 2.4.1). Therefore, to record the radiant temperature of the Earth's surface with high radiometric accuracy a large area of the ground must be sampled and this results in thermal images with a low spatial resolution.

Four Earth resource satellites have, do, or will, carry thermal infrared sensors with a low spatial resolution. They are Landsat 3, Landsat D' and the Heat Capacity Mapping Mission (HCMM) satellite.

### 5.4.10 Heat Capacity Mapping Mission (HCMM) satellite

The HCMM satellite was the first of a small and relatively inexpensive series of NASA's Applications Explorer Mission (AEM) satellites. It was launched in April 1978 and lasted until September 1980. Its orbit was near circular and fairly low, at an altitude of 620 km. The satellite contained a scanning radiometer recording in a visible and near infrared waveband (0.55–1.1 $\mu$m) and a thermal infrared waveband (10.5–12.5 $\mu$m). The orbits of the satellite were arranged to ensure that images were obtained of each scene during times of maximum and minimum surface temperature for the determination of thermal inertia (sect. 2.5.5.3). The radiometer had a very wide scan angle of $\pm 60°$ resulting in a swath width of 720 km (Fig. 5.24). The ground resolution decreased from around 0.6 km at the centre of image swath to around 1 km at the edge of the swath (Price 1978). The satellite did not carry tape recorders and so data were only obtained when the satellite was within range of one of the five receiving stations which were spread between North America, Europe and Australia.

The data from HCMM were intended primarily for conversion to thermal inertia maps for geological mapping (Kahle *et al.* 1981). However, the images have found wider application (Foster *et al.* 1981) for example, in vegetation mapping (Byrne *et al.* 1981; Kalma *et al.* 1983), vegetation stress detection, microclimatology (Carlson 1981; Seguin and Itier 1983), soil moisture mapping (Heilman and Moore 1981), snow melt prediction, the mapping of urban heat islands, and monitoring industrial thermal pollution (NASA 1982e). These data can be purchased from the World Data Centre (Appendix A).

## 5.4.11 Unmanned Earth resources satellites carrying microwave sensors

Environmental scientists are often ready to dismiss microwave images due to the unfamiliarity of the relationships between image tone and the characteristics of the Earth's surface (sect. 2.5.9). This situation is likely to change in the 1980s and 1990s when these relationships are better documented and the advantages of collecting images of the Earth's surface independently of cloud cover are fully realised. To date, only one unmanned Earth resources satellite has carried a synthetic aperture radar (SAR) and that was Seasat. The next unmanned satellite to be launched to carry a SAR will be ERS-1.

## 5.4.12 Seasat

Seasat was an experimental satellite designed by NASA to establish the utility of microwave sensors for remote sensing of the oceans. Images of the land were also obtained giving environmental scientists their first synoptic view of the Earth in microwave wavelengths. Seasat had a circular non Sun-synchronous orbit at an altitude of 800 km, sensing the Earth's surface from 72 °N to 72 °S, orbiting the

**Fig. 5.24** Daytime images of northern England and Scotland taken from the Heat Capacity Mapping Mission (HCMM) satellite on 30.5.1978. Image (a) is in the visible waveband and image (b) is in the thermal infrared waveband. (Courtesy, NASA)

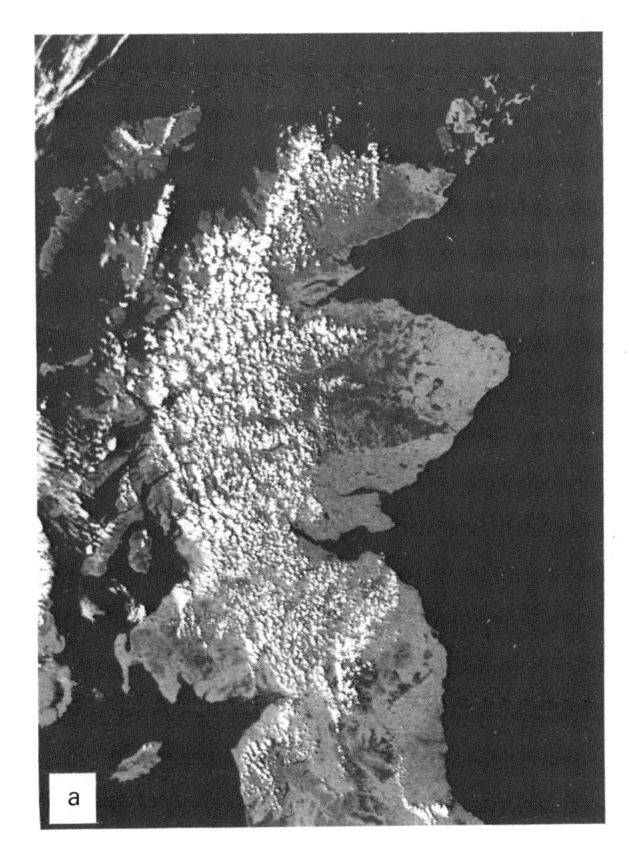


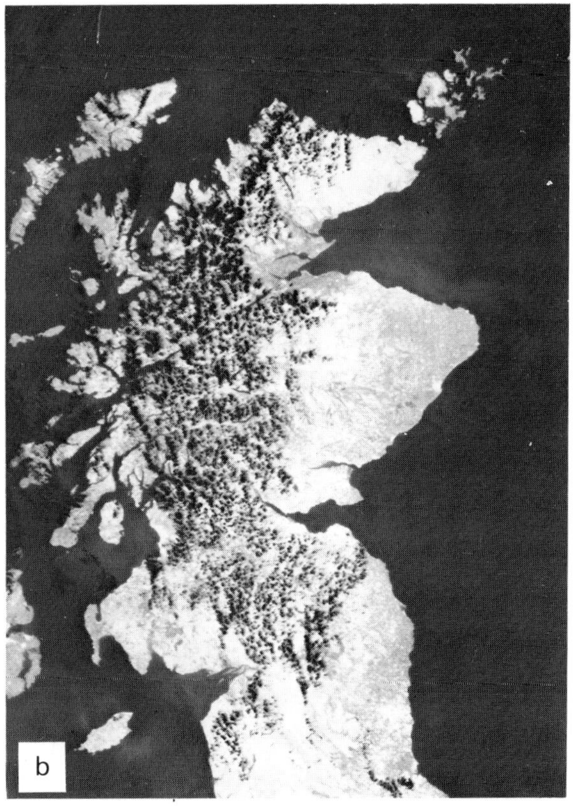

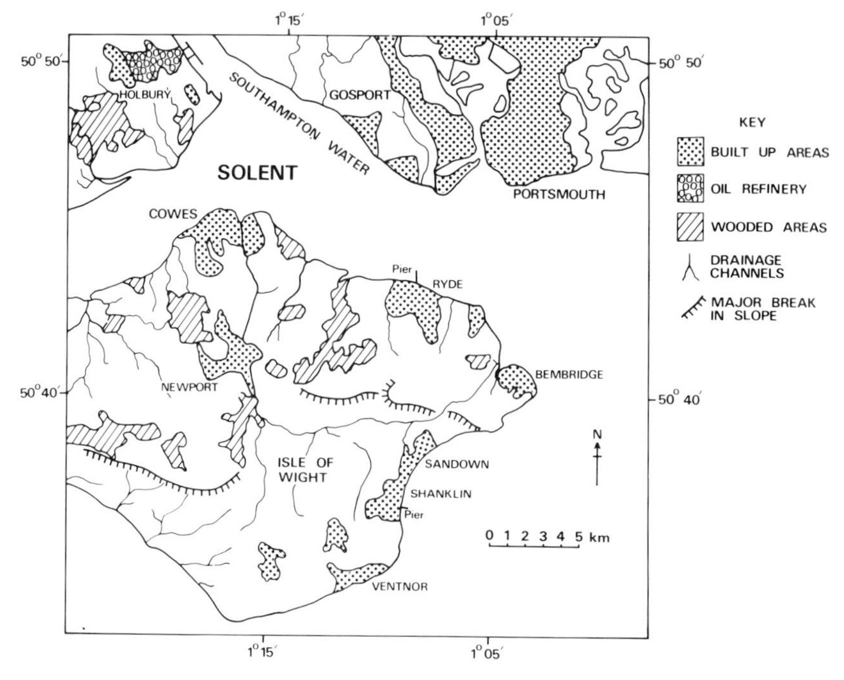


**Fig. 5.25**
A Seasat synthetic aperture radar (SAR) image taken of the Isle of Wight on 2.10.1978 (Lodge 1983). Note the relationship between landcover and backscatter. (Courtesy, Space and New Concepts Department, RAE Farnborough, UK)

Earth 14 times a day and passing over the same area once every 152 days. The satellite carried five sensors, two of which were of potential interest to environmental scientists (Lodge 1981), these being a radiometer and a synthetic aperture radar (SAR). The radiometer recorded in two wavebands, visible (0.47–0.94 $\mu$m) at a spatial resolution of 2 km and thermal infrared (10.5–12.5 $\mu$m) at a spatial resolution of 4 km. The SAR sensor operated at a wavelength of 23.5 cm (L band) with an HH polarisation and produced images with a nominal ground resolution of 25 m over 100 km wide swaths (Fig. 5.25 and 5.26).

**Fig. 5.26**
A comparison between image (i), which is part of an image collected by the synthetic aperture radar (SAR) onboard Seasat on 2.8.1978 and image (ii), which is part of an image collected by the return beam vidicon (RBV) camera onboard Landsat 3 on 10.8.1978. The area is centred on Phoenix Arizona, USA and is bordered by agricultural land and desert. In the radar image the backscatter and therefore the image tone is primarily dependent upon surface roughness, the smooth desert has a dark tone, the rough agricultural lands have a medium tone and the urban areas have a light tone. In the visible RBV image agricultural areas have a dark tone and urban and desert areas a light tone. Individual points that can be compared on both images are (A) ridge, (B) ephemeral drainage, (C) canal, (D) airfield, (E) caravan park, (F) high density housing, (G) agricultural land and (H) wet field. For further details refer to Kozak *et al.* (1981). (Courtesy, NASA)

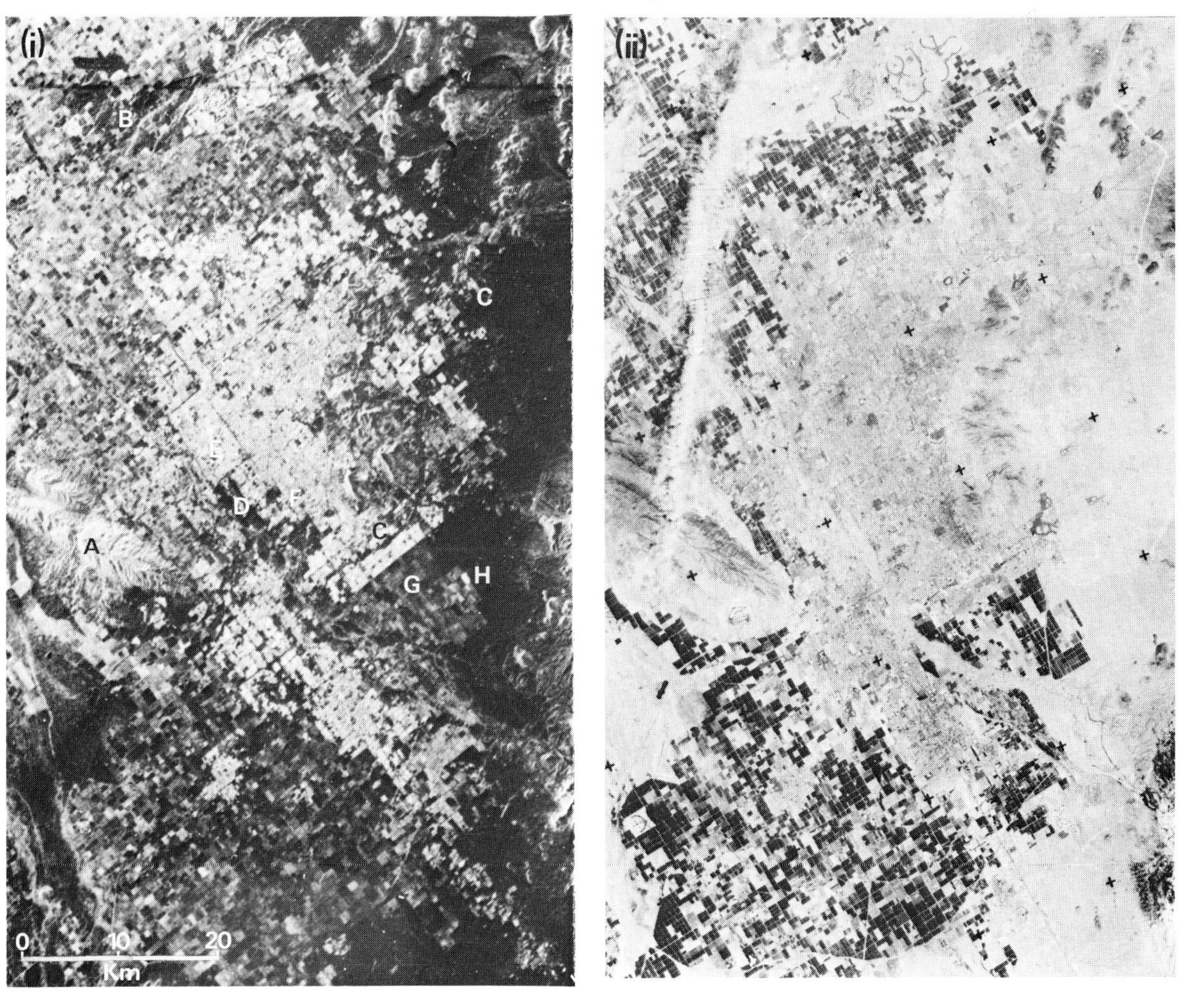

The satellite had only a short life–from June 1978 until it short circuited its electrical system in October 1978. During this time the SAR recorded 100 million square km of imagery, for areas around the receiving stations in the USA, Canada and the UK. The SAR data collected from the Seasat satellite have been used for monitoring the state of the sea (Mattie *et al.* 1980; Beal *et al.* 1981; Darnell and Harriss 1983) and for mapping sea ice (Hall and Rothrock 1981), vegetation (Bryan 1981; Lyon and McCarthy 1981), urban form (Henderson 1982), land cover (Blom and Daily 1981), geology (Frost *et al.* 1983) and a whole host of hydrological features (Walling 1983).

Seasat SAR data can be purchased from either the Environmental Data and Information Service (EDIS) or the National Oceanic and Atmospheric Administration (NOAA), a receiving station, or a national remote sensing centre (Appendix A), (Ford *et al.* 1980; Holt 1981).

## 5.4.13 Earth Resources Satellite (ERS) series

The first remote sensing satellite to be launched by the European Space Agency (ESA) will be the Earth Resources Satellite ERS-1. This is due for launch in 1987 or 1988 and will hopefully be followed by ERS-2 before the end of the decade.

The satellite will be launched into a Sun-synchronous orbit at an altitude of around 700 km with a repeat cycle of three days. Two pay loads have been designed, one for land applications and one for ocean monitoring. As the land applications pay load is similar, at least in concept, to that carried by the French satellite SPOT only the ocean monitoring pay load has so far been authorised. This pay load will carry two remote sensors, one of which is a C band synthetic aperture radar with a 80 km swath width.

So far experiments have been designed to use these data to improve weather and ocean state forecasts, sea surface temperature and phytoplankton measurements; it is hoped that knowledge of these factors will aid fish location and the monitoring of icebergs and oil and chemical pollution (ESA 1981a, 1981b; Anon 1983).

## 5.4.14 Radarsat

This satellite is hopefully to be launched by Canada in 1989. It will carry a C band SAR with a moveable angle of incidence (20 °–45 °) and a nominal ground resolution of 25 m. The major application of data from this SAR will be for mapping ice (Fig. 4.28) especially in the offshore oil drilling areas of northern Canada. As the time factor in the availability of these data is all important, users have been promised a three-hour turn around between data collection and delivery. If this proves to be possible then the door will be open for the speedy monitoring of our environment.

## 5.5 Meteorological satellites

Meteorological satellites have been in use longer and are better developed than Earth resources satellites. A large fillip to the development of these satellites occurred as early as 8 June 1962, when a summary of understanding was signed between the USA and the USSR. The initial tortuous sentence sets the scene:

'Following the exchange of views between Nikita S. Khrushchev, Chairman of the Council of Ministers of the Union of Soviet Socialist Republics and John F. Kennedy, President of the United States of America regarding cooperation in the exploration and use of space for peaceful purposes, the USSR and USA representatives designated for this purpose have discussed in some detail the possibilities of cooperation in meteorology, a world geomagnetic survey and satellite telecommunications' (Blagonravov and Dryden 1962).

The prime use of the data collected by the sensors on board meteorological satellites has always been for the study and prediction of short-term changes in weather, usually on a time scale of a few hours or days. Environmental scientists rarely need these images for meteorological applications and usually wish to use them for studies of climate or terrain (Vette and Vostreys 1977; Barrett 1970, 1974; Barrett and Curtis 1982). The images are often interpreted using methodologies based upon the aerial photographic interpretation techniques discussed in section 3.5. The most useful of these methodologies has proved to be nephanalysis, which is the visual interpretation and classification of cloud type for the production of cloud maps (Blankenship 1962; Harris and Barrett 1978). In recent years an increasing number of images collected by sensors onboard meteorological satellites are being used in digital form by image processors (sect. 6.3.3) for mapping and also the estimation of meteorological parameters like sea surface temperature (Fig. 5.29) or cloud height.

The first satellites to carry sensors for the collection of meteorological data were the USA's Vanguard and Explorer satellites that were launched in 1959. Since then the USA have launched between 40 to 50 meteorological satellites and these can be divided into two groups (Fig. 5.1); polar orbiting satellites and geostationary satellites (Wilkes 1975; Barrett and Martin 1981).

Polar orbiting meteorological satellites occupy relatively low level orbits, usually between 500 km to 1500 km above the surface of the Earth and cross the Earth at very high angles to bring the satellite close to the North and South poles. These satellites are usually Sun-synchronous which means that the satellite moves with the Sun to keep a constant angular relationship between itself and the Sun. As a result a typical polar orbiting meteorological satellite would travel a total of 14 to 15 100-minute orbits each day, passing each point on the Earth's surface at the same time each day and night.

Geostationary meteorological satellites are placed in an orbit at least 35,000 km above the plane of the Equator and advance in the same direction as the rotation of the Earth. The earliest geostationary satellites were the Applications Technology Satellites (ATS) of the early 1970s followed by the Synchronous Meteorological Satellites (SMS) of the late 1970s (Barrett and Hamilton 1982).

Environmental scientists have found greatest utility in data collected by sensors onboard the three polar orbiting satellites of TIROS, NOAA and Nimbus and the geostationary SMS satellites which include GOES, Meteosat and Himawari.

### 5.5.1 TIROS/NOAA series

TIROS and NOAA satellites are operated by the National Oceanic and Atmospheric Administration (NOAA) of the USA (Schwalb 1982). The first generation of these satellites was called the Television and Infrared Observation Satellite (TIROS). This was the first purpose built weather satellite and remained in operation from April 1960 to July 1966. It was modified to give improved image quality and renamed not surprisingly as the Improved TIROS Observational Satel-

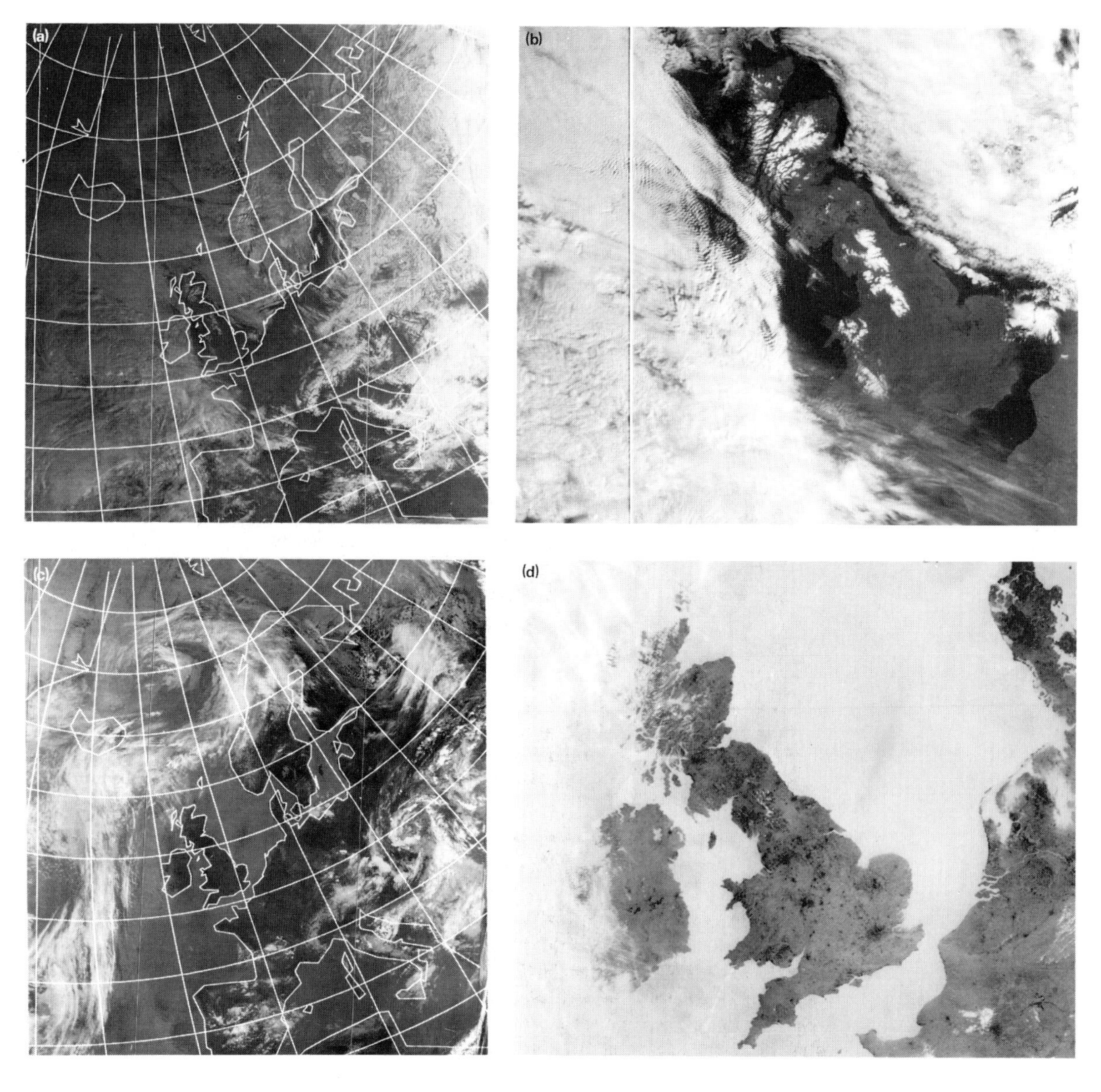

lite (ITOS) for its flights between January 1970 and June 1971. The data from this satellite series made a large impact on meteorology but it was the second and third generation of this series that were of most interest to environmental scientists.

The second generation had satellites called ITOS and NOAA, both of which had a Sun-synchronous orbit at an altitude of 1500 km and carried a Very High Resolution Radiometer (VHRR). This radiometer recorded in a red (0.6–0.7 $\mu$m) waveband and a thermal infrared (10.5–12.5 $\mu$m) waveband and produced images with a very wide swath of 4,400 km at a spatial resolution of 0.9 km (Fig. 5.27). The last satellite in this series was NOAA 5 which was launched in July 1976 and retired in February 1979. Data from the sensors onboard this satellite have been widely used for weather analysis and forecasting. They have also been used for many environmental science applications (Darnell and Harriss 1983), including the mapping of sea surface temperature (Legeckis *et al.* 1980), sea ice (Weeks 1981) and topography (Schneider *et al.* 1979).

**Fig. 5.27**
Visible and thermal infrared band images taken by the VHRR sensor onboard the satellite NOAA 5. Image (a) was taken at 9.24 GMT on 28.2.1977. It shows the distribution of cloud in western Europe over an area 5,625 × 4,400 km in extent. Apart from cirrus cloud over southwest England and waveclouds over Northern Ireland and the western Highlands of Scotland the UK is cloud free. This was due to a ridge of high pressure that extended over much of western Europe. Snow cover can be seen on the Scottish Highlands, the Pennines, North Wales and the Alps. The image has the standard 10° longitude, 5° latitude meteorological grid. Image (b) is a 4 times enlargement of image (a) and covers an area 1,350 × 1,350 km in extent, the image emphasises the difference in cloud heights between the cirrus and lower level clouds, the pattern of the wave clouds and the distribution of snow cover. The line is an image defect. Image (c) was taken at 9.30 GMT on 29.5.1978 and covers the same area as image (a). It shows the location of a stable anticyclone over western Europe that was blocking the eastward movement of Atlantic depressions. The image has the standard 10° longitude, 5° latitude meteorological grid. Image (d) is a 4 times enlargement of image (c) and covers the same area as image (b), (Fotheringham 1979). (Courtesy, University of Dundee, UK)

The third generation of this series have satellites called TIROS-N and NOAA 6 to 13. They have an orbit similar to that of the Landsat satellite which is near polar and Sun-synchronous, at an altitude of 830 km. The repeat time is one day, but as there are always two of these satellites in orbit at any one time, an image of each area on the Earth's surface is collected twice daily (Foote and Draper 1980). These satellites carry an Advanced Very High Resolution Radiometer (AVHRR) which has a spatial resolution of 1.1 km and a swath width of 3,000 km. The AVHRR onboard TIROS-N recorded in four wavebands, visible and near infrared (0.55–0.9 $\mu$m), near infrared (0.725–1.0 $\mu$m), middle infrared (3.55–3.93 $\mu$m) and thermal infrared (10.5–11.5 $\mu$m), (Fig. 5.28). TIROS-N was operational from October 1978 to June 1981 when it was replaced by NOAA 7. During its life, data from its sensors have been used for weather forecasting and a whole host of environmental science applications (Huh and DiRosa 1981; Darnell and Harriss 1983) including monitoring the albedo of the Earth's surface (Hughes and Henderson-Sellers 1982), measuring sea surface temperatures (Wannamaker *et al.* 1980; Saunders *et al.* 1981), mapping sea surface temperature patterns (Saunders *et al.* 1982; McKenzie and Nisbet 1982), the prediction of fish spawning grounds (Lasker *et al.* 1981) and the monitoring of oceanic fronts (Clark and La Violette 1981).

The AVHRR onboard the NOAA satellites recorded in the five wavebands of red (0.58–0.68 $\mu$m), near infrared (0.725–1.1 $\mu$m), middle infrared (3.55–3.93 $\mu$m) and two thermal infrared wavebands (10.5–11.5 $\mu$m and 11.5–12.5 $\mu$m); however, for technical reasons the second thermal waveband (11.5–12.5 $\mu$m) was omitted from NOAA 6 (Baylis 1981). These images are well known to UK readers as they are used to illustrate television weather reports. Despite the relatively low spatial resolution of this AVHRR radiometer, environmental scientists have been quick to exploit the thermal infrared waveband for locating forest fires and the mapping of sea surface temperature (Fig. 5.29), the visible and near infrared wavebands for estimating the moisture content of snow cover (Dozier *et al.* 1981) and the red and near infrared waveband for monitoring vegetation (Townshend and Tucker 1981, and Fig. 6.20). For certain applications the low spatial resolution but rapid repeat time and large

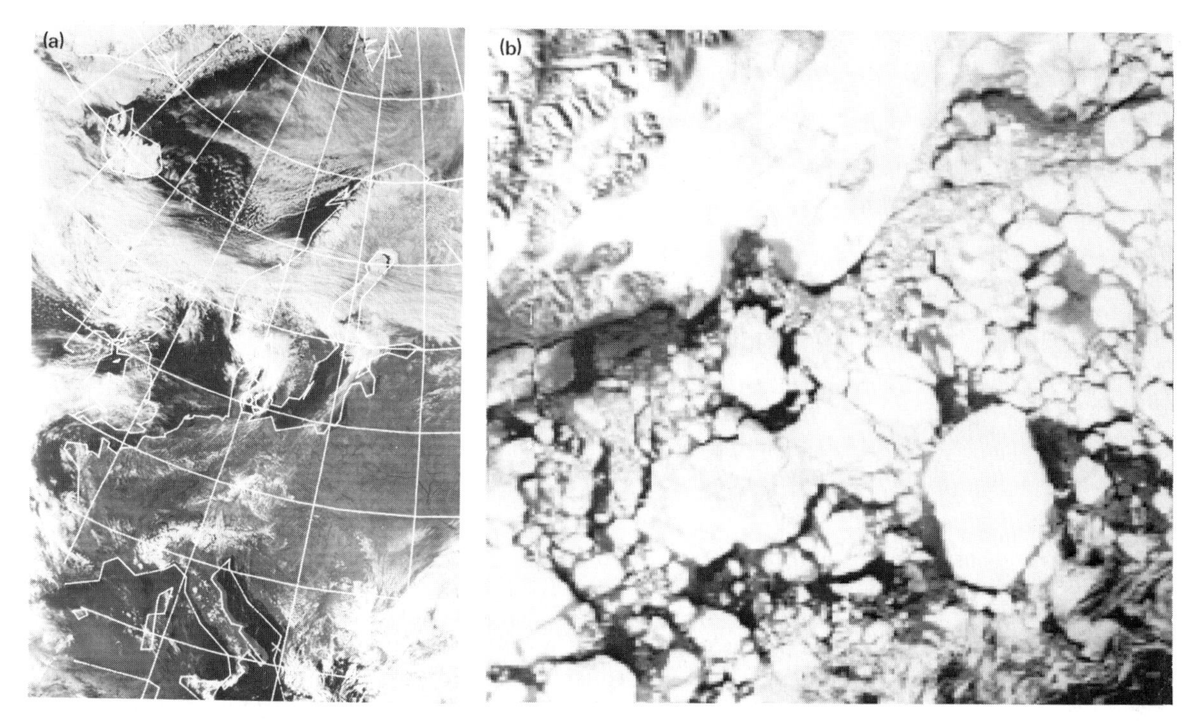


**Fig. 5.28**
Visible band image taken by the AVHRR sensor on board the satellite TIROS-N at 13.44 GMT on 13.4.1979. Image (a) shows the distribution of cloud in western Europe over an area 4,100 × 3,000 km in extent. A band of cirrus cloud stretched over northern Europe and Scandinavia. In the cloud free areas drainage patterns (e.g. around Kiev in the USSR) and snow cover (e.g. in Iceland, the Alps and Northern Norway) can be seen. The image has the standard 10° longitude, 5° latitude meteorological grid. Image (b) is an 18 times enlargement of image (a) and covers an area 275 × 275 km. This image includes part of the Greenland coast (see if you can locate it in image (a)) showing the breakup of a glacier into iceflows and bergs (Fotheringham 1979). (Courtesy, University of Dundee, UK)

aerial coverage of these data can make them superior to data collected by sensors onboard Earth resources satellites like Landsat. For example, to map desertification in the Sahel zone of Africa would require 150 Landsat MSS images but only one AVHRR image.

AVHRR data are readily available for areas within range of a receiving station, in either Virginia, Alaska, France or Scotland (Appendix A).

## 5.5.2 Nimbus series

The Nimbus satellite is the principal meteorological research and development satellite operated by NASA. From 1964 to 1981 Nimbus's 1 to 6 have had orbital altitudes of 1,000 km and carried two sensors, a vidicon camera (sect. 5.4.2) with a spatial resolution of 1 km and a thermal infrared scanner with a spatial resolution of 8 km. Nimbus 7 which is the current satellite in this series was launched into a Sun-synchronous orbit at a height of 910 km in October 1978. It carried a coastal zone colour scanner (CZCS) which is a multispectral scanner designed for oceanographic applications (Gordon and Clark 1980). This scanner records a 1500 km wide swath at a spatial resolution of 800 m in the six wavebands of blue (0.43–0.45 μm), green (0.51–0.53 μm and 0.54–0.56 μm), red (0.66–0.68 μm), near infrared (0.7–0.8 μm) and thermal infrared (10.5–12.5 μm), (Baylis 1981). Like the data collected by TIROS-N and NOAA 6 and 7 they have been used for a variety of environmental studies (Singh 1982).

Complete catalogues of Nimbus sensor data can be obtained from the World Data Centre (Appendix A) and local coverage can be obtained from a receiving station, which in the case of the UK is in Scotland (Appendix A).

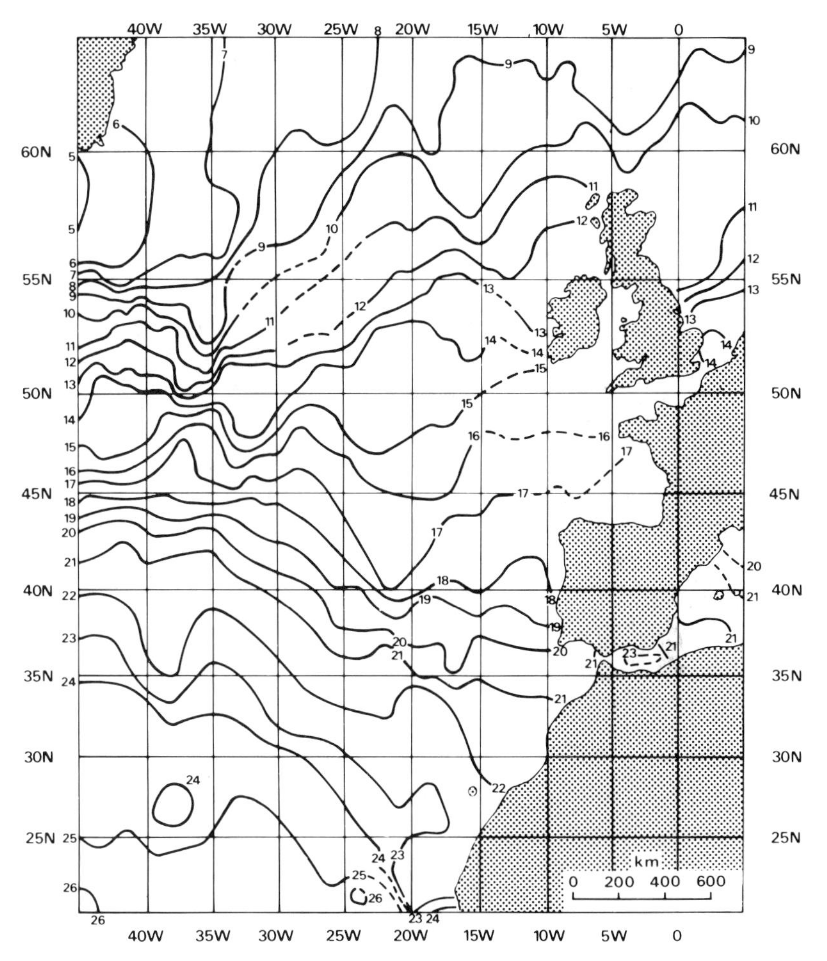


**Fig. 5.29**
A map of sea surface temperature (°C) produced from a NOAA AVHRR image of 30 October 1979 (Dismachek *et al.* 1980). (Courtesy, NOAA)

## 5.5.3 Synchronous Meteorological Satellites (SMS)

The sensors onboard a total of six geostationary satellites are intended to provide coverage of the Earth between 60 °N to 60 °S. These satellites will sit over the Equator at an altitude of around 36,000 km while their sensors record images of an entire Hemisphere. To date four of these satellites are operational; they are the Geostationary Operational Environmental Satellites (GOES) west and east, Meteosat and Himawari (Wiesnet and Matson 1983).

### 5.5.3.1 GOES west and east

The GOES are operated by NOAA (Bristor 1975; NOAA/NASA 1980). GOES west covers the western Americas and the Pacific and GOES east covers the eastern Americas and the Atlantic (Fig. 5.30). They have sensors that record visible (0.66–0.7 $\mu$m) and thermal infrared (10.5–12.6 $\mu$m) wavelengths (Hamilton 1981). As the repeat frequency

**Fig. 5.30**
Image taken from the satellite GOES east in the visible waveband on 12.3.1980. Image (a) is the whole scene and image (b) is an enlargement. (Courtesy, NOAA)

of the image is only limited by the time it takes to scan and relay the image, they can be produced every half an hour for transmission to 'user-stations' for visual (sect. 3.5) or digital interpretation (sect. 6.3). These 'user-stations' are often very remote from NOAA headquarters in the USA and include for example, the British Antarctic Survey in the Antarctic, Trans-Canada Pipelines Ltd in the Arctic and numerous environmental companies and research institutes in Canada, Latin America and Australasia. These images have been used in meteorology (Purdom 1976) for the monitoring of tornadoes (Hung and Smith 1982); in climatology for the identification and monitoring of anomalously low surface temperatures (Maddox and Reynolds 1980) and the identification of clouds (Parikh and Ball 1980). They have also been used in forestry for the assessment of soil moisture to estimate of fire risk (Waters 1976); in agriculture, for determining the thermal regime of drained soils (Chen *et al.* 1979) and in oceanography for the monitoring of sea surface temperatures (Legeckis *et al.* 1980). For further examples refer to Darnell and Harriss (1983). GOES data can be obtained from NOAA (Appendix A).

### 5.5.3.2 Meteosat

This satellite is operated by the European Space Agency and is located on the Greenwich Meridian over West Africa. Meteosat-1 was launched in November 1977 and failed in November 1979. Meteosat-2 was

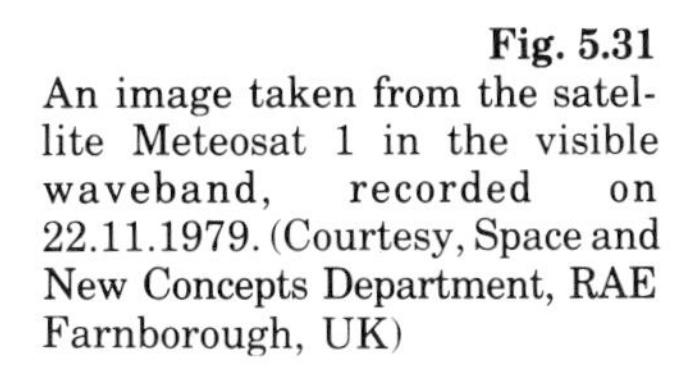

**Fig. 5.31**
An image taken from the satellite Meteosat 1 in the visible waveband, recorded on 22.11.1979. (Courtesy, Space and New Concepts Department, RAE Farnborough, UK)

launched in June 1981 and is still operational. Like GOES its sensors record images every half an hour but do so in three wavebands, in the visible and near infrared (0.4–1.1 μm), (Fig. 5.31), in the water vapour absorption region of the middle infrared (5.7–7.1 μm) and the thermal infrared (10.5–12.5 μm). The water vapour waveband records atmospheric moisture and is used in conjunction with the thermal infrared waveband for atmospheric modelling and weather forecasting (Eyre 1981). The spatial resolution varies with waveband and latitude. At the Equator the spatial resolution is 2.5 km in the visible and near infrared waveband and 5.0 km in the water vapour and thermal infrared waveband. In northern Europe this decreases to 4 km in the visible and near infrared waveband and 8 km in the water vapour and thermal infrared waveband (Mason 1981; Battrick 1981).

Data from the Meteosat sensors have been used for many meteorological applications (Barrett and Hamilton 1982; Darnell and Harriss 1983) for example studies of synoptic climatology (Endlich and Wolf 1981; Hunt *et al.* 1981), sea temperature measurements (Domain *et al.* 1980; Saunders *et al.* 1981), land surface temperature measurements (Browning 1982; Bardinet *et al.* 1982) and wind vector determinations (Eigenwillig and Fischer 1982). Meteosat sensor data have also been used for terrestrial applications, for example land use mapping in Africa (Gomber 1980; Maes *et al.* 1982) and the monitoring of natural disasters (Robson *et al.* 1982).

This imagery can be purchased from the Meteosat Data Service in Germany or a national remote sensing centre (Appendix A).

#### 5.5.3.3 Himawari

The Himawari, or 'sunflower satellite' as it is commonly called, is the Japanese satellite in the SMS grouping. It was launched in April 1978 and carried a similar sensor to that used on GOES west and east except that the sensor can be used in both low and high spatial resolution modes. The low resolution mode is 4 km in visible wavelengths and 7 km in thermal infrared wavelengths and the high resolution mode is 2.6 km in visible wavelengths and 5 km in thermal infrared wavelengths (Bell 1981; Barrett and Hamilton 1982). The applications of the data collected by the sensors onboard the Himawari are much the same as for the GOES and Meteosat satellites, as is discussed in Darnell and Harriss (1983).

## 5.6 Military and USSR satellites

The majority of remotely sensed images collected from satellite borne sensors are not available to Western environmental scientists, either because they are taken for military purposes or because they are taken by the USSR for Eastern use only. Unfortunately myths have developed around these images and their properties have been exaggerated. To enable you to make up your own mind on their likely characteristics, military and USSR satellite programs will be discussed briefly.

## 5.6.1 Military satellites

Since the 'space-age' began, over 1,600 satellites have been launched (King-Hele *et al.* 1981), 60% of which were for military purposes and of these around half carry imaging sensors for application in reconnaissance or meteorology.

## 5.6.2 Military satellites used for reconnaissance

In the 1980s we can expect around 40 military reconnaissance satellites to be launched each year, 35 by the USSR and five by the USA (SIPRI 1981). As the satellites launched by the USA are well documented (SIPRI 1978, 1982) they will be discussed first.

The USA has collected remotely sensed data from four separate types of reconnaissance satellite; these are the 'area-survey' satellite, the 'close-look' satellite, 'Big Bird' and more recently 'KH 11'. They all carry an imaging sensor and all have the ability to manoeuvre in space to place themselves in a particular orbit at a particular time – something Earth resources satellites have never done.

The 'area-survey' satellites were used during the 1960s and early 1970s to obtain broad photographic coverage of large areas. Each satellite had a short life of around a month and carried cameras that took photographs with a wide angle lens. The film was automatically processed onboard the satellite, scanned by a digitiser and the data transmitted to Earth. These satellites burnt up on re-entry and so were not re-used (SIPRI 1978). The 'close-look' satellite employs cameras to obtain high resolution photographs of areas of interest, either as identified by the 'area-survey' satellite or prompted by political events. As it carries a camera with a focal length of around 6 m, uses a very high resolution film and orbits at a low altitude of around 150 km, its images are expected to have a spatial resolution of between 0.25 m and 0.5 m (SIPRI 1978; Baker 1981). Once a roll of film has been exposed it is ejected in a capsule that is either retrieved in mid-air by an aircraft, or is picked up at sea (Colvocoresses 1975). The lifetime of these satellites was only five days in the early 1970s but it is now nearer three months.

The satellite Big Bird combines the reconnaissance attributes of both the 'area-survey' and 'close-look' satellites as it carries sensors for the collection of both high and low resolution images. Big Bird was launched in the early 1970s and is today the main USA reconnaissance satellite (SIPRI 1981). The sensors are primarily photographic and as with the 'close-look' satellite, the film is returned to Earth by capsule. Big Bird can also act as a platform for other sensors and these have included a high resolution thermal infrared scanner for detecting thermal plumes produced by nuclear powered submarines and a synthetic aperture radar (SAR) for tracking ship movements.

The satellite series KH 11 which was first launched in 1977, is used by the military but operated by the CIA. These satellites fly on a relatively fixed orbit at an altitude of 250 km and carry a multispectral scanner similar to the Thematic Mapper sensor carried by Landsats 4 and D′. Like Big Bird sensors, KH 11 sensors can record at two spatial resolutions, the higher of which is a spatial resolution of 5 m. Of interest to environmental scientists is the ability of the sensor onboard the satellite to automatically compare a past with a present image, as a means of detecting change (SIPRI 1980, 1981, 1982).

The Soviet program of reconnaissance satellites is less clear, as it is based on an ever-developing series of Cosmos satellites which have been launched in their hundreds since the early 1960s. Like the USA's 'close-look' satellites they carry long focal length cameras and today some also carry a SAR for ship location.

During important world events both the USA and USSR place their reconnaissance satellites over the region of interest; this was the case during the Middle East War of 1973, during India's first nuclear test in 1974, during the Turkish invasion of Cyprus in 1973, when the Argentinians invaded the Falkland Islands in 1982, and during the conflict between Iran and Iraq from 1980 onwards (SIPRI 1978, 1979, 1980, 1981, 1982).

To judge the value of such data it is worthwhile reflecting on the comments of the former US President Jimmy Carter concerning the holding of the USA hostages in Iran, 'Our satellite photographs of the embassy compound and the surrounding area kept us abreast of any changes in the general habits and overall composition of the terrorists' details. We could, for instance, identify individual cars and trucks that went inside the compound each day', (Carter 1982).

The military reconnaissance satellite programs of countries other than the USA or USSR are small but interesting. For example, the Chinese launched a joint reconnaissance and Earth resources satellite named Chinasat 1 in July 1975. This had a very low orbit and carried sensors similar to the Landsat multispectral scanner (SIPRI 1978). After 1985 these satellites will carry multispectral linear array sensors similar to those carried by the French satellite SPOT (NASA 1982d). The Indians launched a joint reconnaissance and Earth resources satellite named Bhaskara 1 in June 1979. This carries low resolution RBV cameras similar in design to those carried on Landsats 1 and 2. Bhaskara 2 was launched in November 1981 and there are more Bhaskaras in the pipeline. In 1986 the French will be launching their reconnaissance satellite which will have characteristics similar to the USA's 'close-look' satellite.

### 5.6.3 Military satellites used for meteorology

Two types of meteorological data are required for military planning. These are low resolution images of large areas for weather forecasting and high resolution images of small areas for military planning or further reconnaissance. The USA military obtain their low resolution images in visible and thermal infrared wavelengths from sensors onboard the Defence Meteorological Satellite Program (DMSP) satellite. Some of these images have been made available for civilian use since April 1973 and have proved to be of value to the environmental sciences (Croft 1981; Darnell and Harriss 1983). For example, thermal infrared images have been used for monitoring volcanic activity and natural gas fields (Brandli 1978), day-time visible images have been used in climatological studies (Harris 1977; Carleton 1982) and night-time visible images have been used for monitoring urban population and energy utilisation patterns (Welch 1980).

The USA military obtain their high resolution meteorological data from sensors onboard the two or three military meteorological satellites launched by the USA each year, or from sensors onboard the UK and French military meteorological satellites which were first launched in the early 1970s.

### 5.6.4 USSR satellites

The USSR has launched seven satellite series for the collection of remotely sensed data; they are three series of manned satellites named Vostok, Voshkod and Soyuz, one series of manned space stations named Salut and three unmanned satellites named Cosmos, Molinya and Meteor (Fig. 5.1).

### 5.6.5 Manned USSR satellites

In the early 1960s, six 1-man Vostok satellites and two 3-man Voshkod satellites paralleled the flights of Mercury (sect. 5.3.2) and in all probability photographs were taken of the Earth's surface. The later Soyuz satellite series which had been gradually developing since the mid-1960s carried several remote sensing devices. The better known examples are Soyuz 9 from which photographs were taken for the production of photomaps of the Caspian Sea area and Soyuz 22 from which around 2,000 multiband photographs were taken using the East German MKF-6 multiband camera (Fig. 5.32), (Zickler 1977) for geological mapping and vegetation assessment (Nikolaev 1981).

### 5.6.6 Manned USSR experimental satellite stations

As the first small step on the Moon was taken by an American, Soviet interest moved away from such a direct space race and greater emphasis was placed on the manned space stations called Salut. These stations enable experiments to be conducted in space and provide platforms for military reconnaissance and remote sensing of Earth resources. For example, a considerable number of stereoscopic and multispectral photographs were taken from Salut 1 for environmental research (Nikolaev 1981), from Salut 3 for the mapping of previously unmapped areas (Ince 1981) and from Saluts 4, 6 and 7 for a wide range of geological, agricultural and cartographic applications (Doyle 1982), (Fig. 5.32).

### 5.6.7 Unmanned USSR satellites

The best known unmanned USSR satellite series is Cosmos. This is because around 80 to 90 are launched each year and around 1,200 have been launched to date (Borrowman 1982). They have been used in a wide range of applications from weather forecasting in the late 1960s to military reconnaissance throughout the 1970s and early 1980s.

Engineering details of the Cosmos satellites remain unknown in the West. However, it is suspected that the well reported bout of military

**Fig. 5.32**
Soviet space photography. The camera is the multiband camera MFK 6. The image is of the Pamir-Alai region of Siberia and was taken with the MFK 6 camera from Soyuz 22. (Courtesy, VEB, Carl Zeiss, Jena, West Germany)

reconnaissance undertaken by these satellites in the mid 1970s was performed with photographic sensors, probably MKF-6 multiband cameras (NASA 1982d). Since the mid 1970s nuclear powered microwave sensors have been used on some of the Cosmos satellites. The nuclear powered Cosmos 953 that fell onto Canada in 1978 carried a microwave sensor as did the well reported Cosmos 1076 and 1402.

The satellite Molniya is the specialised meteorological satellite series that relieved Cosmos of its meteorological role in the mid-1960s. However, its life was short, being replaced by the more sophisticated Meteor satellite in 1969. The first Meteor satellites carried an infrared scanner (8–12 $\mu$m) for ocean and climatic studies (Vlasov *et al.* 1981). The more recent Meteor satellites have carried a number of sensors which are suitable for both Earth resources and meteorological applications. For example, the Meteor satellite launched in June 1980 had specifications similar to Western Earth resources satellites as it had a near polar orbit, an altitude of 600 km and carried five sensors, one of which had a spatial resolution of 80 m in eight wavebands and one of which had a spatial resolution of 30 m in three wavebands (Sidorenko 1981).

## 5.7 Recommended Reading

**Allison, L. J., Schnapf, A.** (1983) Meteorological satellites: *In* Colwell, R. N. (ed.) *Manual of Remote Sensing.* (2nd edn). American Society of Photogrammetry, Falls Church, Virginia, pp. 651–79.

**Barrett, E. C.** and **Hamilton, M. G.** (1982) The use of geostationary satellite data in environmental science, *Progress in Physical Geography*, **6**: 159–214.

**Beal, R.C., Deleonibus, P.S.** and **Katz, I.** (eds) (1981) *Spaceborne Synthetic Aperture Radar for Oceanography*. Johns Hopkins Press, Baltimore, Maryland.

**Carter, W. D., Rowan, L. C.** and **Weill, G.** (eds) (1983) *Remote Sensing and Mineral Exploration–1982*. Advances in Space Research Series (COSPAR Volume 3). Pergamon Press, Oxford; New York.

**Deutsch, M., Wiesnet, D. R.** and **Rango, A.** (1981) *Satellite Hydrology*. American Water Resources Association, Minneapolis.

**Elachi, C.** (1983) Microwave and infrared satellite remote sensors: *In* Colwell, R. N. (ed.) *Manual of Remote Sensing* (2nd edn). American Society of Photogrammetry, Falls, Church, Virginia, pp. 571–650.

**Houghton, J.T., Cook, A.H.** and **Charnock, H.** (1983) *The Study of the Land Surface from Satellites*. The Royal Society of London (first published in *Philosophical Transactions of the Royal Society, Series A*, **309**, pp. 241–464).

**Lulla, K.** (1983) The Landsat satellites and selected aspects of physical geography, *Progress in Physical Geography*, **7**: 1–45.

**Sheffield, C.** (1983) *Man on Earth*. Sidgwick and Jackson, London.

**Short, N. M.** (1982) *The Landsat Tutorial Workbook*. National Aeronautics and Space Administration Reference Publication 1078, Washington DC.

**Slater, P.N.** (1980) *Remote Sensing: Optics and Optical Systems*. Addison-Wesley Reading, Massachusetts; London, pp. 438–576.

**Townshend, J. R.G., Gayler, J. R., Hardy, J. R., Jackson, M. J.** and **Baker, J. R.** (1983) Preliminary analysis of Landsat-4 Thematic Mapper products. *International Journal of Remote Sensing*, **4**: 817–28.

# Chapter 6 Image processing

*'Every picture tells a story,'*
(Advertisement for Sloane's Backache and Kidney Oils, c. 1907)

## 6.1 Introduction

The term image is rather broad, as by definition it refers to a representation of something else and includes both statues and paintings! Within the confines of remote sensing the term image refers to a continuous or discrete record of a two dimensional view. An aerial photograph is an example of a continuous image where detail is displayed by means of a continuous signal that we can see and interpret. A Landsat multispectral scanning system (MSS) image is an example of a discrete image where detail is held in discrete digital units that we cannot see but we can handle quantitatively. These divisions are not fixed for each image type, as the ability to transfer images between the continuous and discrete states is a basic requirement of many image processing procedures. For example, an environmental scientist may wish to convert a continuous thermal infrared linescanner image (sect. 4.3) into a discrete image for digital analysis and then convert it back to a continuous image for visual interpretation.

As all remotely sensed images are but a poor representation of the real world, environmental scientists have turned their attention to both continuous image processing and discrete image processing as a means of maximising the value of particular images to particular applications.

## 6.2 Continuous image processing

Continuous image processing can be used to correct, enhance and classify all types of continuous imagery. The five most popular image correction and enhancement techniques (Fig. 6.1) are first, selective enlargement to facilitate the observation of detail within a scene or to isolate a particular study area or feature. Second, contrast modification as a means of optimising the range of image grey tones; this can involve increasing image contrast to aid in the location of subtle tonal changes or decreasing image contrast to reveal detail in areas of very dark and very light tone. Third, the enhancement of tonal edges for

| Equipment \ Technique | Image enhancement | | | | | | Image classification |
|---|---|---|---|---|---|---|---|
| | Selective enlargement | Contrast modification Increase | Contrast modification Reduce | Edge enhancement | Colour/false colour composite production | Directional and spatial filtering | Density slicing |
| Photographic enlarger | ■ | ■ | ■ | ■ | ■ | □ | ■ |
| Diazo colour printer | □ | ■ | □ | □ | ■ | □ | ■ |
| Electronic dodger | □ | □ | ■ | □ | □ | □ | □ |
| Scanning microdensitometer | ■ | ■ | ■ | □ | □ | □ | ■ |
| Analogue image processor | ■ | ■ | ■ | ■ | □ | □ | ■ |
| Multi-additive viewer | □ | ■ | ■ | □ | ■ | □ | □ |
| Optical processor | □ | □ | □ | □ | □ | ■ | □ |

□ Equipment not useful for particular application ■ Equipment useful for particular application

**Fig. 6.1** Techniques and equipment for continuous image processing.

the mapping of geological faults, breaks of terrain slope and textural boundaries. Fourth, the production of colour composites as a means of manipulating and displaying several multiband images of a scene as one colour image. Fifth, directional and spatial filtering to both remove regular features in the image, like scan lines and cloud bands, and to enhance the remaining detail.

The classification technique most frequently used is that of density slicing where specific tones within a remotely sensed image are isolated, density level by density level. This has proved to be particularly useful for the location of density boundaries (Ranz and Schneider 1971) and for visually correlating non-contiguous areas of similar tone (Rohde and Olson 1970; Frazee *et al.* 1972).

Seven instruments are used for continuous image processing (Fig. 6.1); they are a photographic enlarger, diazo colour printer, electronic dodger, scanning microdensitometer, analogue image processor, multi-additive viewer and optical processor.

## 6.2.1 Photographic enlarger

The photographic enlarger is used for five tasks. First, the enlargement of images, as illustrated in Fig. 6.2. Second, the alteration of image contrast by printing low contrast images onto what photographers call hard (high contrast) paper and by printing high contrast images onto what photographers call soft (low contrast) paper (Horder 1976; Sabins, 1978). Third, the enhancement of tonal edges by printing a positive image while it is in register with a blurred negative image of itself; this both suppresses low spatial frequencies within a scene and enhances high spatial frequencies within a scene. Fourth, the production of colour composites by exposing a colour film three times; the first time with an image illuminated with blue light, the second time with an image illuminated by green light and the third time with an image illuminated by red light. Unfortunately, the technical problems associated with this technique make it unsuitable for routine use. Fifth, the slicing of image density by printing a remotely sensed image onto 'Agfacontour' film. This very unusual film has two D log E curves (sect. 3.3.1) and therefore performs in the same way as two films, one

**Fig. 6.2**
(A) Black and white panchromatic aerial photograph of the Neepsend Road area of Walkley, Sheffield, UK. At this scale, details of gardens can be seen, vehicle types identified and if you look carefully in the north west corner of the photograph you can locate where the polluted river Loxley enters the River Don (Courtesy, Meridian Airmaps Ltd., UK). (B) A small portion of the photograph has been enlarged × 25, from the paper print. Note that as the row of terraced houses is on the edge of the aerial photograph, walls as well as roofs can be seen, giving the appearance of an oblique view.

acting as a mask for the other. As the gamma (sect. 3.3.1.1) of the two D log E curves is high, the density range of the resultant image is small. By using a series of increasing exposures it is possible to move the D log E curves along the exposure axis, exposure by exposure, to produce an image with sharp breaks in tone at given density levels. This has proved to be an inexpensive yet very effective means of density slicing remotely sensed imagery (Nielson 1972).

## 6.2.2 Diazo colour printer

A diazo colour printer uses diazo graphic arts film (diazochrome) for the production of positive, single colour, contact, transparency prints from single black and white remotely sensed images (Sabins 1978). This film, which is composed of an acetate or polyester base embedded with light sensitive salts and single colour couplers, is put face to face with a transparent remotely sensed image before being slipped inside the diazo colour printer. A mercury vapour lamp then exposes the two films causing the salts in the diazo graphic arts film to be destroyed, at a rate that is inversely proportional to the density of silver grains in the remotely sensed image. The diazo graphic arts film is then moved into a different section of the diazo colour printer where it is developed by heated ammonia vapours.

This instrument can be used for the production of colour composites or for the density slicing of single images. To produce a colour composite the above procedure is repeated using a different colour for each multiband image and the resultant three images are then sandwiched together. For example, to produce a false colour composite of a Landsat MSS scene, MSS 4 (green) would be printed in yellow, MSS 5 (red) would be printed in magenta and MSS 6 or 7 (near infrared) would be printed in cyan, prior to them being sandwiched together and viewed on a light table. To produce a density sliced image a separate colour print is formed for each of three development times. For example, low image densities could be recorded on a yellow image, medium image densities could be recorded on a magenta image and high image densities could be recorded on a cyan image prior to them being sandwiched together (Skaley 1980). It is interesting to note that diazo colour graphics film can also be used in the field without the aid of a diazo colour printer by exposing the film in the sunlight and developing it in a box containing ammonia soaked cotton wool, so not all image processing is expensive!

### 6.2.3 Electronic dodger

An electronic dodger is used to decrease image contrast as a means of exposing the detail that is hidden in areas of shadow and highlight. This is done while producing prints from a negative image, using an illumination source with a strength that varies in direct proportion to the density of each area of the image. As a result high density areas are illuminated strongly and low density areas are illuminated weakly. Unfortunately; the penalty for such processing is a loss of detail in the middle density range (Genderen 1975).

### 6.2.4 Scanning microdensitometer

A scanning microdensitometer automatically measures the density of all points on a film transparency. This information can either be converted to digital data for discrete image processing (sect. 6.3) or can be plotted mechanically (Buckley 1971; Owen-Jones 1977).

### 6.2.5 Analogue image processor

The input to an analogue image processor is usually a television camera and the output is usually one, or several television monitors. The most simple version of this system comprises only a camera and a television monitor; the operator controls the degree of image enlargement by the image to camera distance and the contrast and colour balance by adjusting the two little knobs on the television monitor, (Williams and Goodman 1980). The two systems that are commonly employed in remote sensing are black and white image analysers that were designed for use in metallurgy and colour video processors that were designed for use in remote sensing. The black and white image analyser (Fig. 6.3) enables the operator to enlarge, alter the contrast, density slice and measure the areas of all or part of an image (Wignall 1977). Colour video processors have similar abilities to the black and white image analyser (Schlosser 1974). In addition they have the ability to edge enhance and to density slice in colour, as is illustrated in Plate 5 and discussed in Estes and Senger (1971) and Townshend (1981b).

**Fig. 6.3**
The console of a 'Quantimet' black and white image analyser used for continuous image processing. (Courtesy, Cambridge Instruments, UK)

## 6.2.6 Multi-additive viewer

Multi-additive viewers provide the interactive flexibility that is necessary for the speedy production of optimised colour composites. A multi-additive viewer holds several photographic positive images in a single plane. Each image is illuminated by two sources: a brightness lamp which controls image brightness and which shines through a filter to give the image colour and a saturation lamp which controls the degree of image colour saturation (Yost and Wenderoth 1967). Manipulation of these three colour variables of brightness, colour and saturation enables the operator to optimise an image for a particular application (Fig. 6.4), (Sapp 1971).

## 6.2.7 Optical processor

An optical processor provides a means of investigating the spatial distribution of image tone (Rosenfeld 1969; Duda and Hart 1973). Optical processing is based on the observation that when intense and spectrally pure (coherent) electromagnetic radiation from a laser is passed through a film transparency it is scattered in a manner which is dependent upon the spatial distribution of tones in that transparency. The transformation of the amplitude and phase of the light waves as they pass through the film transparency can be described mathematically by the Fourier transform (Goodman 1978), which is an elegant summary of the spatial patterns within an image in terms of the sine and cosine curves of the image frequencies (Townshend 1981b). To visualise this Fourier transformation it is necessary to insert a lens behind the film transparency, as this focuses the radiation

**Fig. 6.4**
A multi-additive viewer made by Clydes, UK. The illumination from the light sources, to the far left of the instrument, that passes through filters in the filter wheels and four images, is collected by the optics at the centre of the instrument and is fused for back projection onto the horizontal screen.

**Fig. 6.5**
Optical processing. Procedure (a) illustrates a system for the formation of a diffraction pattern. Procedure (b) illustrates a system for the spatial or directional filtration of diffraction patterns. In both procedures a camera can be included to record the end product.

at a point (Fig. 6.5a) and produces a spread of light called the diffraction pattern. This pattern has directionality and intensity; the directionality is orthogonal to the directionality in the film transparency and

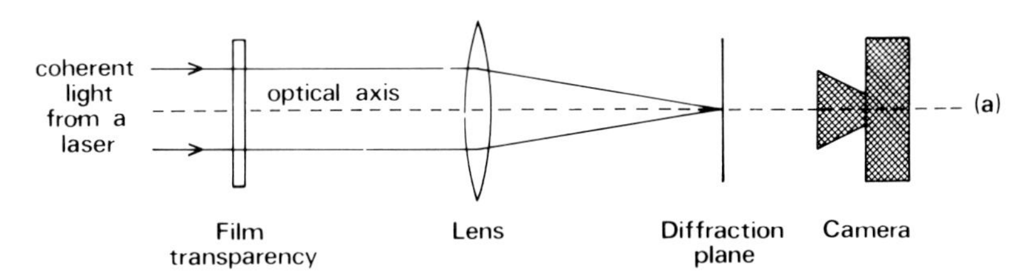


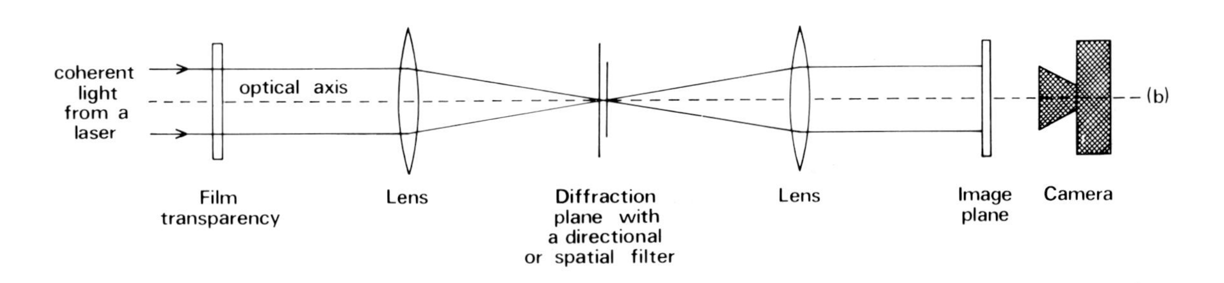

the intensity is related to the spatial frequency of that directionality. This diffraction pattern has provided a means of identifying and quantifying trends within remotely sensed images (Pincus 1969; McCullagh 1981). However, today these tasks are more easily handled by digital image processors (sect. 6.3.3).

Informative as the diffraction pattern is, environmental scientists have found greater utility in the partial reconstruction of the original image from the diffraction pattern (Steiner and Salerno 1975). This is achieved by reversing the procedure by which the diffraction pattern was produced and slipping a filter or two into the optical path to prevent the transmission of certain spatial detail (Harnett *et al.* 1978), (Fig. 6.5b). The three most popular types of filter, the 'bow tie', 'annular' and 'full stop' are illustrated in Fig. 6.6. In image (a) the north west/south east trending fault lines are removed by inserting a north east/south west trending 'bow tie' filter into the optical path. In image (b) the north east/south west trending fault lines are removed, this time by inserting a north west/south east trending 'bow tie' filter into the optical path. In image (c) the west/east trending scan lines with a high spatial frequency are removed by inserting an 'annular' or 'high pass' filter into the optical path and in image (d) the west/east trending cloud ridges with a low spatial frequency are removed by inserting a 'full stop' or 'low pass' filter into the optical path. As can be seen from these examples the removal of certain detail enhances the remaining detail. This has been put to good effect by environmental scientists (Holdermann *et al.* 1978), for example, in the mapping of archaeological sites (Chevallier *et al.* 1970), the removal of parallel lines of sand dunes from bare rock surface (Barnett and Harnett 1977) and the enhancement of terrain lineation (Bauer *et al.* 1967; Ulaby and McNaughton 1975).

**Fig. 6.6**
The modification of images by the spatial or directional filtration of image diffraction patterns.

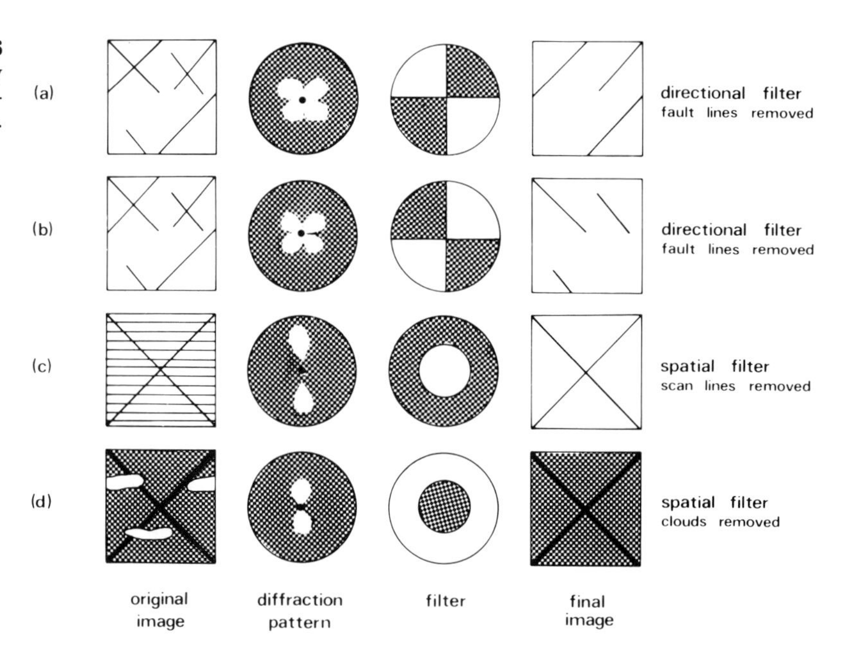

## 6.3 Discrete image processing

A discrete image comprises a number of individual picture elements called pixels, each one of which has an intensity value and an address in two dimensional image space. The intensity value of a pixel which is recorded by a digital number or DN, is dependent upon the level of electromagnetic energy received by the sensor from the Earth's surface and the number of intensity levels that have been used to describe the intensity range of the image. Many remotely sensed images are given an 8-bit ($2^8$) intensity range which stretches from 0 for low radiance or radar return to 255 for high radiance or radar return.

The address of the pixel in two dimensional image space is given in distances along rows and down columns. For example, in Fig. 5.9 the pixel in the first column and row has an address of 1,1 whereas the pixel in the last column and row has an address of 2,340 , 3,240.

There are three stages in the processing of a discrete image. First, the images which are stored on computer compatible tapes (CCTs) are read into a computer, second, the computer manipulates these data and third, the results of the manipulations are displayed. These three stages of image reading, processing and display can be performed using a mainframe computer, a micro-computer with graphics or a purpose built digital image processor.

### 6.3.1 Mainframe computer

The advantage of a mainframe computer is its ability to read and manipulate large remotely sensed data sets. The disadvantages are that the data cannot be handled interactively, output is usually limited to the standard overstrike capability on a lineprinter and specialised image processing software is not usually available. Despite these limitations a mainframe computer has proved to be of particular value in three areas of application; first, for the extraction of DN from CCTs, second, for the production of simple density sliced images (Fig. 6.7), (Curran 1980a; Jensen and Hodgson 1983; Ford *et al.* 1983) and third, for educational purposes (Mather 1980a, 1980b; Williams *et al.* 1981; Gurney and Templeman 1981).

### 6.3.2 Micro-computer with graphics

The use of micro-computer based image processing systems is increasing rapidly. This has been helped by the ready availability of Landsat MSS subscenes on 'floppy' disk, a decrease in micro-computer prices and the introduction of graphics packages that are suitable for basic image processing (Boulter 1979; Short 1982). In comparison with purpose-built digital image processors these systems are relatively slow and lack sophistication and flexibility but at less than one tenth the cost of their big brothers their future in this market is assured (Fig. 6.8).

**Fig. 6.7**
The use of the lineprinter to display a Landsat MSS image. Image (a) is a × 85 photographic enlargement of a Landsat MSS 7 image of part of Oxfordshire and Berkshire, UK (Courtesy, NASA). Image (b) is a lineprinter output of the Henley area and image (c) is a lineprinter output of the Reading area. (Courtesy, Dr C. Gurney Washington DC)

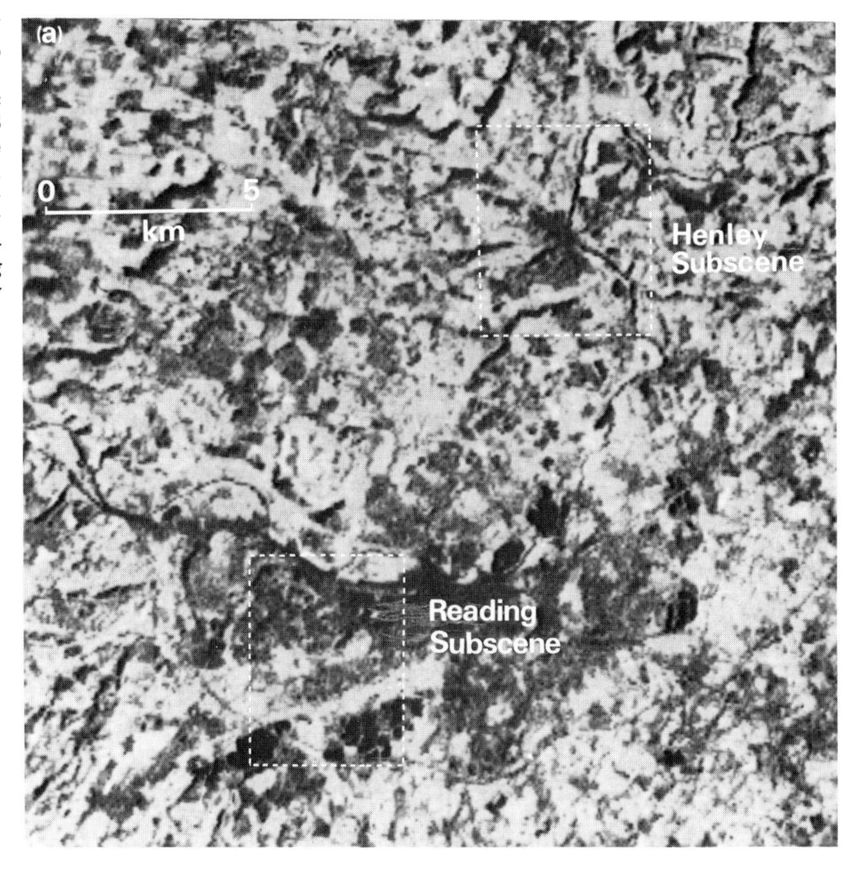


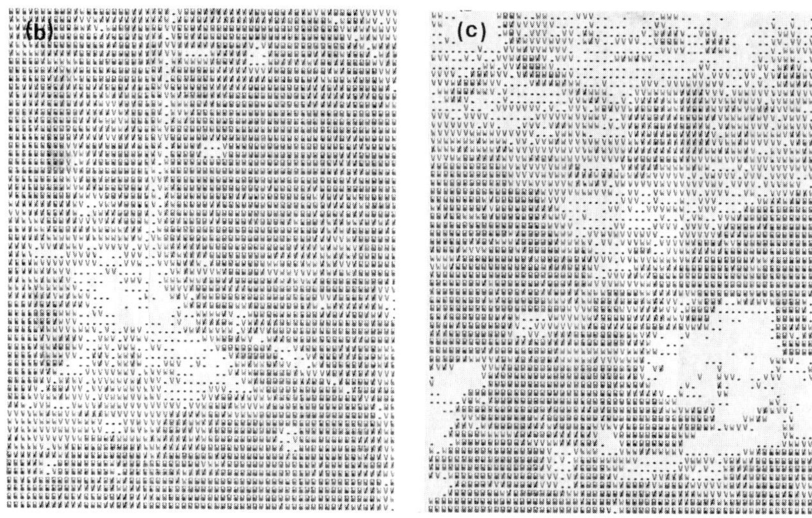


CLASS LIMITS AND SYMBOLS ARE AS FOLLOWS
CLASSES ARE LESS THAN NUMBER STATED
SYMBOL FOR THE CLASS IS GIVEN UNDERNEATH THE LIMIT

| 1 | 21 | 31 | 41 | 51 | 101 |
|---|---|---|---|---|---|
| | . | V | W | M | B |

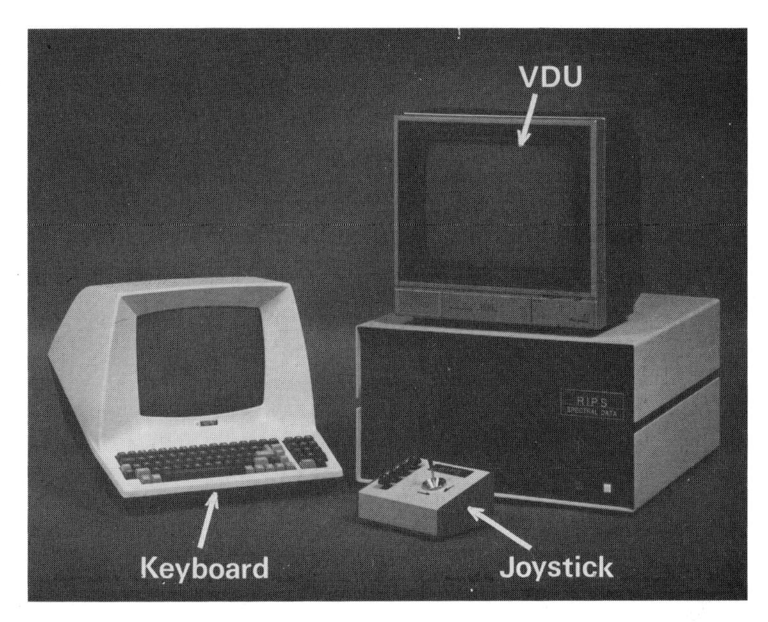


**Fig. 6.8**
A micro-computer with graphics. (Courtesy, Spectral Data Corporation, USA)

## 6.3.3 Digital image processor

The current healthy state of remote sensing is in part due to the rapid development of speedy and flexible interactive digital image processing systems. This equipment has been greeted with glee by the remote sensing community who are purchasing them at a rate, which in the UK alone has increased five-fold over the years 1980 to 1983. The success of the digital image processor hinges on its potential ability to read almost anything put before it, to interactively process vast amounts of data and to display the results in a number of ways (Landgrebe 1976; Moik 1980; Mulder 1980), (Fig. 6.9).

### 6.3.3.1 Image reading

The three reading devices used by a digital image processing system are first, a tape reader to read CCTs of remotely sensed data; second, a table digitiser to input auxiliary information like topography

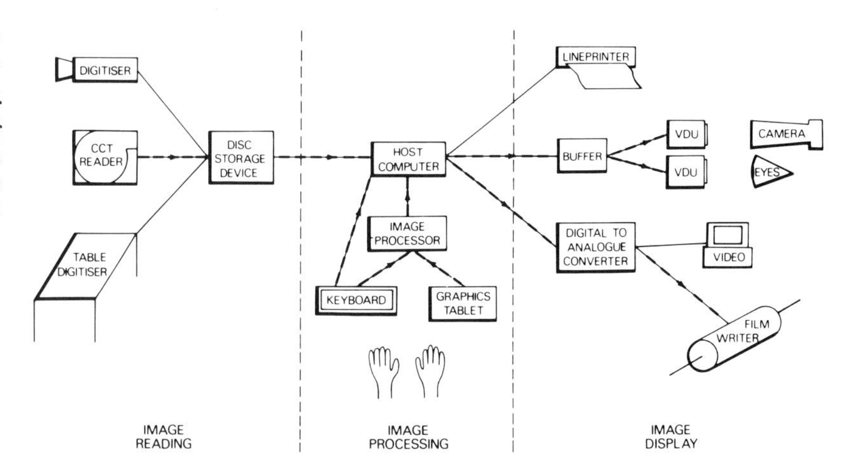


**Fig. 6.9**
The components of a large digital image processing system, comprising three methods of data input and five methods of data output. The dashed lines with arrows indicate a probable processing route for digital satellite sensor data.

political boundaries and census data (Fig. 3.29) and third, a digitiser to convert continuous images like photographs into discrete images.

Three types of digitiser are in widespread use; these are the 'drum', the 'flat bed' and the 'flying spot' (Sabins 1978). The drum is the most popular digitiser for reasons of cost and speed rather than accuracy. It comprises a rotating drum with a hole in one side over which the image is fixed. As the drum spins it moves along its long axis enabling the detector within the drum to measure image transmittance (sect. 3.3.1). The analogue signal produced by the detector is sampled at fixed time intervals to yield discrete data that are suitable for digital image processing. The flat bed digitiser is slow and expensive but very accurate. It works on the same principle as the drum digitiser, only the film is held on a flat bed which moves in $x$ and $y$ directions between a fixed light source and detector. The flying spot digitiser is very fast and accurate and although expensive has a cost commensurate with the other peripherals of a digital image processor (Fig. 6.9). It operates in the same way as a television camera and speedily scans the original images in both the $x$ and $y$ directions (Gonzalez and Wintz 1977; Castleman 1979).

**Fig. 6.10**
Four makes of digital image processor: (A) IDP 3000, made by Plessey Radar Ltd., UK. (Courtesy, Space and New Concepts Department, RAE Farnborough, UK. British Crown copyright reserved); (B) GEMS, made by Computer Aided Design Centre, UK (Courtesy, Space and New Concepts Department, RAE Farnborough, UK. British Crown copyright reserved); (C) I$^2$S model 75 made by International Imaging Systems, USA; (D) Dipix, Aries II made by Dipix Ltd, Canada.

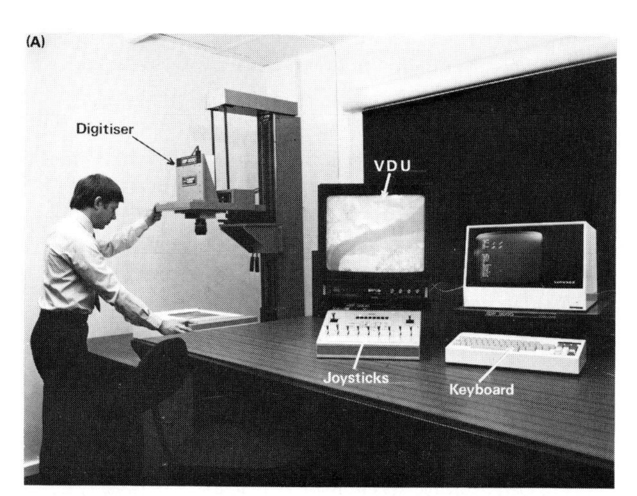


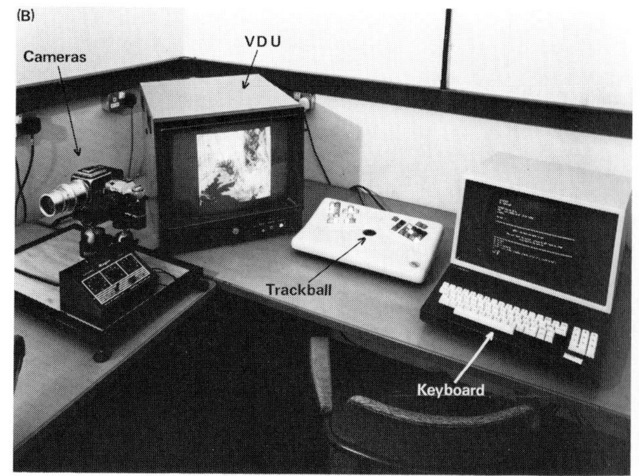


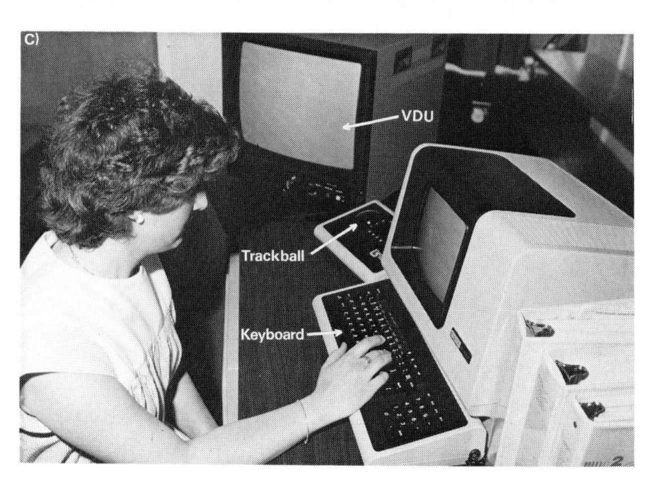

### 6.3.3.2 Interactive image processing

A digital image processor is a cunning blend of computing hardware and software which is designed to undertake many parallel computing operations under the management of a host computer (Andrews 1970; Balston and Custance 1979; Strome and Goodenough 1979). The workings of a digital image processor are transparent to the operator who specifies tasks by a combination of a keyboard, into which commands are typed, (Fig. 6.9) and either a pen guided graphics tablet or a finger guided control panel (Fig. 6.10), (Haralick 1977). To make matters even easier many image processors use menu-based instruction cards which appear on the visual display unit (VDU) or keyboard screen to direct the user through a processing sequence. For example, if the user chooses 'density slice' (sect. 6.3.7.1), a menu card would appear on the VDU or keyboard screen to give a choice of density levels and colours.

### 6.3.3.3 Image display

Four types of image display are commonly used. These are a line-printer, a video tape recorder/monitor, a film writer and a visual display unit (VDU), (Fig. 6.9). The two most important of these are the VDU and the film writer.

One of the first tasks performed by an operator is to transfer some of the data to the VDU. The way in which this is achieved is dictated by the amount of detail required. For example, the digitised photograph in Fig. 6.11, which is composed of 512 × 512 square pixels, can be displayed in a recognisable form by using only one pixel in 12 in both the $x$ and $y$ directions. To display a Landsat MSS image which is composed of 2,540 × 3,240 rectangular pixels also involves image sampling, as only one in six or seven pixels in the $x$ direction and one

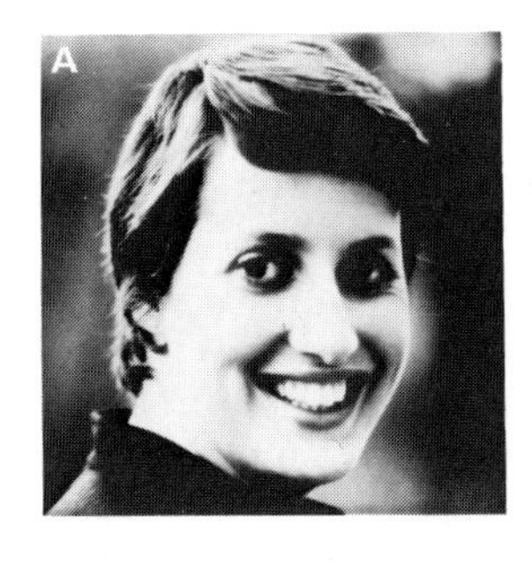


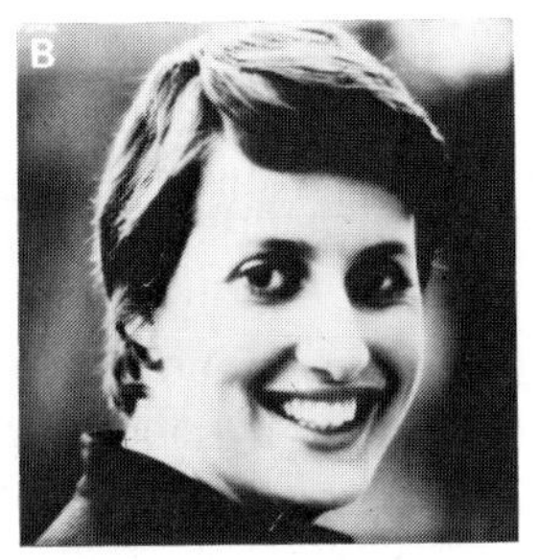


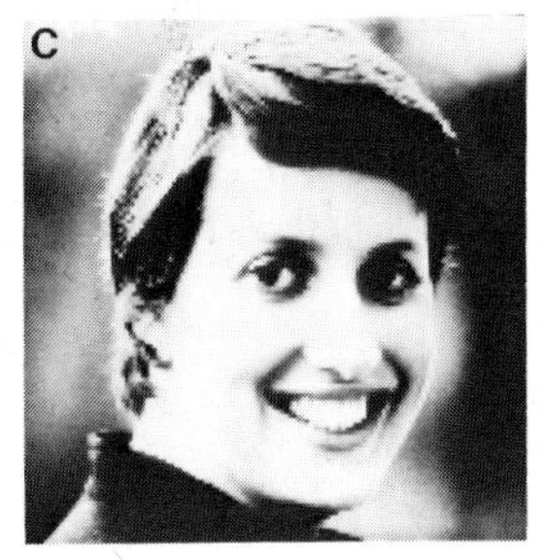


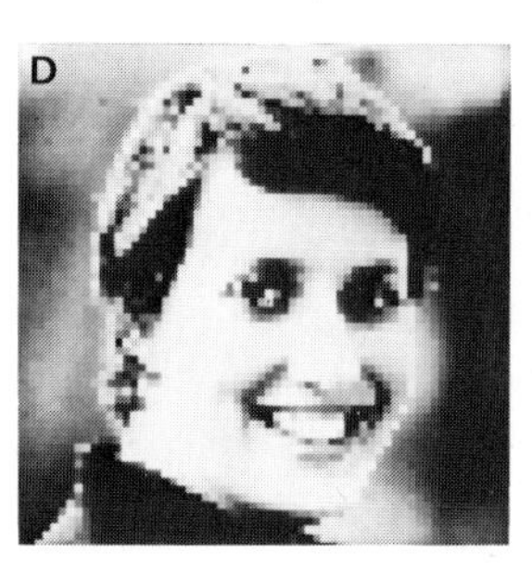

**Fig. 6.11**
The effect of pixel number on the display, of a digitised photograph (Plate 2): (A) Original image 512 × 512 ($26.2 \times 10^4$) pixels; (B) 256 × 256 ($6.6 \times 10^4$) pixels; (C) 128 × 128 ($1.6 \times 10^4$) pixels; (D) 64 × 64 ($0.4 \times 10^4$) pixels; (E) 42 × 42 ($0.2 \times 10^4$) pixels.

**Fig. 6.12**
Methods for the display of a Landsat MSS image on a VDU.

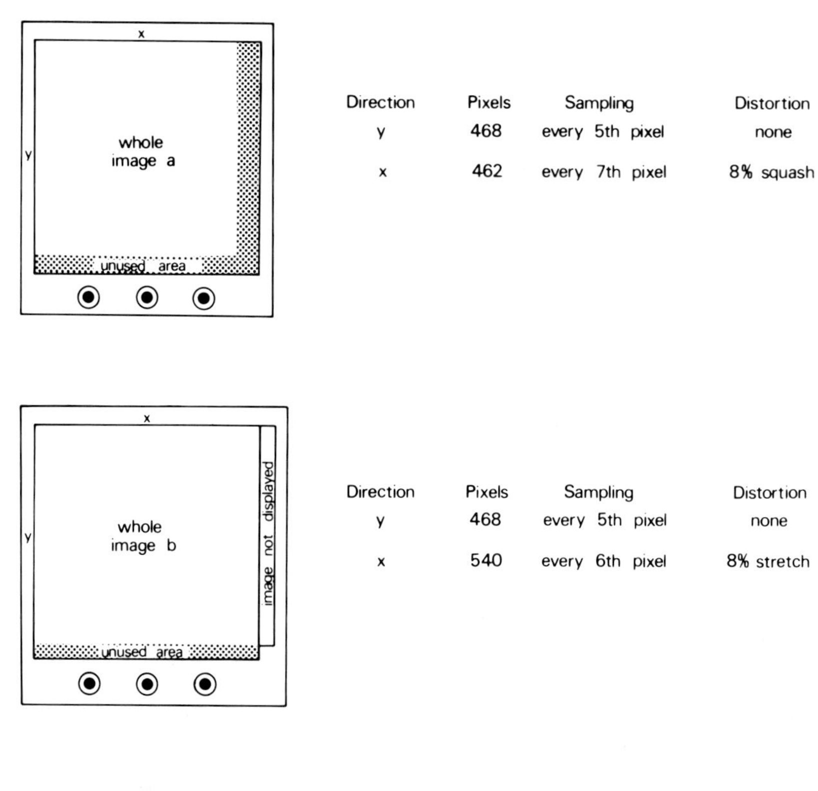


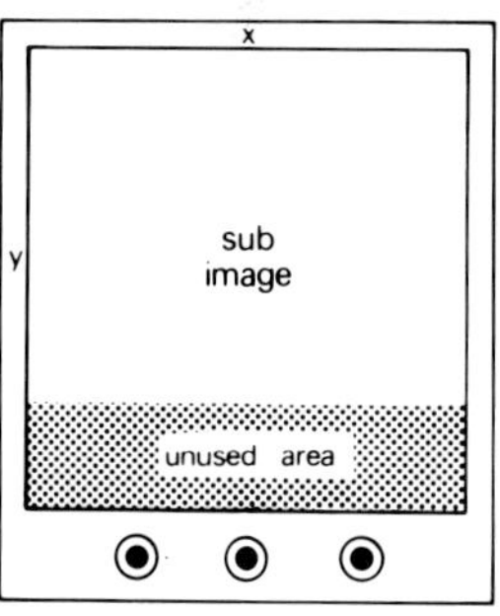


| Direction | Pixels | Sampling | Distortion |
|---|---|---|---|
| y | 394 | various | none |
| x | 512 | various | 30% stretch |

in five pixels in the $y$ direction can be displayed; however, this degree of sampling does decrease with image enlargement (Fig. 6.12).

Film writers are used to produce high quality final images for interpretation and publication. As the drum revolves a scan line of data is exposed onto photographic film by a light source, the intensity of which is modulated by the DN of individual pixels. When a scan line is complete the drum advances and the process is repeated until a transparency suitable for the production of photographic prints has been produced (Fig. 6.13).

To put all of these image processing steps together a typical processing pathway for satellite sensor images is shown in Fig. 6.9. A CCT is read into the disk storage device and all or part of these data are called into the host computer by the operator controlled keyboard. The image is then processed, under instruction from both the operator controlled keyboard and the graphics tablet, as the data moves back-

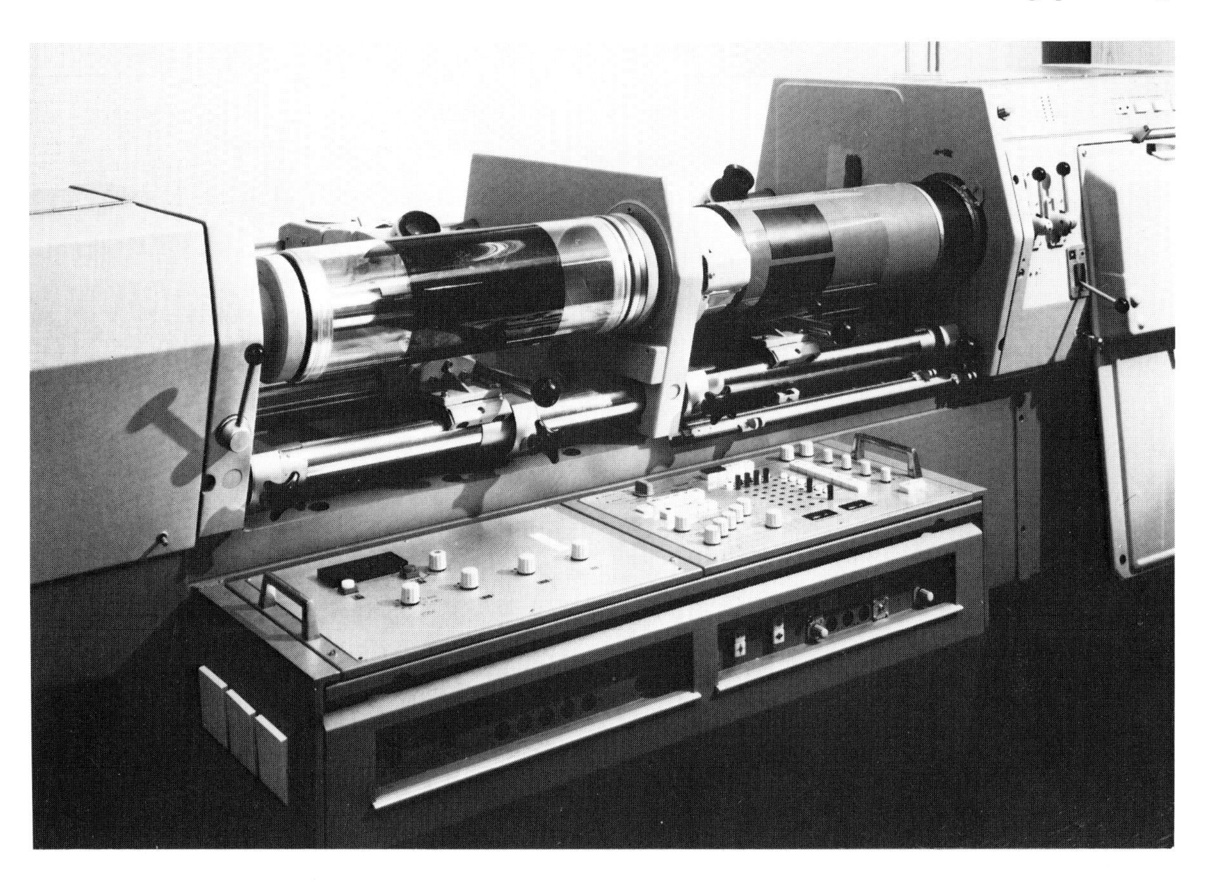

**Fig. 6.13**
Hell DC-300 laser film scanner. (Courtesy, Eurosense, Belgium)

wards and forwards between the host computer and image processor. The results of each image processing step are displayed on a VDU screen to be observed and photographed by the operator. When the processing is complete the image is fed to the digital to analogue converter and then to the film writer.

### 6.3.4 Image processing operations

There are many ways of processing discrete images and environmental scientists have become particulary attracted to six of them. They are, image restoration and correction, image enhancement, image compression, colour display, image classification, and the development of geographic information systems.

### 6.3.5 Image restoration and correction

This is the first stage in any image processing sequence and includes first, the restoration of the image by the removal of effects whose magnitude is known like the non-linear response of a detector and the curvature of the Earth. Second, the correction of the image by the suppression of effects whose magnitudes are not known, like atmospheric scatter and sensor wobble (Andrews 1974).

For ease of discussion image restoration and correction techniques will be discussed in relation to first, the radiometry of the image and second, the geometry of the image.

### 6.3.5.1 Radiometric restoration and correction

Radiometric correction involves the re-arrangement of the DN in an image so that all areas of the image have the same linear relationship between the DN and either radiance or backscatter. This is achieved by two methods, first, the correction of the output from each detector and second, the manipulation of the output from all detectors (Moik 1980).

### 6.3.5.2 Correcting the output from a detector

Each detector within a sensor has a curvilinear response to radiance or backscatter; when the form of this response is known it can be used to transform the output of a detector from a curvilinear to a linear response (Moik 1980). This procedure is computationally trivial as it requires only the addition or subtraction of DN from each DN in the scene. For example, in Fig. 6.14, DN of 184 in the original image would be displayed as 160 in the corrected image. It is easy to correct Landsat MSS data because the relationship between the original and derived linear response is so well known for each of the 24 detectors (6 per waveband) that a look-up-table can be incorporated into image

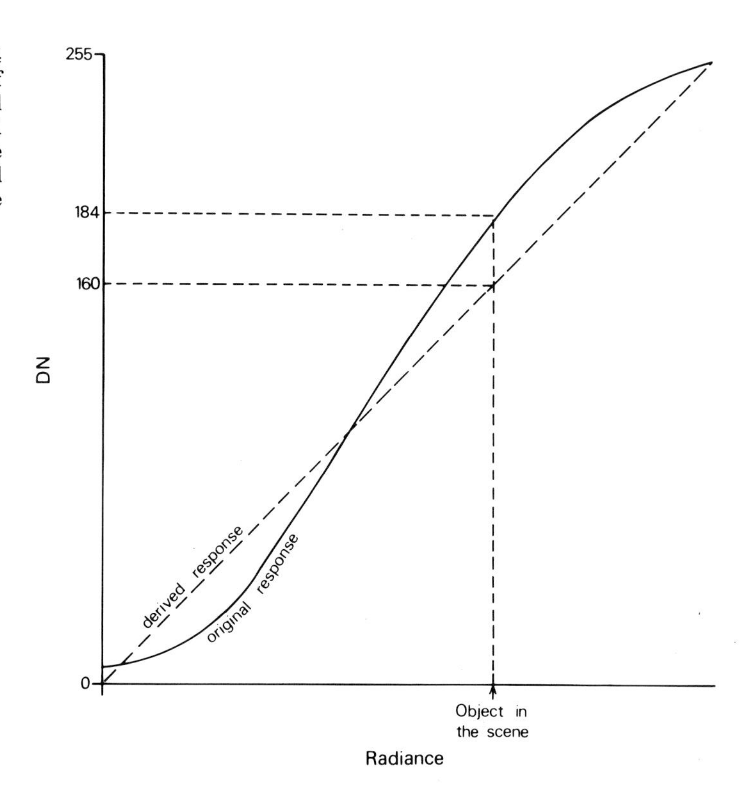


**Fig. 6.14**
Radiometric correction of remotely sensed data. The solid line indicates the original response of the detector and the dashed line indicates the derived and corrected response of the detector.

processing software. One of the most complex radiometric corrections is for Landsat RBV data where image shading and other anomalies result in a spatially variable relationship between the original and derived linear response. For this reason it is customary to segment Landsat RBV images into 'light' and 'dark' regions prior to performing any radiometric correction (Bernstein 1976).

Whatever the data set, no radiometric correction is perfect and so the response of each detector remains unique and to some extent curvilinear, simply because the original detector responses have changed since they were measured in the laboratory. This is evident for Landsat MSS data where the detectors are so unbalanced that even after radiometric correction many images suffer a complaint called sixth line banding, where every sixth scan line has a different response to radiance (Fig. 5.9). To correct these and any images for which response curves are not available it is usual to manipulate the output from all the detectors at the same time (Short 1982). Three of these so called 'cosmetic' manipulations are presented in Table 6.1. As these cosmetics have severe side-effects on the health of the image they are best avoided.

**Table 6.1**
Three methods for the removal of the sixth line banding effect on Landsat MSS images.

| *Method* | *Effect* | *Side effect* |
|---|---|---|
| Low pass (smoothing) filter (sect. 6.3.6.8) | Removes stripes which are high frequency features in the image | Smooths image and lowers spatial resolution |
| Directional (vertical) filter (sect. 6.3.6.10) | Removes stripes which are horizontal features in the image | Emphasises all vertical and suppresses all horizontal detail in image |
| Histogram matching | Correct the mean and variance of each scan lines DN to the mean and variance of a random scan line | Alters considerably the relationship between DN and radiance |

*Source*: Condit and Chavez 1979.

### 6.3.5.3 Skylight and haze suppression

The effect of skylight and haze is negatively related to wavelength and positively related to the distance from a point directly below the sensor. In an attempt to remove the wavelength dependent effect, the assumption can be made that the longest waveband recorded by a particular sensor is unaffected by skylight or haze and that this waveband can be used as a yardstick with which to judge the effect of skylight and haze in the other wavebands. For example, in areas of dark shadow the DN for Landsat MSS 7 is typically zero therefore, as a first approximation the DN above zero in the other three Landsat MSS wavebands can be attributed to skylight and haze and can be subtracted from all the DN. For example, Sabins (1978) and Condit and Chavez (1979) have noted that for Landsat (1, 2 and 3) MSS data, recorded on seemingly haze-free days, when the DN of shadow in MSS 7 is zero, the DN in MSS 6 will be around three, the DN in MSS 5 will be around seven and the DN in MSS 4 will be around eleven.

To correct for the off-vertical viewing effects workers have either applied a correction based on the path length from the ground to the sensor, or as is more often the case, they have omitted all data at the image edge.

### 6.3.5.4 Shade suppression

Meteorological satellite images, especially those collected by sensors onboard geostationary satellites (sect. 5.5.3) cover such vast areas of the globe that irradiance will vary markedly across the scene. To overcome this influence, the DN of each pixel can be increased in direct proportion to the distance from the Sun (Condit and Chavez 1979). For example, to correct for shade on a Meteosat sensor image (Fig. 5.31) the pixels comprising the UK have their DN increased by an amount greater than those for West Africa.

### 6.3.5.5 Geometric restoration and correction

The aim of geometric restoration and correction is to make an image conform to a pre-arranged scheme. This can involve a large number of image manipulations, like the correction of predictable sampling errors that arise from changes in aircraft or satellite altitude, or the alteration of image projection to produce a stereo pair of images (Batson *et al.* 1976), (Fig. 6.15), or the distortion of an image to make it fit onto another image or map (Haralick 1976; Hardy 1978; Moik 1980).

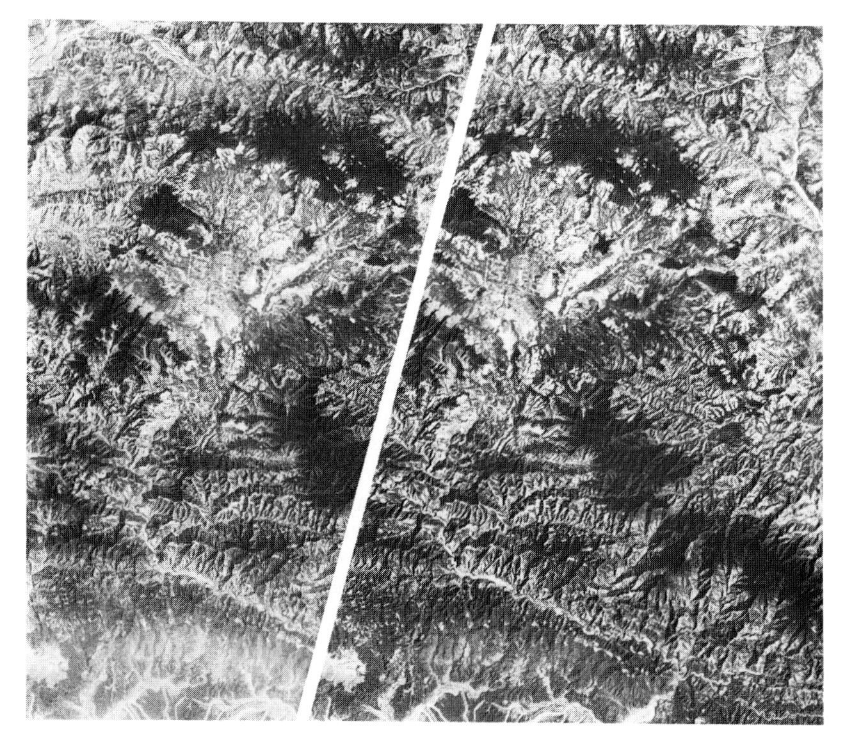

**Fig. 6.15**
False colour stereoscopic pair of Landsat MSS images of the Kathmandu Valley in Nepal, here reproduced in black and white. (Courtesy, Environmental Research Institute of Michigan, USA)

These geometric corrections can be achieved either by changing the location of lines of pixels in the image or by completely resampling the image (Naraghi *et al.* 1983). To change the location of lines of pixels in the image is a relatively simple procedure and is very useful for removing the predictable systematic distortion in an image. To cite two examples, first, to ensure complete ground coverage, airborne multispectral scanners (sect. 4.2) usually overlay their scan lines in the direction of flight. The geometric effect of this oversampling can be removed by simply discarding scan lines. Second, the images collected by satellite borne linescanners and linear array sensors are systematically distorted as the Earth rotates ever eastwards during image acquisition, as is illustrated in Fig. 5.13. This effect can be removed by rectifying each scan line during image formation. As the majority of these systematic distortions are removed before the environmental scientist lays hands on the data, they will not be discussed further, the reader being referred to Sabins (1978) and Moik (1980).

Image resampling involves the reformation of an image onto a new base. This is achieved by using features that are common to both the image and the new base (Bernstein 1976). These features, which are termed ground control points (GCPs), are chosen to be in sharp contrast to their surroundings and are often road intersections, field boundaries, the edges of waterbodies and airport runways (Bernstein 1976; Benny 1983). The GCPs are located on the image by their $x$ and $y$ co-ordinates and on the new base by their latitude and longitude. The functional relationships ($f_1$ and $f_2$) between image $x$ and $y$ and latitude and longitude are determined by least squares regression, as in formula [6.1].

$$x = f_1 \text{ (latitude and longitude)}$$

$$y = f_2 \text{ (latitude and longitude)} \qquad [6.1]$$

To use these two equations to geometrically transform an image involves four stages. First, a geometrically correct geographical grid is defined in terms of latitude and longitude. Second, the computer proceeds through each cell in this geographical grid and at each cell the computer transforms the latitude and longitude values (formula 6.1) into values of $x$ and $y$ which become the new address of an image pixel. Third, the computer visits this address in the image and transfers the appropriate DN from the nearest pixel to this address, to its new home in the geographical grid (Fig. 6.16). Fourth, this process is repeated until the geographical grid is full at which point the image has been geometrically corrected, usually to an accuracy of $\pm 1$ pixel (Bernstein 1976). This procedure is often available on image processing software with the user being asked to stipulate the method by which a nearby pixel is chosen in stage three. The methods commonly available are nearest neighbour, bilinear interpolation and cubic convolution. Details of these techniques are given in Table 6.2, Rosenfeld and Kak (1976), Timon (1980), Moik (1980) and Bernstein (1983).

As these methods are both time consuming and expensive, environmental scientists only apply such corrections when it is absolutely necessary; for many applications the simple superimposition of an image grid or coastline plot is all that is needed, as is demonstrated Fig. 5.27.

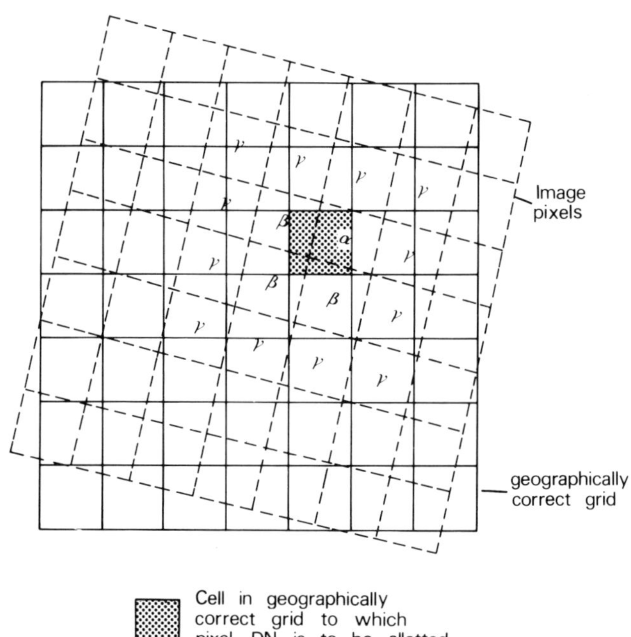


**Fig. 6.16**
Resampling procedure used to geometrically correct an image. (Modified from Lillesand and Kiefer 1979)

| *Method* | *Technique* | *Example pixels to be transferred in Figure 6.16* | *Relative computing time* | *Advantages* | *Disadvantages* |
|---|---|---|---|---|---|
| Nearest Neighbour | Transfer DN of nearest pixel | α | 1 | Simple to compute. DN unaltered. | Image disjointed. Pixels offset by up to half a pixel. |
| Bilinear Interpolation | Transfer proximity weighted average of nearest 4 pixels | α and β | 10 | Smooth image. Geometrically accurate. | DN altered |
| Cubic Convolution | Transfer evaluated weight of nearest 16 pixels | α, β and γ | 20 | Very smooth image | Complex to compute. DN altered |

**Table 6.2**
A comparison between three methods of resampling remotely sensed images.

*Sources*: Bernstein 1976; Lillesand and Kiefer 1979; Moik 1980; Short 1982.

## 6.3.6 Image enhancement

Image enhancement involves the improvement of an image in the context of a particular application. The most popular image enhancements are contrast stretching, band to band ratio and subtraction, digital filtering, data compression and colour display.

### 6.3.6.1 Contrast stretch

Remote sensing instrumentation is designed to record the intensity of electromagnetic radiation whether it be radiance or backscatter over a very wide range of conditions. For example, the Landsat MSS and TM sensors have to record the radiance from both poorly illuminated Arctic oceans and well illuminated deserts and as a result only a relatively small proportion of their measurement scale will be used for any particular scene. If these pixels were to be displayed in their raw state the image would have such a low contrast that it would be difficult to differentiate between objects with a slightly different DN. To overcome this problem the measured DN of the pixels can be stretched onto a new longer display scale by taking pixels with a fairly low DN on the measurement scale and pulling them to the very low end of a display scale and by taking pixels with a fairly high DN on the measurement scale and pulling them to the very high end of a display scale, thus altering all absolute DNs (Rosenfeld and Kak 1976; Ronde *et al.* 1978; Condit and Chavez 1979).

Of the many methods available for stretching image contrast the two most popular are the semi-automatic and the interactive manual.

### 6.3.6.2 Semi-automatic contrast stretch

The minimum and maximum DN on the measurement scale are established either by observing a histogram of DN frequency (Fig. 6.17) or more likely by density slicing the image (sect. 6.3.7.1). These minimum and maximum values are then automatically stretched to become 0 and 255 respectively and the intervening pixels are arranged between these two extremes. The method by which this is achieved depends on the users objective. A linear stretch is used when equal weight is to be given to all DN regardless of their frequency of occurrence and a histogram equalisation stretch is used when it is felt necessary to weight the DN by their frequency of occurrence.

An example of the more popular linear stretch is illustrated in Fig. 6.17; a pixel with a DN of 51 has been given the DN of 0, a pixel with a DN of 178 has been given the value of 255 and all other pixels have been stretched in between. Equal weight has been given to all DN with the result that 80% of the pixels, that is those which occur within a DN range of 105–147 on the measurement scale are spread over only a third of the display scale within a DN range of 110–195. If a histogram equalisation stretch had been performed, 80% of the pixels, that is those which occur within a DN range of 105–147 on the measurement scale, would have been spread over 80% of the display scale. The resultant image would have greater overall contrast but smaller features would be distinguished less easily.

### 6.3.6.3 Manual contrast stretch

This involves interactive manipulation of the transfer function which converts the measurement scale to the display scale and this usually assumes an *a priori* understanding of the scene under study (Steiner and Salerno 1975; Sabins 1978). Fig. 6.18 illustrates the transfer function used for the linear contrast stretch of Fig. 6.17, where the measurement scale of 51–178 DN has been stretched to a display scale of 0–255. If it were known that for the scene under study, pixels with a

**Fig. 6.17**
Effect of a linear contrast stretch. The original data on the upper histogram have DN in the range 51–178. After a linear contrast stretch the data are displayed on a much larger scale of 0–255.

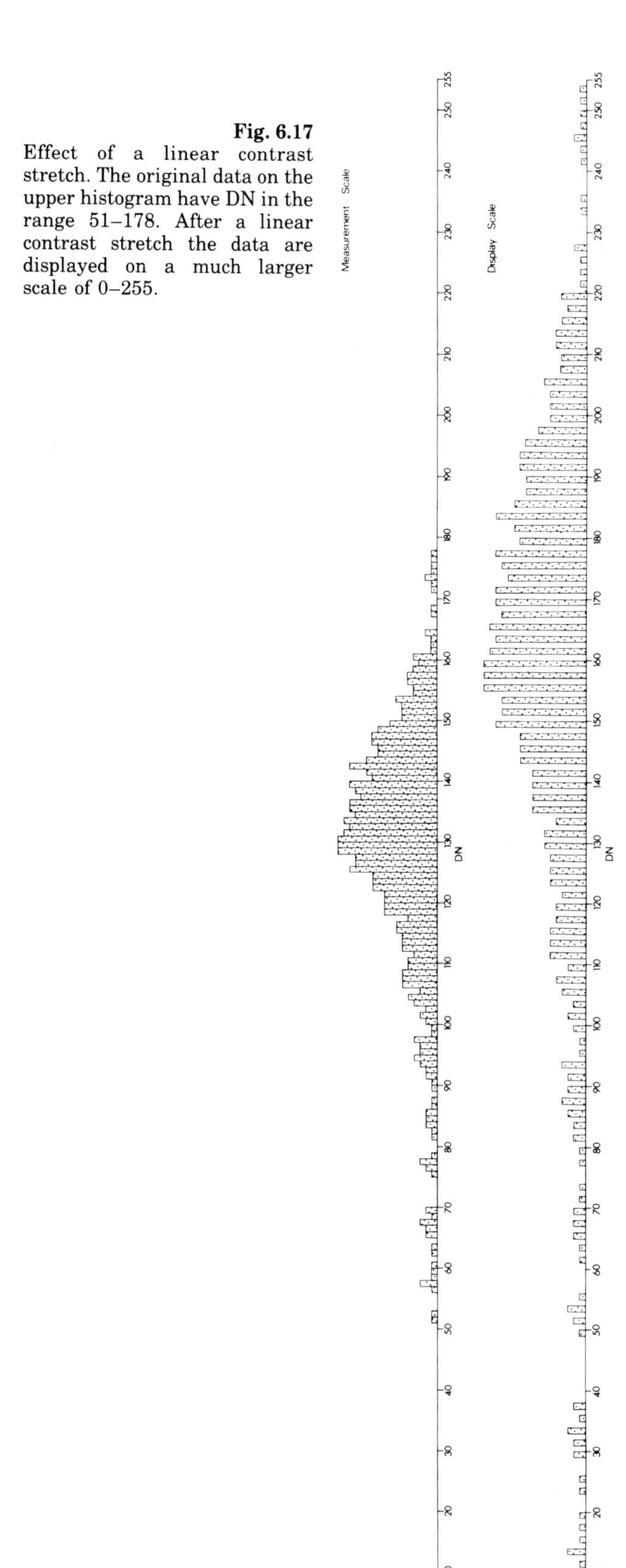

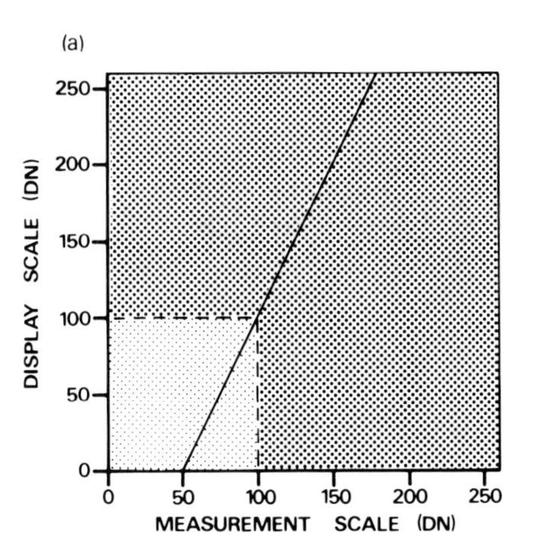


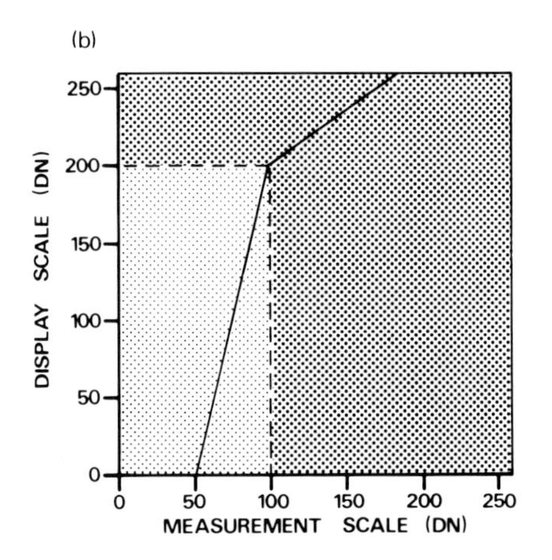


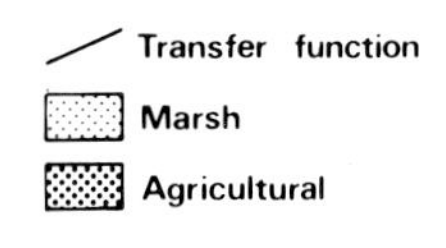


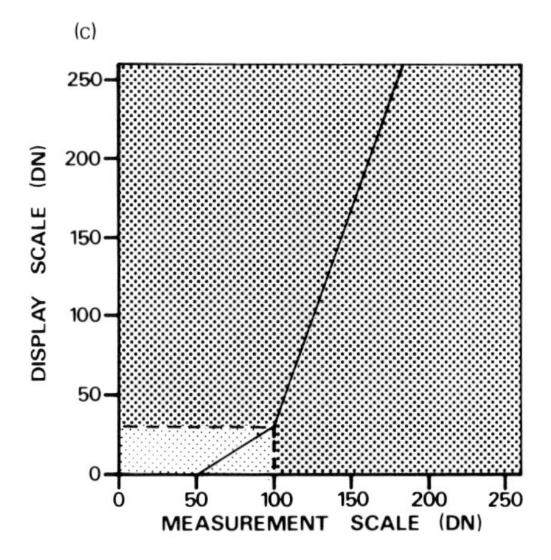


**Fig. 6.18**
Manual contrast stretch of a scene with two land covers of marsh and agriculture. Graph (a) displays the measurement scale via a linear transform. Graphs (b) and (c) display the measurement scale via a two-part linear transform, in (b) 'marsh' pixels are emphasised at the expense of 'agricultural' pixels and in (c) 'agricultural' pixels are emphasised at the expense of 'marsh' pixels.

DN of less than 100 represented marsh land and pixels with a DN greater than 100 represented agricultural land then specific transform functions could be used to preferentially stretch one of these scene components. For example, a user with a hydrological or botanical bent may wish to expand the display scale for 'marsh'. Such a transform is illustrated in Fig. 6.18, where marsh pixels are spread over the display scale range from 0–200 DN, relegating agricultural pixels to the display scale range of 201–255 DN. Conversely an agricultural user may wish to expand the display scale for agricultural land. Such a transform is illustrated in Fig. 6.18, where marsh pixels are relegated to a display scale range of 0–30 DN while agricultural pixels have the display scale range of 31–255 DN.

Other manual stretches include fitting the data to particular functions, for example logarithmic or gausian as shown in Fig. 5.9 and the allocation of particular DN ranges in the display image for particular DN ranges in the input image (Steiner and Salerno 1975; Gonzalez and Wintz 1977; Sabins 1978).

### 6.3.6.4 Band to band ratio and subtraction

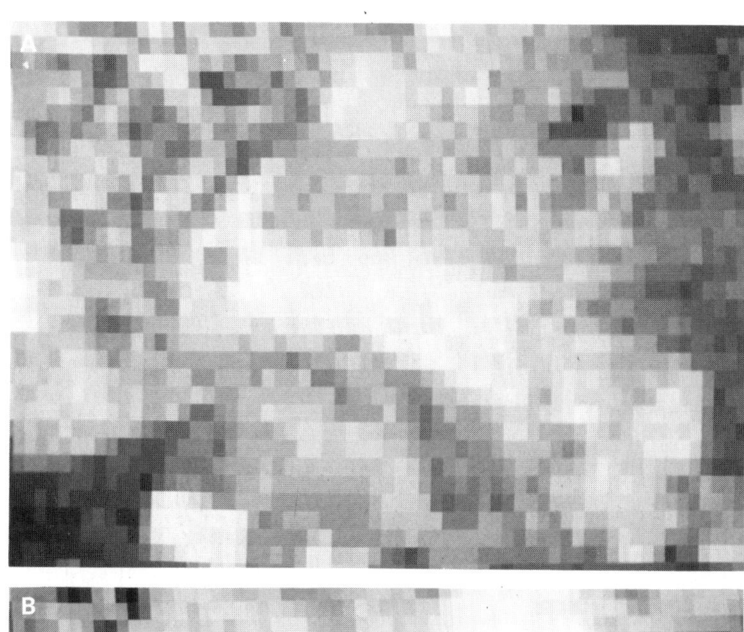


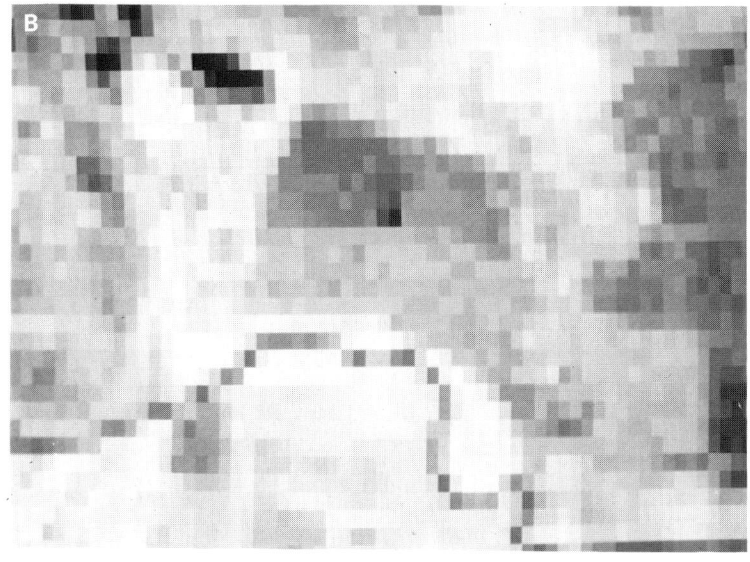

**Fig. 6.19**
Landsat MSS images, enlarged × 3,700 of part of the Thames valley in Oxfordshire, UK. Image (A) is MSS 5 (red); image (B) is MSS 7 (near infrared); image (C) is a ratio of MSS 7 to MSS 5; image (D) is a panchromatic aerial photograph of the same area. (Image (D) courtesy, Clyde Surveys Ltd, UK)

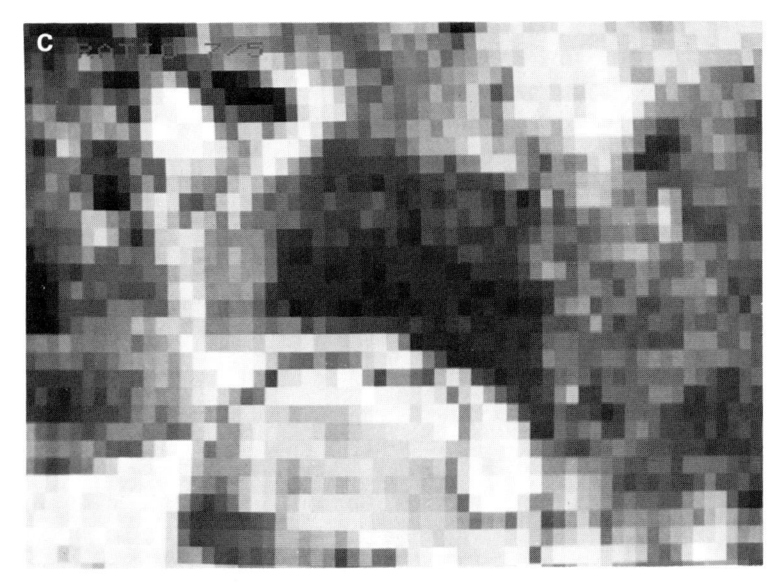

It is often useful to display, in image form, the difference in the amount of electromagnetic radiation received by a sensor in different wavebands that are recorded at the same time, or to display the same wavebands that are recorded at different times. Such an image is achieved by ratioing or subtracting the DN of pixels in one image with the corresponding pixels in another image and contrast stretching the result. This ratio or difference image will suppress detail common to the two images (Sabins 1978; Condit and Chavez 1979; Townshend 1981b) and will enhance detail that is different between the two images. For example, when working with Landsat MSS images, workers have used ratios of MSS 4 (green) to MSS 5 (red), for the location of red soils and iron ore deposits (Vincent 1973) and MSS 6 (near infrared) to MSS 7 (near infrared), to minimise the influence of vegetation cover (Siegal and Goetz 1977). Perhaps the most useful ratio has been MSS 5 (red) to MSS 7 (near infrared) as this ratio is positively related to vegetation amount.

For example, Fig. 6.19 shows much enlarged atmospherically corrected Landsat MSS images of an agricultural area. In MSS 5 (red) the woodland and pasture have a radiance which is lower than for the bare field. In MSS 7 (near infrared) the woodland and pasture have a radiance which is higher than for the bare field. In the MSS 7/MSS 5 ratio image the bare field is shown to have a much smaller radiance difference between MSS 7 and MSS 5 than either the pasture or the woodland, as would be expected from discussions in section 2.5.2. Ratio images of near infrared and red wavebands are employed to provide indications of vegetation amount using data from many different types of sensor (Curran 1983c), (Plate 3). For example, the red to near infrared ratio image created using Advanced Very High Resolution Radiometer (AVHRR) data from the satellite NOAA 7 (Fig. 6.20) is currently sold by NOAA as an aid to the monitoring of vegetation amount (Short 1982; Norwine and Greegor 1983).

A difference image derived from two images recorded at different times can be used for monitoring change. The procedure is slightly more complicated than the multiwaveband case, as the images must first be geometrically corrected by resampling one image directly onto the other (sect. 6.3.5.5). This technique has proved to be of particular value in the monitoring of short-term events like flood, fire, cloud movement and crop growth or long-term changes like urban development, deforestation or desertification (Moik 1980; Nelson and Grebowsky 1982).

### 6.3.6.5 Spatial filtering

Filtering provides a means of improving images by suppressing or enhancing certain spatial frequencies, directions and textures (Rosenfeld and Kak 1976). These procedures can be undertaken on the transformed image frequency data in what is known as the 'frequency domain' or, on the image data itself, in what is known as the 'spatial domain'.

### 6.3.6.6 Spatial filtering in the frequency domain

The image is converted using a frequency transform like the Fourier

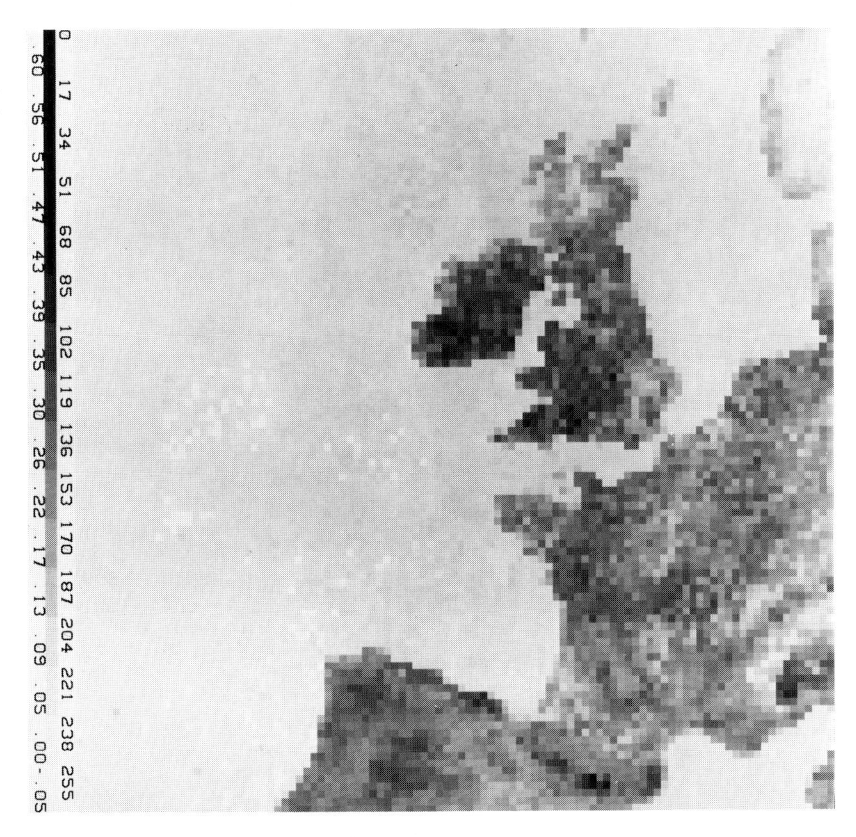


**Fig. 6.20**
Normalised vegetation index (near infrared – red/near infrared + red) image for northwest Europe. This image was produced using 7 days of spatially degraded AVHRR data from 21.3.1983 to 27.3.1983. (Courtesy, NOAA)

or Hadamard into a diffraction pattern (Steiner and Salerno 1975). In the discussion of diffraction patterns in section 6.2.7 it was noted that as low frequencies in the scene (e.g. major landforms) usually occurred near to the centre of the diffraction pattern and high frequencies in the scene (e.g. urban areas) usually occurred near to the edge of the diffraction pattern the manipulation of their location could be a means of enhancement. For example, to improve the radiometric properties of a thermal infrared linescanner image it may be desirable to suppress high frequency noise by removing some of the higher frequency pixels at the edge of the diffraction pattern. Likewise to improve the spatial resolution of a thermal infrared linescanner image it may be desirable to enhance some of these high frequency pixels by suppressing some of the lower frequency pixels nearer to the centre of the diffraction pattern.

As filtering in the frequency domain is both abstract and expensive it is used with lower frequency (!) than filtering in the spatial domain.

### 6.3.6.7 Spatial filtering in the spatial domain

This involves passing a filter of N × N pixels over an image and converting the central pixel of this so called 'kernel' to some measurement of the total array *en route*. As this technique involves the rolling together of all DN within the pixel array it is often known by the descriptive term of convolution (Kim and Strintzis 1980). There are many breeds of filter in use. The four most popular are the low pass filter, high pass filter, directional filter and textural filter (Holdermann *et al.* 1978).

### 6.3.6.8 Low pass filter

A low pass filter enhances low spatial frequencies – features that are larger than the array used, at the expense of high spatial frequencies – features that are smaller than the array used. As such it has earned the nickname of the 'smoothing' or 'defocusing' filter.

This filter replaces the central pixel in the array with the mean of the total array as it passes over the image. (Fig. 6.21). It has proved to be particularly useful for a number of operations like the smoothing out of over enlarged discrete images to make them appear continuous, the suppression of banding that is common to both Landsat MSS and

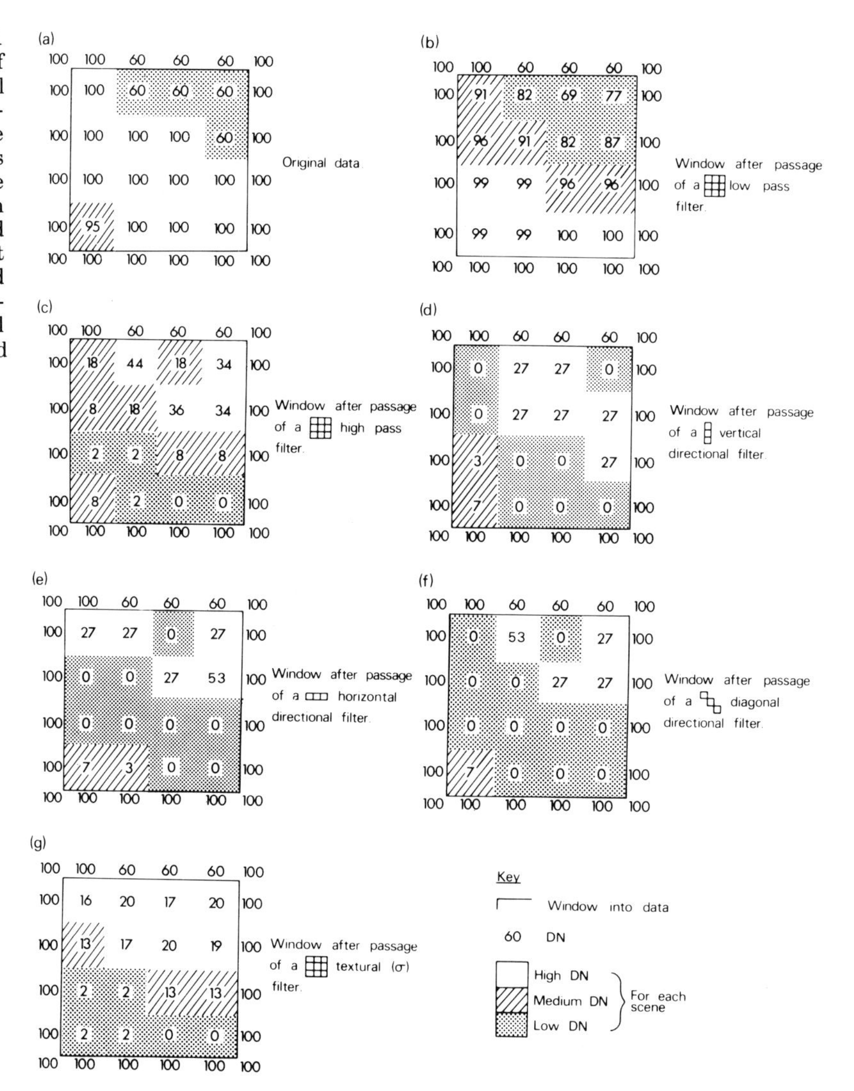


**Fig. 6.21**
An illustration of the effects of spatial filtering on a 16-pixel matrix. Matrix (a) is the original data; matrix (b) shows the smoothing effect of a low pass filter; matrix (c) shows the edge enhancement effect of a high pass filter; matrices (d), (e) and (f) show the edge enhancement effect of directional filters and matrix (g) shows the enhancement of roughness by a textural filter, where σ is the standard deviation

TM images and the suppression of 'speckle' in radar images (sect. 4.4.5), (Lee 1981).

As the size of the filter array is positively related to the smoothness and spatial resolution of the final image, filters with an array larger than 9 × 9 pixels are infrequently used, however they are discussed in Chavez and Bauer (1982) and illustrated in Fig. 6.22.

### 6.3.6.9 High pass filter

A high pass filter enhances high spatial frequencies – features that are smaller than the array used, at the expense of low spatial frequen-

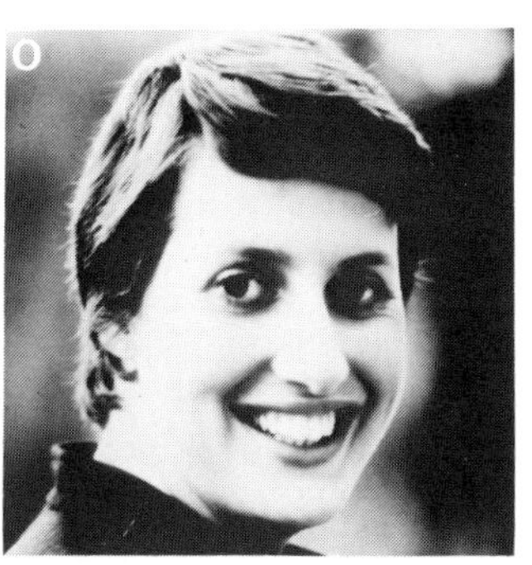


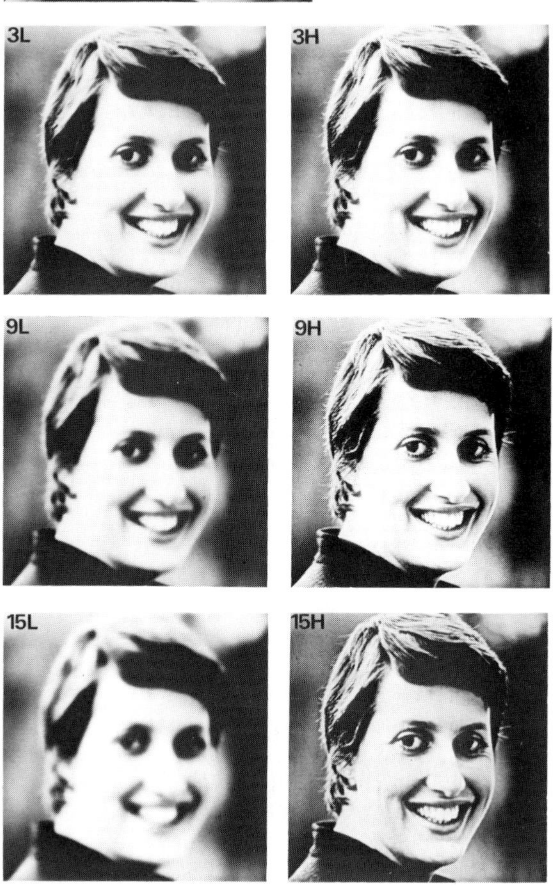


**Fig. 6.22**
The effect of low pass (smoothing) and high pass (edge enhancing) spatial filters on a digitised photograph (Plate 2). (O) original, (3L) effect of a 3 × 3 low pass filter, (3H) effect of a 3 × 3 high pass filter, (9L) effect of a 9 × 9 low pass filter, (9H) effect of a 9 × 9 high pass filter, (15L) effect of a 15 × 15 low pass filter and (15H) effect of a 15 × 15 high pass filter.

cies, – features that are larger than the array used. As such it has earned the nickname of the 'sharpening' or 'edge enhancement' filter (Short 1982).

This filter multiplies the absolute difference between the mean of the pixel array and the central pixel by an arbitary number, as it passes over the image (Fig. 6.21). It has proved to be particularly useful for both cosmetic and enhancement purposes. Perhaps its most useful cosmetic purpose is in the restoration of high frequency information which has been lost as a result of discrete sampling. This loss occurs because sub-pixel sized features with a high spatial frequency are integrated into one DN. For example in Fig. 6.21, if 100 DN represented grass, 60 DN represented trees and 95 DN represented a tree surrounded by grass then the single tree pixel would have a DN that is only 5% lower than the neighbouring grass pixels. After the passage of a high pass filter with a 3 × 3 pixel array and using a multiplicative constant of 2 (Fig. 6.21) the single tree pixel has a DN that is two and a half times higher than the neighbouring grass pixels and therefore the detectability of this low frequency object has been enhanced. Perhaps the most useful enhancement application is the increasing of contrast between edges and their surroundings as the interpretability of an image is improved if the image appears 'sharper' (Fig. 6.22). This has proved to be of great value in the mapping of all kinds of terrestrial, aquatic and atmospheric boundaries (Hartl 1976).

### 6.3.6.10 Directional filter

Directional filters are either high or low pass filters that enhance given spatial frequencies in their direction of travel (Fig. 6.23). In the environmental sciences high pass spatial filters are the most popular kind of directional filter as a result of their ability to enhance high spatial frequencies (edges) in a particular direction. Such a filter, which is often one by three or more pixels in size, multiplies the difference between the mean of the pixel array and the central pixel by an arbitary number as it passes over the image. This has proved to be of particular value in the enhancement of features with a preferred orientation including geological lineations, oceanic waves and the man-made features of roads and urban boundaries. In Fig. 6.21 the results of vertical, horizontal and diagonal high pass directional filtering are illustrated.

### 6.3.6.11 Textural filter

Image texture whether it be smooth, rough, fluffy or knobbly is an important discriminating characteristic of an image region, but to enhance this for the interpreter, or to use it as a characteristic for image classification it must be given a value at a point (Haralick and Shanmugam 1974; Hartl 1976; Thomas 1977; Stanley 1977).

Digital filters have been used to derive these values. The earliest examples involve a three by three pixel array which substitutes the central pixel of the array with a simple measure of DN variance (range, kurtosis, standard deviation) as it passes over the image (Haralick *et al.* 1973; Logan *et al.* 1979). This is done with the assump-

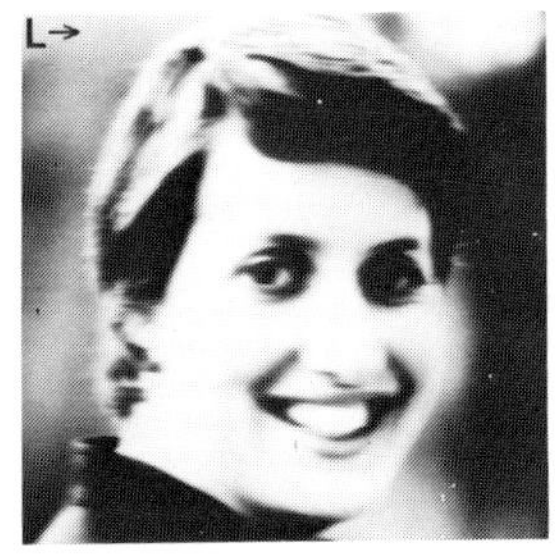


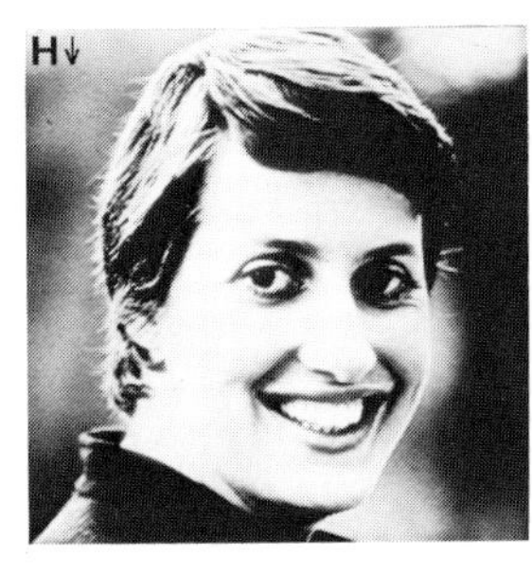


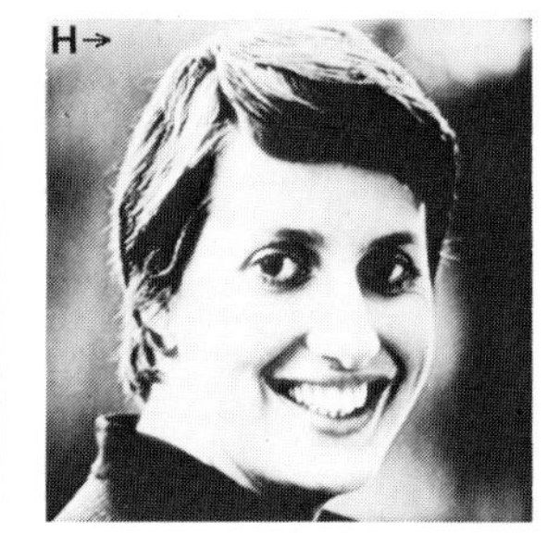


**Fig. 6.23**
The effect of directional, low pass and high pass spatial filters on a digitised photograph (Plate 2). (O) original, (L↓) effect of a 1 × 15 low pass filter that smooths the image from top to bottom (L→) the effect of a 15 × 1 low pass filter that smooths the image from side to side (H↓) the effect of a 1 × 15 high pass filter that edge enhances the image from top to bottom and (H→) the effect of a 15 × 1 high pass filter that edge enhances the image from side to side.

tion that the region with the highest DN variance would have the roughest texture, Fig. 6.21. Today far more complex textural filters are available (Wechler and Citron 1980; Irons and Petersen 1981) and have been used in many applications including geological mapping (Thomas *et al.* 1981), terrain analysis (Weszka *et al.* 1976; Shih and Schowengerdt 1983), land cover mapping (Hsu 1978; Irons and Petersen 1981), forest mapping (Logan *et al.* 1979), urban classification (Jensen 1981) and the differentiation of sea ice (Gersen and Rosenfeld 1975) and clouds (Harris 1977).

## 6.3.6.12 Data compression

Environmental scientists can be faced with the Herculean task of trying to comprehend many multispectral images of a scene at the same time (Fig. 4.7). This problem can be eased if the information which is particular to each image is combined into a new image by means of a statistical transformation. The most popular techniques for the transformation of multispectral images are the statistical techniques of principal components analysis (PCA) and canonical analysis (Short 1982). Both of these techniques aim to replace the original wavebands which describe the data with new orthogonal axes that better describe the particular scene under study.

For example, in Landsat MSS data it is noted that the statistical variance (or information) is divided between all four wavebands. A hypothetical scatter plot of the two most dissimilar wavebands of red (MSS 5) and near infrared (MSS 6 or 7) illustrates that while these two bands are correlated and therefore hold some redundant information, neither the red nor the near infrared image would be a suitable ambassador for the whole data set (Fig. 6.24). To overcome this problem PCA creates new axes called band axes along the lines of maximum variance within the data. Therefore once the pixels have been located by their new co-ordinate system (Fig. 6.24) a band axis A image would contain more information than any other band axis image (Meijerink and Donker 1978).

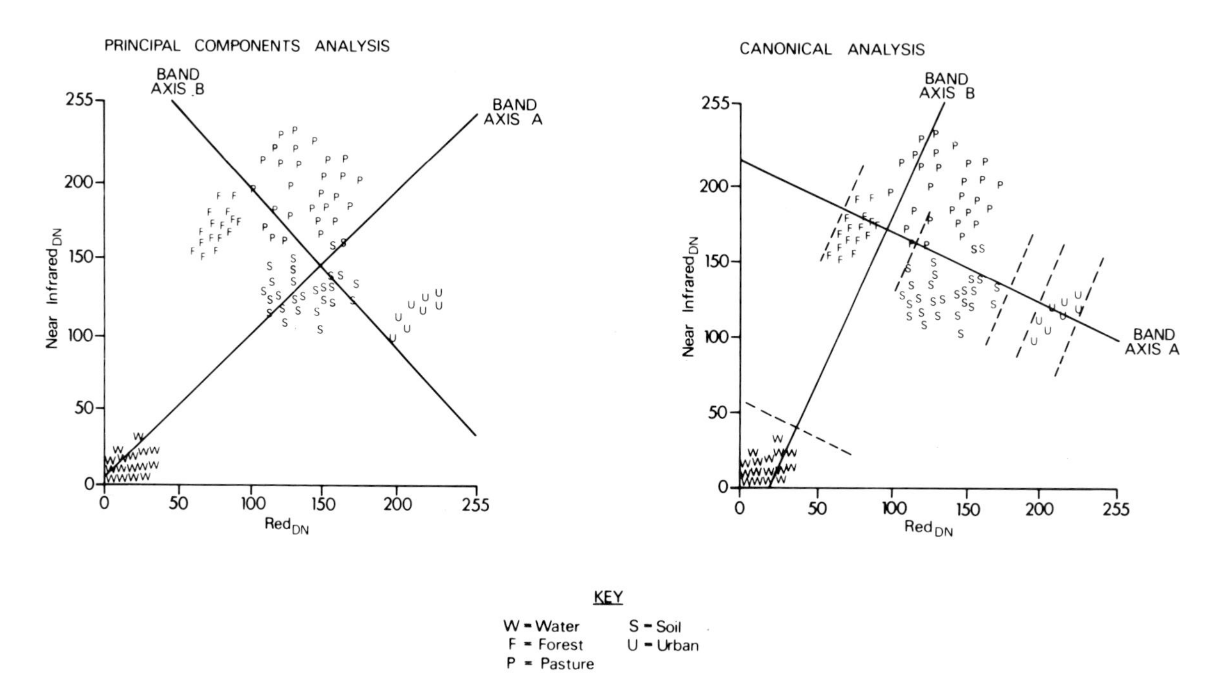


**Fig. 6.24**
The use of principal components analysis and canonical analysis to compress spectral data onto new orthogonal axes.

This transformation is illustrated for a Landsat MSS scene of south western England where Fig. 5.9 are the original four wavebands of the Landsat MSS scene and Fig. 6.25 are the transformed four bands of the Landsat MSS scene (Davidson 1980). Around 90% of the total variance has been compressed into band axis A making it a very effective single band composite of the multispectral data (Abboteen 1978). The remaining three images contain detail that would otherwise be obscured by the more dominant features in the scene. For example in the band axis B image, woodlands, estuarine sediment and sixth line banding are enhanced. To avoid the loss of this information some workers have combined the band axes A, B and C together in a false colour composite as is discussed in Lillesand and Kiefer (1979), Moik (1980) and Byrne *et al.* (1980).

Canonical analysis, like PCA, compresses the information component of several images into one image and this maximises the differ-

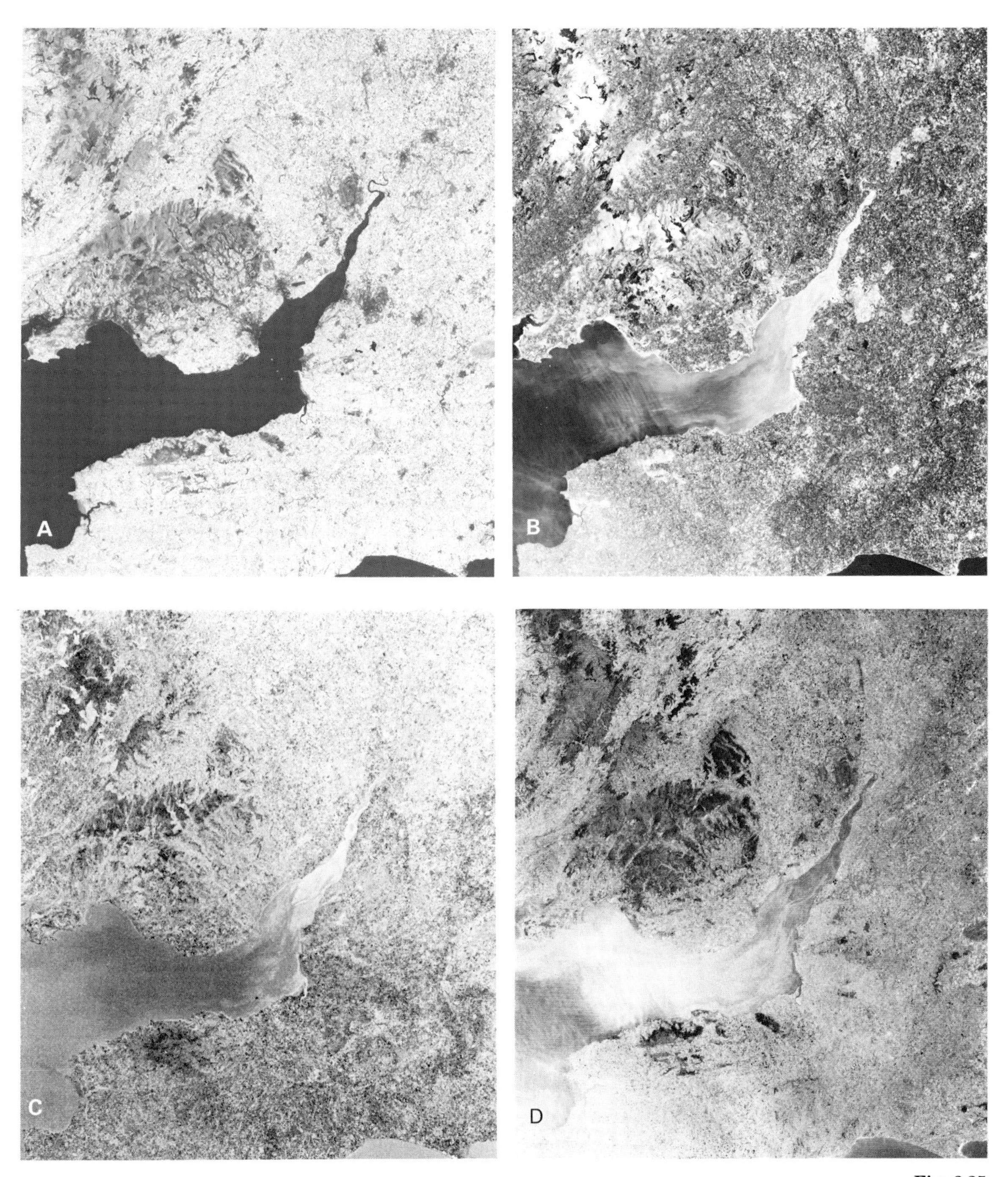


**Fig. 6.25**
Landsat MSS images of southwestern England following a principal components transform. Image (A) is the product of band axis A and carries around 90% of the total variance. Image (B) is the product of band axis B and carries around 8% of the total variance. Image (C) is the product of band axis C and carries around 1% of the total variance and image (D) is the product of band axis D and also carries around 1% of the total variance. The original images are in Fig. 5.9. (Courtesy, Space and New Concepts Department, RAE Farnborough, UK)

ence in DN between the major scene components. Figure 6.24 illustrates how canonical axes would be placed through a hypothetical data set (Johnston 1980). Note that while there is overlap between the classes of land cover in both red and near infrared wavelengths the overlap is reduced when viewed from the vantage of the new band axes A and B (Short 1982). As with PCA, each pixel has a new set of DN for each axis and so each axis can be displayed as an image (Lillesand and Kiefer 1979).

PCA and canonical analyses are not very popular among environmental scientists due to their heavy use of computing time and the destruction of the relationship between object radiance or backscatter and image DN, as discussed in Chapter 2. However, it is likely that the use of compression techniques will increase, if only for the handling of seven waveband Landsat TM data, multitemporal data and integrated multisensor data sets.

#### 6.3.6.13 Colour display

As our eyes can perceive more colours than shades of grey (sect. 3.3.5) the use of coloured images can dramatically increase the amount of information that can be displayed (Billingsley *et al.* 1970). Images are usually coloured using 'pseudocolour' on single images and 'normal' or false colour on several images.

Pseudocolour involves the replacement of each grey level in the image with a colour which separates the small grey scale differences that could not be distinguished by the human eye. This technique is relatively easy to achieve on many image processing systems which readily divide the density levels of an image over a continuum of colours (Moik 1980).

'Normal' and false colour composites are used to display three images of a scene. The procedure is to contrast stretch and composite the images and then colour each waveband with a primary colour (Eyton 1983). The variations in the spectral response of objects in these three wavebands result in colour differences which aid in the object's identification (Plates 3 and 4). For example, to produce the standard Landsat MSS false colour composite on the cover of this book, MSS 4 (green) was displayed as blue, MSS 5 (red) was displayed as green and MSS 7 (near infrared) was displayed as red. As with optical superimposition there is some slight loss of spatial information due to misregistration, although this is unlikely to be greater than ±0.5 pixel (Moik 1980).

### 6.3.7 Image classification

In the same way it is possible to classify a geographic region by its geology, flora or population (Grigg 1965), it is possible to classify a geographic region by its remotely sensed spectral response (Robinove 1981; Merchant 1982). The two most popular methods of doing this are a density slice of one waveband or a supervised classification of several wavebands (Hajic and Simonett 1976; Short 1982).

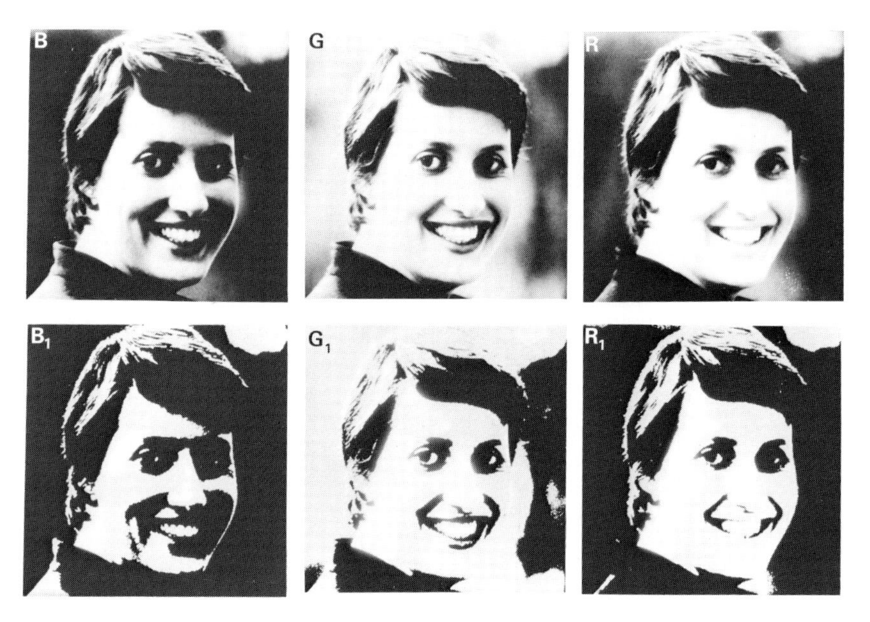


**Fig. 6.26**
Density slicing multispectral images. The three images (B) blue, (G) green and (R) red are derived from the digitised colour photograph in Plate 2 and have 256 density levels from 0–255 DN. The three images $B_1$, $G_1$ and $R_1$ are density sliced versions of B, G and R and have only 3 density levels of black (0–85 DN), grey (86–107 DN) and white (171–255 DN).

## 6.3.7.1 Density slicing

Density slicing involves the grouping of image regions with similar DN (Simon 1978), (Plate 3) either automatically or interactively. The effect of automatic thresholding is shown in Fig. 6.26 where the 255 grey levels of the original images have been divided into three equal classes of black (0–85 DN), grey (86–170 DN) and white (171–255 DN).

A preferable method is to choose thresholds on the basis of breaks or troughs in the scene histogram. For example in Fig. 6.28 the red image of the training area (Fig. 6.27) has three clear breaks in the histogram at 50, 100 and 195 DN enabling an unambiguous four class density slice into urban, pasture and soil, forest and water. The near infrared image of this training area had only one clear break at around 60 DN and two troughs at 150 and 200 DN enabling an ambiguous four class density slice which separates pasture and water but groups pasture, soil, forest and urban.

## 6.3.7.2 Supervised classification

Contrary to popular belief the computer classification of remotely sensed images is not the difficult part of the image classification procedure. A computer operator armed with an image processor and a CCT of remotely sensed date could in minutes, churn out a myriad of colourful image classifications (Merchant 1982). The results would have little environmental meaning because the computer operator will have restricted analysis to the spectral domain on the assumption that the product will be valid in the spatial domain. To obtain meaningful and accurate image classifications there is a need for the environmental scientist to take the computer operators seat and interact with the image data by supervising the classification sequence (Schmidt 1975). This will involve the careful choice of wavebands, the location

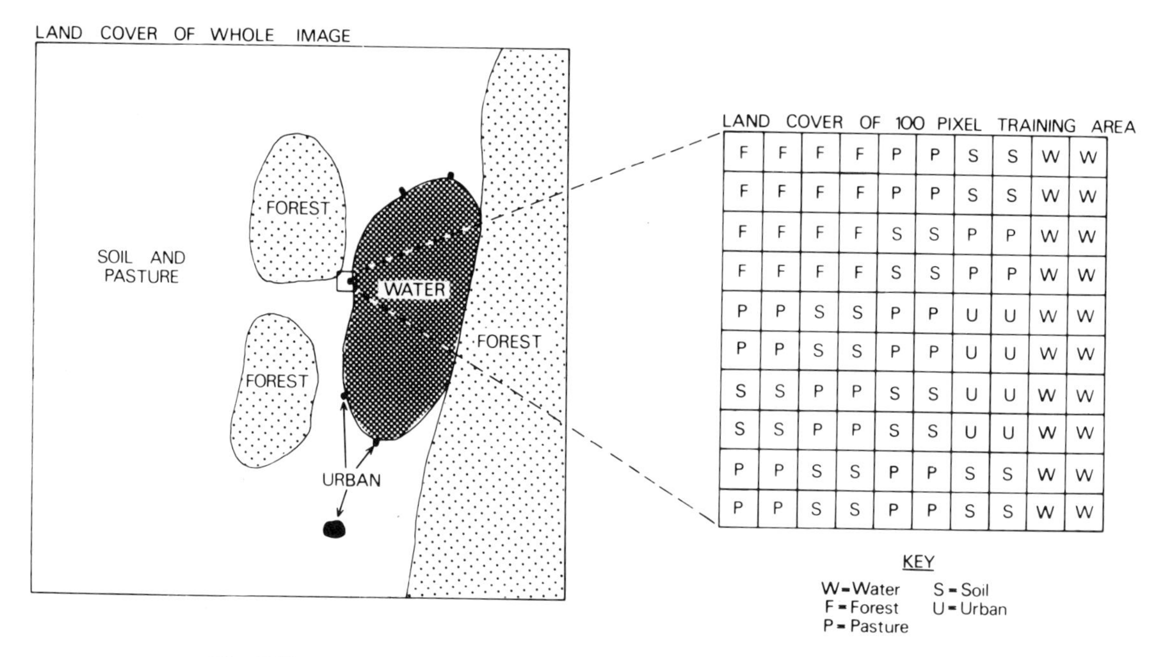


**Fig. 6.27**
A training area 10 × 10 pixels in size, extracted from a hypothetical Landsat MSS image.

of small but representative training areas (Fig. 6.27), the determination of the relationship between object type and DN in the chosen wavebands, the extrapolation of these relationships to the whole image data set and the display and accuracy assessment of the resultant images. To illustrate these stages the land cover of an area recorded on two wavebands of an imaginary Landsat MSS image will be classified with the aid of one training area (Fig. 6.27 and 6.28). This illustration although simple, could be used for any application with any multiband imagery, in any number of wavebands and with any number of training areas.

### 6.3.7.3 Waveband selection

There is a great temptation to load every waveband of data into the image processor and simply press the buttons for the result. This is very wasteful, as the use of highly correlated wavebands will increase computing time rather than classification accuracy (Moik 1980). It is preferable to determine the degree of interband correlation and to use only wavebands that are poorly correlated to each other.

### 6.3.7.4 Training

The aim of training is to obtain sets of spectral data that can be used to determine decision rules for the classification of each pixel in the whole image data set (Merchant 1982), (Fig. 6.29). The environmental scientist has considerable freedom in choosing these classes, providing two guidelines are adhered to (Haralick 1976). First, the training data for each class must be representative of all data for that class and second, the training data for each class must come close to fitting the

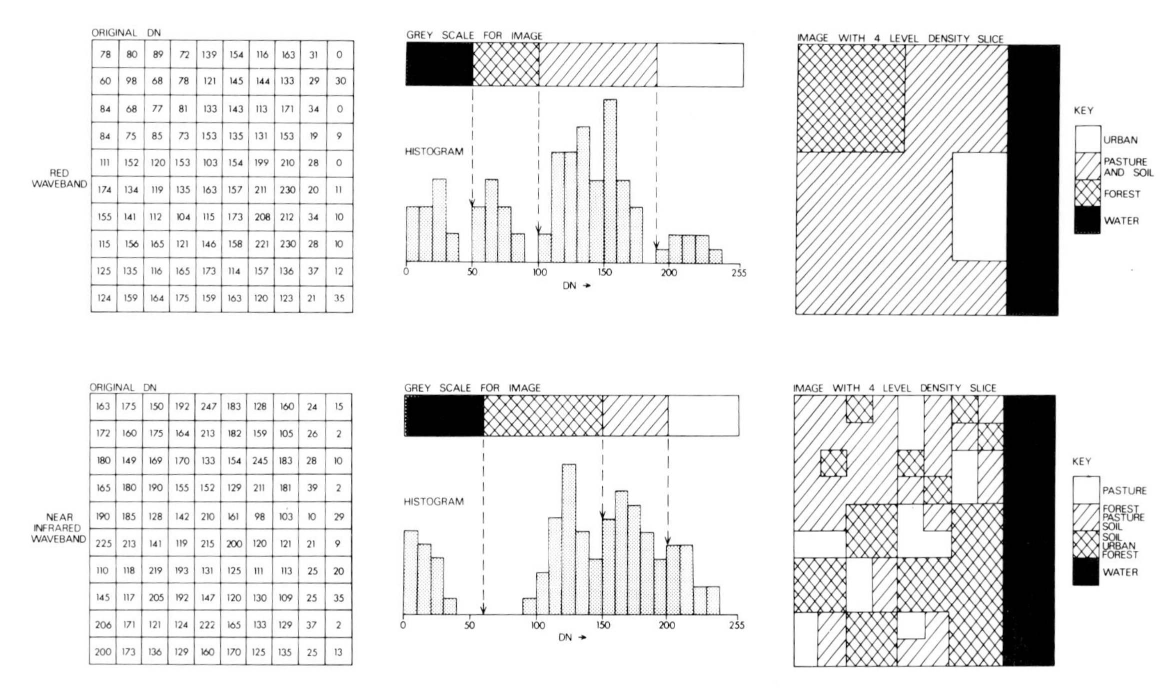


ORIGINAL DN — RED WAVEBAND

| 78 | 80 | 89 | 72 | 139 | 154 | 116 | 163 | 31 | 0 |
|---|---|---|---|---|---|---|---|---|---|
| 60 | 98 | 68 | 78 | 121 | 145 | 144 | 133 | 29 | 30 |
| 84 | 68 | 77 | 81 | 133 | 143 | 113 | 171 | 34 | 0 |
| 84 | 75 | 85 | 73 | 153 | 135 | 131 | 153 | 19 | 9 |
| 111 | 152 | 120 | 153 | 103 | 154 | 199 | 210 | 28 | 0 |
| 174 | 134 | 119 | 135 | 163 | 157 | 211 | 230 | 20 | 11 |
| 155 | 141 | 112 | 104 | 115 | 173 | 208 | 212 | 34 | 10 |
| 115 | 156 | 165 | 121 | 146 | 158 | 221 | 230 | 28 | 10 |
| 125 | 135 | 116 | 165 | 173 | 114 | 157 | 136 | 37 | 12 |
| 124 | 159 | 164 | 175 | 159 | 163 | 120 | 123 | 21 | 35 |

ORIGINAL DN — NEAR INFRARED WAVEBAND

| 163 | 175 | 150 | 192 | 247 | 183 | 128 | 160 | 24 | 15 |
|---|---|---|---|---|---|---|---|---|---|
| 172 | 160 | 175 | 164 | 213 | 182 | 159 | 105 | 26 | 2 |
| 180 | 149 | 169 | 170 | 133 | 154 | 245 | 183 | 28 | 10 |
| 165 | 180 | 190 | 155 | 152 | 129 | 211 | 181 | 39 | 2 |
| 190 | 185 | 128 | 142 | 210 | 161 | 98 | 103 | 10 | 29 |
| 225 | 213 | 141 | 119 | 215 | 200 | 120 | 121 | 21 | 9 |
| 110 | 118 | 219 | 193 | 131 | 125 | 111 | 113 | 25 | 20 |
| 145 | 117 | 205 | 192 | 147 | 120 | 130 | 109 | 25 | 35 |
| 206 | 171 | 121 | 124 | 222 | 165 | 133 | 129 | 37 | 2 |
| 200 | 173 | 136 | 129 | 160 | 170 | 125 | 135 | 25 | 13 |

**Fig. 6.28**
The spectral characteristics of the training area extracted from a hypothetical Landsat MSS image (Fig. 6.27). The spatial distribution of DN, a frequency histogram of DN and a four level density slice of DN are illustrated for red (MSS 5) and near infrared (MSS 6 or 7) wavebands.

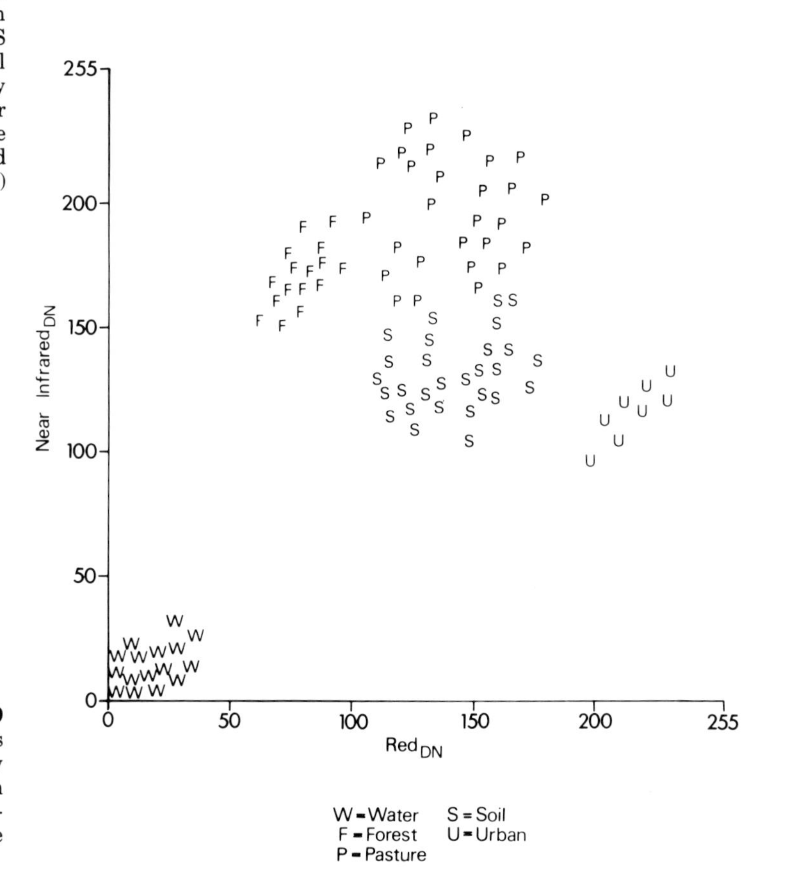


**Fig. 6.29**
A scatterplot of 100 pixels labelled by the land cover they represent. The pixels are from the training area of a hypothetical Landsat MSS image (Fig. 6.27).

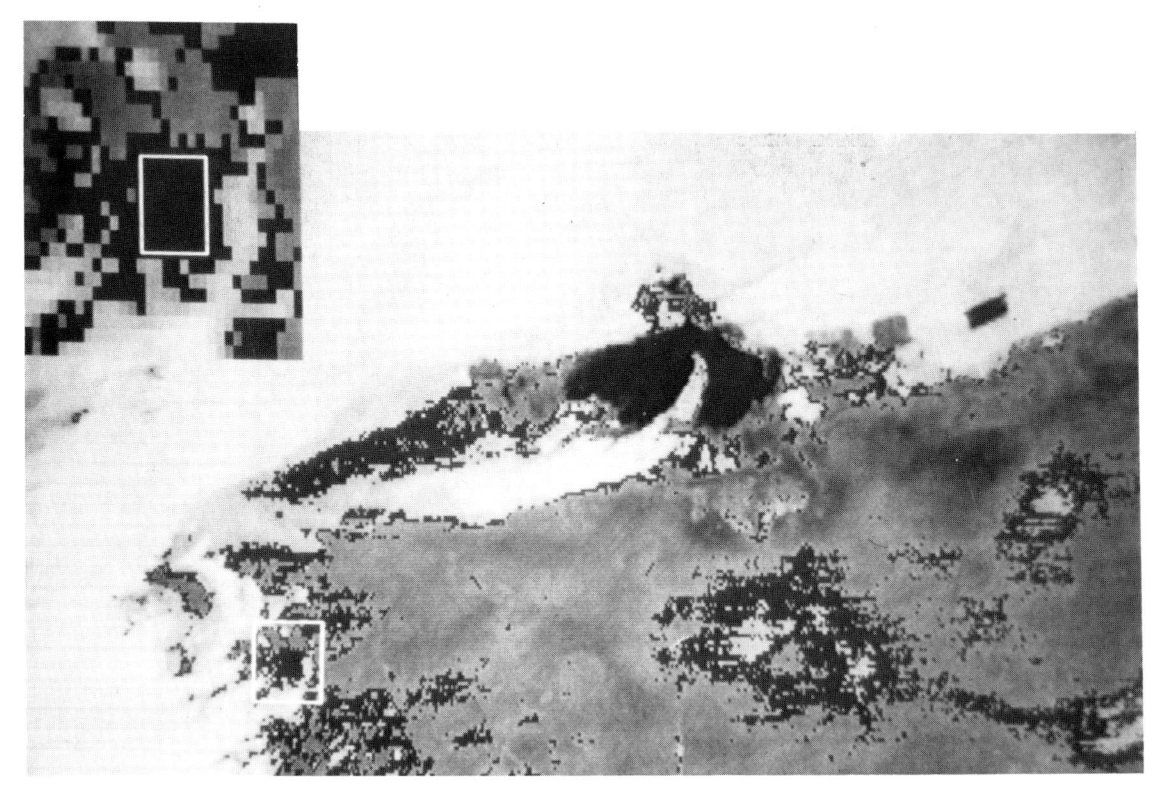

**Fig. 6.30**
Defining a training area on an enlarged (× 70) Landsat MSS image of southern Tunisia. An area for which land cover is known has been enlarged and placed in the top left of the VDU. The spectral response of a particular homogeneous land cover within the training area has been used in a parallelepiped classification to classify areas with similar spectral response over the whole image. (Courtesy, T. Munday, Durham)

distributional assumptions on which any decision rules are based. For example, to comply with the first guideline a 'soil' class (Fig. 6.29) must include all of the light and dark toned soil types in the scene. To comply with the second guideline a 'pasture' class (Fig. 6.29) which has a bimodal distribution (Fig. 6.31) should be divided into its two components. If this were not done the data would be unsuitable for classification using techniques such as maximum likelihood (sect. 6.3.7.8), which assumes data normality. It is necessary therefore, to observe the frequency distribution of each image class (Fig. 6.31); some image processors tabulate this information on the VDU while others output the data on a lineprinter.

As part of the training procedure it is advisable to instigate an unsupervised classification of the whole data set to determine if the number of classes chosen during training bear anything more than scant relation to the number of statistically separable spectral classes in the whole data set. This can be achieved by dividing the image data into its natural groupings, on the assumption that similar objects will have similar DN and more importantly dissimilar objects will have dissimilar DN. Such groupings can be achieved using several readily available statistical packages of which 'cluster analysis' has proved to be the most popular. This works by locating the desired number of cluster centres in the waveband to waveband measurement space (Fig. 6.29) and continues to move the cluster centres until the clusters have maximum statistical separability (Fu 1976; Hartl 1976). The results of this classification are dependent upon the number of classes that are initially chosen. For example, if only three classes had been chosen to

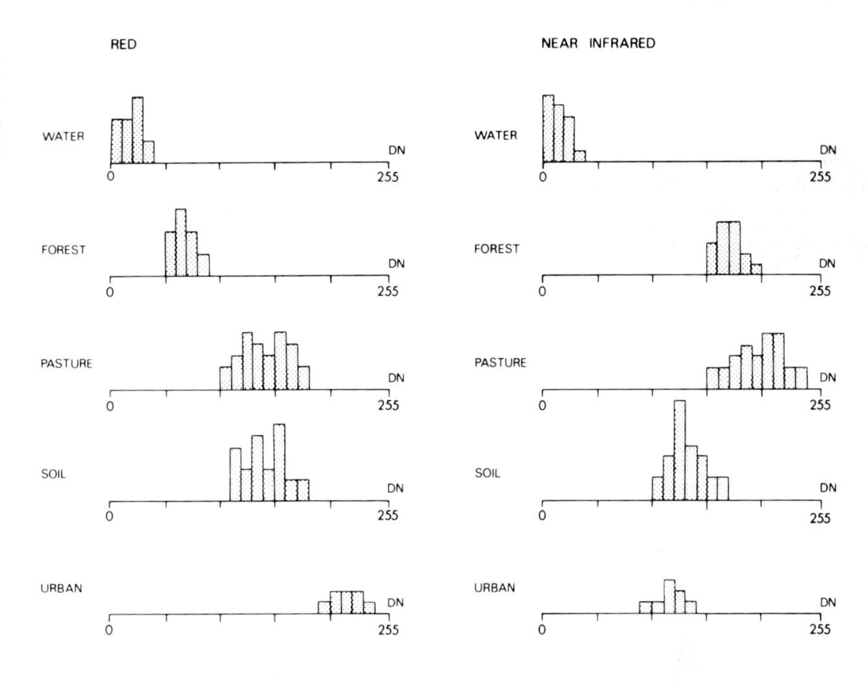


**Fig. 6.31**
Frequency histograms of DN for the training area of a hypothetical Landsat MSS image (Fig. 6.27) displayed by land cover and waveband.

classify the hypothetical Landsat MSS image in Fig. 6.27, the resulting classification (Fig. 6.32) would have been 1, water; 2, forest, pasture and soil and 3, urban, which is rather limiting for those with an agricultural interest but may be ideal for a planner who wishes to record the rural/urban boundary. If seven classes had been chosen then the resulting classification (Fig. 6.32) would have been 1, water; 2, forest; 3, shaded pasture; 4, unshaded pasture; 5, dark toned soil; 6, light toned soil; and 7, urban. This may be rather too detailed for land cover mapping where two classes within a field boundary are not desired but would be suitable for the production of a simple soil map.

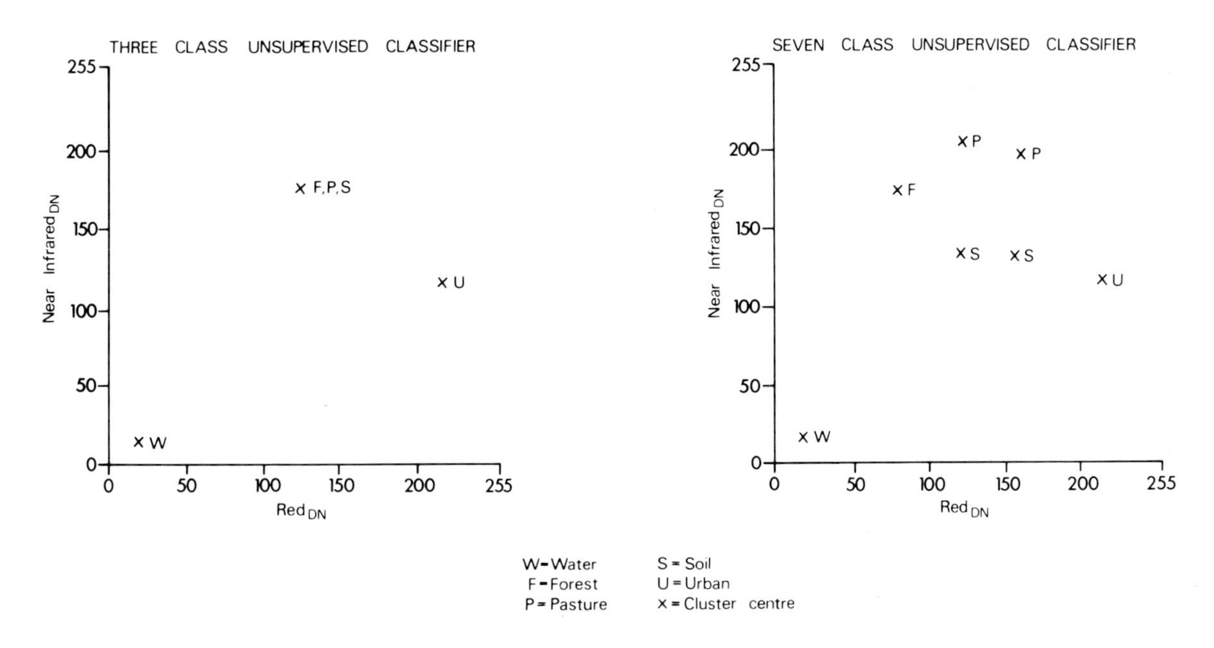


**Fig. 6.32**
The results of an unsupervised 3 and 7 class classification of a hypothetical Landsat MSS image (Fig. 6.27).

Although this is really an exploratory technique, it has been used to classify whole images, with interesting results (Armstrong 1977; Mayer and Fox 1981; Wharton and Turner 1981; Justice and Townshend 1982).

### 6.3.7.5 Extrapolation of the results of training over the whole data set

There are several classifiers that can be used to extrapolate the results of training over the whole data set (Lee *et al.* 1977; Hixson *et al.* 1980). The three most popular are, the minimum distance to means classifier, the parallelepiped classifier and the maximum likelihood classifier (Lillesand and Kiefer 1979).

### 6.3.7.6 Minimum distance to means classifier

This is the most simple and therefore cheapest classifier to compute as it comprises just three simple tasks. First, the mean DN of a class in the training data is calculated for all wavebands, this is termed the mean vector. Second, the pixels to be classified in the whole data set are allotted the class of their nearest mean vector; for example in Fig. 6.33 pixel Pi would be classified as 'soil'. Third, a data boundary is located around the mean vectors such that if a pixel falls outside of this boundary, as has pixel Pii in Fig. 6.33 then it will be classified as 'unknown'.

The limitation of this classifier is its insensitivity to variance in the spectral properties of each class. For example, as pixel Piii in Fig. 6.33 is nearest to the 'forest' mean vector it would be classified as 'forest' even though the class 'pasture' would seem to be more suitable (see Fig. 6.29).

### 6.3.7.7 Parallelepiped classifier

The parallelepiped or 'box' classifier is currently the most popular classifier for remote sensing applications as it is both fast and efficient (Rose and Rosendahl 1983), (Fig. 6.30). It operates in the same way as a simultaneous density slice (sect. 6.3.7.1) in all wavebands. For example, if the aim was to identify the class 'soil' (Fig. 6.29) in the whole data set, a density slice in the red waveband between 100–180 DN would produce an unsatisfactory composite class of 'pasture and soil' and a density slice in the near infrared waveband between 60–160 DN would produce an unsatisfactory composite class of 'urban, forest and soil'. However, a parallelepiped classification between 100–180 DN in the red waveband and between 60–160 DN in the near infrared waveband would produce a highly discriminatory 'soil' class (Fig. 6.33).

These boxes can be placed around each class in the training data and the pixels in the total data set can be classified by the box into which they fall. For example, in Fig. 6.33 pixel Piv falls into the 'urban' box and would be classified as 'urban' while pixel Pv falls between boxes and would be classified as 'unknown'.

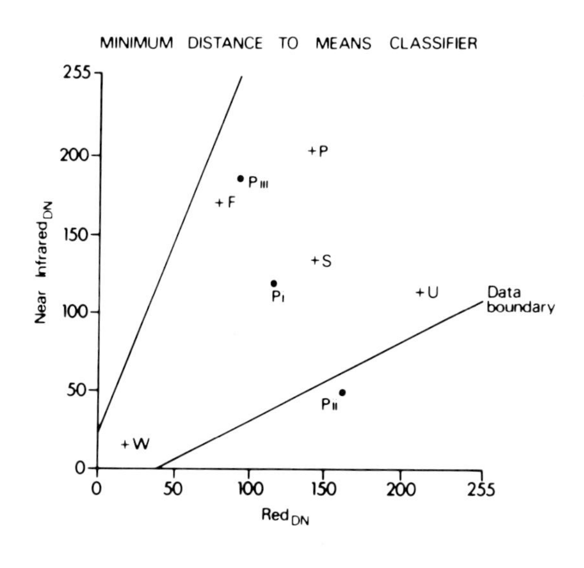


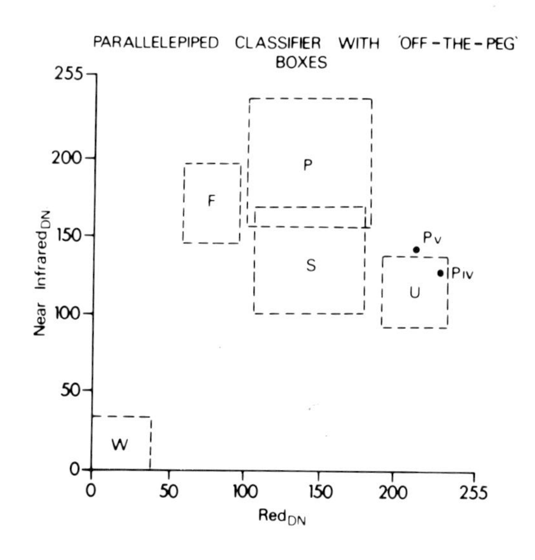


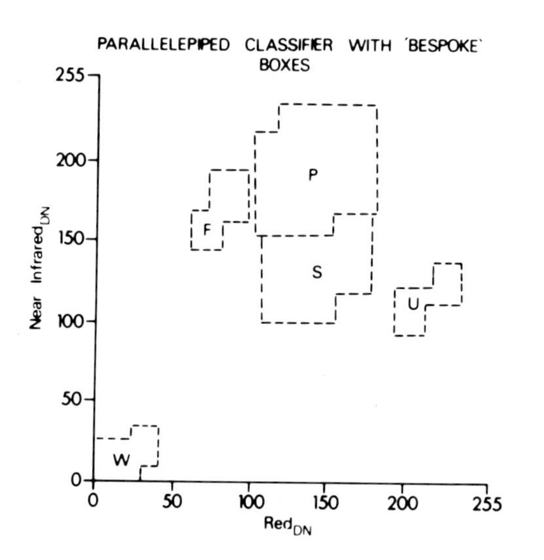


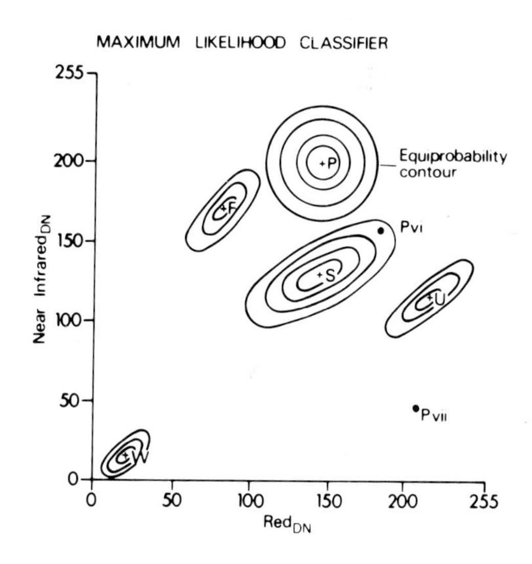


KEY

W = Water
F = Forest
P = Pasture
S = Soil
U = Urban
+ = Mean vector
P$_{I}$ = Pixel number

**Fig. 6.33**
Four classification strategies used in supervised classification to extrapolate training data over the whole data set.

Problems arise when the boxes overlap, as is the case for the classes 'soil' and 'pasture' in Fig. 6.33. These overlaps can occur when spectrally similar classes have been chosen or when there is a high degree of correlation between the spectral properties of objects in different wavebands. This second effect results in elongated bands of pixels that do not fit neatly into boxes (Fig. 6.29). To avoid such problems the discerning user spurns 'off the peg' boxes and creates 'bespoke' boxes, by tailoring the 'off-the-peg' boxes until they fit. A fine set of bespoke boxes are illustrated in Fig. 6.33; note the pleasing fit especially in the area of overlap between the classes of 'pasture' and 'soil'.

### 6.3.7.8 Maximum likelihood classifier

This is the most expensive and usually the most accurate classifier (Tomlins 1981). It works by first, calculating the mean vector (sect. 6.3.7.6), variance and correlation for each land cover class in the training data, on the usually valid assumption that the data for each class are normally distributed (Castleman 1979). With this information the spread of pixels around each mean vector can be described using a probability function, as illustrated in Fig. 6.33. Pixels from the whole data set are allocated to the class with which they have the highest probability of membership which in the case of pixel Pvi in Fig. 6.33 is the class 'soil'. As every spectral response has a probability, however low, of representing a class, no pixels are left out in the cold. For example, the lone pixel Pvii in Fig. 6.33 would be classified as 'urban'.

To improve the accuracy of this classifier for particular applications it is possible to interactively weight each class (Hajick and Simonett 1976; Strahler 1980). Such manipulation is justified in two situations. First, when the training set does not represent the proportion of the classes in the total scene. For example, as the class 'urban' was probably over-represented in the training set (Fig. 6.27) it would be permissible to reduce the probability of classifying 'urban' pixels in the whole data set by reducing the 'urban' weighting. Second, when all pixels in a given class must be classified as belonging to a given class, even though this would mean large errors of commission. For example, if all forest must be recorded the weighting of the 'forest' class could be increased.

To improve the speed and therefore reduce the cost of this classifier, it is possible to restrict the search area for each class using information gained during training (Eppler 1974). Once defined, these restricted search areas can be incorporated as 'look-up' tables for implementation on similar data sets (Brooner *et al.* 1971; Moik 1980).

### 6.3.7.9 Classification display

Once the operator is satisfied that spectrally sensible classes have been chosen by a classifier then the results can be extrapolated over the whole scene (Fig. 6.34). This extrapolation is usually undertaken interactively, class by class to give the operator a chance to reconsider any spatially anomalous classes (Fig. 6.30).

### 6.3.7.10 Classification accuracy

Once the objects in the scene have been classified there is need for a *post facto* exercise to determine the accuracy of the classification. The method chosen is dependent upon the availability of ground data. The most simple case would be where every image pixel can be labelled as belonging to an object class. As this is often unrealistic the image is usually sampled and a representative number of pixels are labelled (Benson *et al.* 1971). As with all spatial sampling the user must compromise on when to sample, where to sample, what to sample and the area and number of samples.

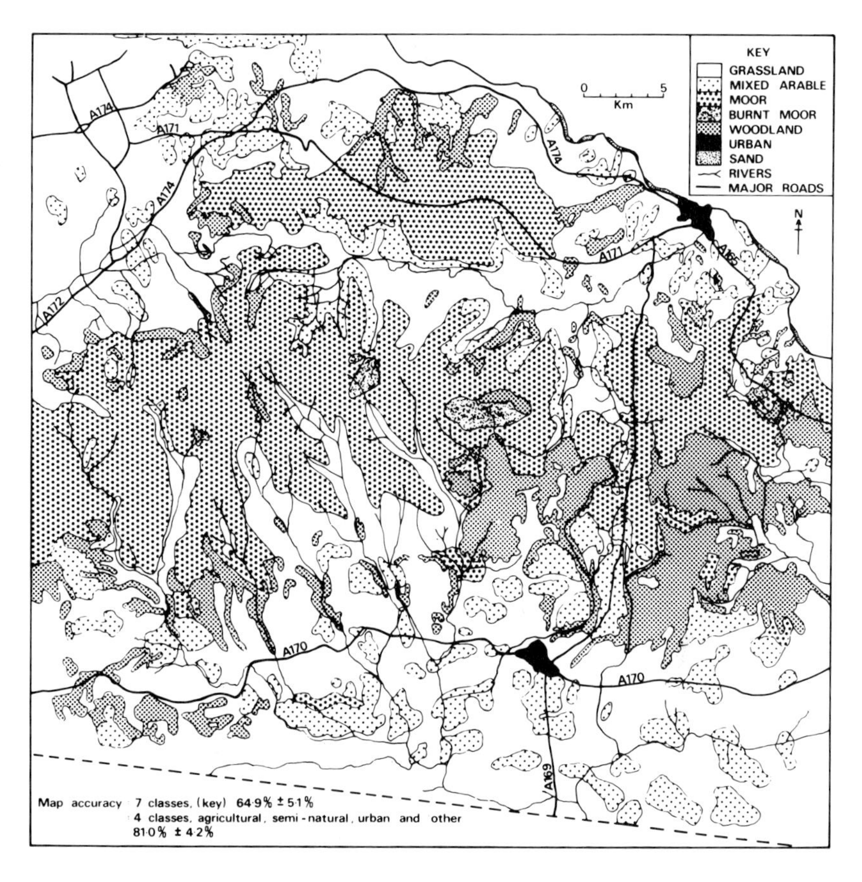


**Fig. 6.34**
Land cover map of the North York Moors, UK, constructed from a supervised parallelepiped classification of a Landsat MSS image. Note cover image. (Courtesy, S. Birch, G. Foody, S. Laffoley and F. Wells, Sheffield University)

## 6.3.7.11 When to sample

Ideally all sample points should be recorded at the time of sensing. This is very expensive as the small army of staff that are required to record a representative number of samples will probably spend most of their time waiting for a suitable overpass day. Such high cost has been justified for the recording of rapidly changing phenomena like soil moisture and marine sediment but can rarely be justified for the recording of less transient phenomena. When mapping land cover it is usually adequate to sample immediately after the confirmation of an overpass and in the case of geological mapping, sampling often takes place several years after the overpass was completed (Bonn 1976; Jackson *et al.* 1976; Justice and Townshend 1981).

## 6.3.7.12 Where to sample

The two sampling methods that are commonly used for the collection of ground data in support of remote sensing missions are purposive sampling and probability sampling (Rosenfield *et al.* 1982). Purposive sampling can vary from the detailed measurement of carefully located sites to 'observation while moving at a rapid pace on adjacent roads' (Myers 1975). The three advantages of such sampling are first, it utilises the field knowledge of the environmental scientist; second, the sites can be located easily on remotely sensed images by their relation to a road or transect line and third, the method is very fast. The one usually overriding disadvantage is that the data are biased and therefore have little statistical validity.

Probability sampling involves the objective selection of sample points over the entire study area (Justice and Townshend 1981). As it is necessary to generate an adequate sample size for the smallest classes a stratified sample is preferred, with an equal number of sample points in each class (Hixson *et al.* 1981; Rosenfield 1982; Rosenfield *et al.* 1982; Card 1982).

### 6.3.7.13 Sample number

As a general rule at least 50 sample points per class are required (Hay 1979). This sample number is only feasible where the object under study changes little over time, the field season is long and site access is easy. As this is rarely the case the accuracy of image classifications can be calculated using a much smaller sample size, providing the results are treated as only approximations.

### 6.3.7.14 Sample area

The area of a sample site should be related to the spatial resolution and geometric accuracy of the remotely sensed imagery (Justice and Townshend 1981). Formula [6.2] indicates the minimum area of a sample site (A) in relation to the ground diameter of a pixel (PD) in metres and the geometric accuracy of a pixel (PG) in pixel units.

$$A = [PD\,[1 + 2\,PG]]^2 \qquad [6.2]$$

For example, the area of the sample sites that are required to sample an airborne multispectral scanner image with pixels that are 2 m in diameter on the ground (Fig. 4.7) and are located to an accuracy of ± 0.5 pixel is

$$A = [2\,[1 + (2 \times 0.5)]]^2$$
$$A = 16\ m^2$$

### 6.3.7.15 What to sample

The features of the environment that are measured by an environmental scientist are discipline dependent. As it is beyond the scope of this book to discuss all of the measurement strategies that could be used, attention will be focused on the collection of data for land cover accuracy determination. These measurements include both land cover data that are input directly to the accuracy calculations, discussed in section 6.3.7.16 and observable and measurable field characteristics, that are used as an aid to the understanding of mis-classification (Smedes 1975; Lintz *et al.* 1976).

Land cover data include the land cover classes identified during training. In the example used in section 6.3.7.5 these would be water, forest, pasture, soil and urban. The observable field characteristics may include such features as slope angle, slope aspect, soil colour, soil wetness and any unusual characteristics. The measurable field characteristics may involve such features as the coverage of vegetation, the soil moisture content and the radiometric properties of each land cover (Milton 1980; Dozier and Strahler 1983). The fact that similar objects reflect or backscatter radiation in a similar way is a probabilistic statement and is not a law. To understand the departures from the probable by using these observable and measurable field characteristics is a useful route to the improvement of classification accuracy.

### 6.3.7.16 Calculating classification accuracy

The seemingly logical approach to the calculation of classification accuracy would be first, to identify say 50 points within a class on the classified image and second, to check the identity of these points in the field. This would be done on the assumption that if 45 of the 50 points had been classified correctly then the class accuracy would be 90%. Unfortunately this assumption is not valid as the result of 45 out of 50 could have been obtained by chance from both a near perfect or a very poor classification. To use the observation that 45 out of 50 points were correctly classified as a measure of classification accuracy

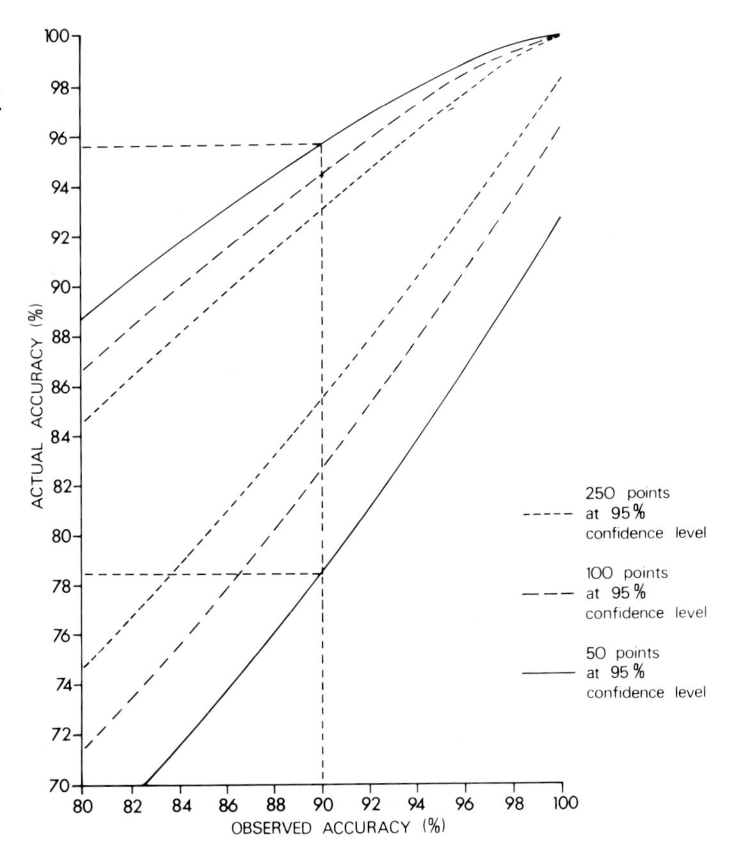


**Fig. 6.35** The relationship between observed and actual accuracy. (Data from Hord and Brooner 1976)

there is a need to delve into probability theory. (Hay 1979). This probability which can be calculated using a binomial expansion, or read directly from a table (Hord and Brooner 1976) or graph (Fig. 6.35), indicates that at the 95% confidence interval a result of 45 out of 50 could be obtained when the actual accuracy of the classification was as low as 78.6% or as high as 95.7%. Therefore the accuracy of this class is not necessarily 90% but is 78.6%–95.7%. An illustration of this approach is presented in Table 6.3 as an example of a five class classification of the hypothetical image shown in Fig. 6.27. Further examples are to be found in Genderen *et al.* (1978), Ginevan (1979); and Todd and Gehring (1980).

When attempting to estimate the accuracy of classified satellite sensor images using ground data derived from collateral sources like maps and aerial photographs, it is usual to have very large sample sizes (N), less than 80% correct observations (p%) and more than 20% incorrect observation (q%). Under these conditions the standard error (SE) can be used to describe the classification accuracy (Hay 1979), formula [6.3].

$$SE = \sqrt{\frac{p\%\ q\%}{N}} \qquad [6.3]$$

For example, if 70% of 1,000 field samples had been correctly classified then the standard error would be:

$$SE = \sqrt{\frac{70 \times 30}{1{,}000}} \quad = 1.5\%$$

The 70% correct observations can be taken to be a mean value with a standard error of ± 1.5% at the 68% significance level and 3% at the 95% significance level. The classification accuracy range at the 95% confidence interval is therefore 67%–73%. For further details refer to Campbell (1981) and Arnoff (1982).

## 6.3.7.17 Increasing classification accuracy

The three techniques for increasing classification accuracy involve decreasing the spatial resolution of the remotely sensed data, decreasing the number of classes and increasing the information available to the classifier.

Paradoxically classification accuracy is negatively related to the spatial resolution of the remotely sensed data. This is particularly evident for land covers with high spatial frequencies. For example, an urban area which can be represented by a few pixels with similar DN in an image with a low spatial resolution, would be represented by many pixels with dissimilar DN in an image with high spatial resolution (Sadowski and Sarno 1976; Townshend 1981b; Forshaw *et al.* 1983).

Classification accuracy is negatively related to the number of classes, as with only one class the accuracy will be 100 per cent and with an infinite number of classes the accuracy will be near to zero. The relationship between class number and accuracy can be seen in

**Table 6.3**
Calculating the classification accuracy of a five class classification. After the method of Hay (1979).

Raw data

| *Observed classes (in field)* | *Predicted classes (on image)* | | | | | |
|---|---|---|---|---|---|---|
| | *Water* | *Forest* | *Pasture* | *Soil* | *Urban* | *Total* |
| Water | 48 | 2 | 0 | 0 | 0 | 50 |
| Forest | 2 | 41 | 5 | 2 | 0 | 50 |
| Pasture | 0 | 5 | 40 | 5 | 0 | 50 |
| Soil | 0 | 0 | 2 | 42 | 6 | 50 |
| Urban | 0 | 0 | 0 | 6 | 44 | 50 |
| TOTAL | 50 | 48 | 47 | 55 | 50 | 250 |

Classification accuracy by binomial expansion

| | *Proportion correct* | *Class accuracy by binomial expansion* |
|---|---|---|
| Water | 48/50 | 86.5–98.9 |
| Forest | 41/50 | 69.2–90.2 |
| Pasture | 40/50 | 69.0–88.8 |
| Soil | 42/50 | 71.5–91.7 |
| Urban | 44/50 | 76.2–94.4 |
| TOTAL | 215/250 | 81.2–89.8 |

relation to the supervised (parallelepiped) classification of a Landsat MSS image of an upland area (Fig. 6.34). The rather ambitious seven class classification gave an accuracy of only 64.9 ± 5.1% which was increased to 81.0 ± 4.2% when only four classes were used. The addition of extra information during classification can increase accuracy (Hutchinson 1982; Richards *et al.* 1982). Two possible additions are first, the spatial measures of texture and context (Starr and Mackworth 1978; Landgrebe 1980; Thomas 1980). and second, data from other data bases. Texture can be introduced using the output from a textural filter, Harris (1980), (sect. 6.3.6.11) and context can be introduced by increasing the probability of the class chosen for the *previously* classified pixel, in a maximum likelihood classification (sect. 6.3.7.8), (Campbell 1981). This weighting can help to increase classification accuracy where there are either, abrupt environmental changes that leave pixels astride spectrally contrasting classes, or, where a class like clouds is spectrally variable or where there are localised and confusing changes in spectral response like weeds in a field (Swain *et al.* 1981; Gurney 1982; Gurney and Townshend 1983).

The introduction of data from other data bases is a little more difficult as it requires the registration of remotely sensed data with other data sources (sect. 6.4). A good example of this approach is the use of topography to separate spectrally similar forest types.

## 6.4 Geographic information systems

To determine the degree of spatial correlation between elements in the landscape, environmental scientists compare thematic maps, ground survey information and remotely sensed images of the same area. To aid interpretation, these spatial data sources can be superimposed. For example, land tenure boundaries can be overlaid on aerial photographs, or spatial data sources can be cartographically combined for example by printing geological or soil maps onto a topographic map base. The environmental scientist can also try to increase the spatial data set by glancing backwards and forwards between several spatial data sources in the hope that the brain will be the synthesiser. Unfortunately, this approach fails, for as Short (1982) quaintly puts it 'The eye becomes confused and the mind tends to "boggle" when more than two or three maps are visually compared.' In the 1970s the problem of spatial data combination was overcome by the rapid increase in computer memory and power that made it possible to unite any spatial data that can be referenced by geographic co-ordinates (Estes 1982). The three processing steps for this operation being (i) data encoding, (ii) data management and (iii) data manipulation (Tomlinson 1968, 1972; Knapp and Rider 1979; Richason 1982).

## 6.4.1 Data encoding

Before spatial data can be handled they need to be broken down into small polygonal or grid units. Polygons can be tailored to fit the Earth's surface as, unlike grids, they can follow the boundaries of terrain (Arnberg 1981), soils (Burrough 1980), fields (Short 1982) and political units (Teicholz 1980). As each polygon can have a unique shape, its storage, manipulation and retrieval is both complex and expensive (Calkins and Tomlinson 1977). However, this has not deterred its use (Mitchell *et al.* 1977).

The grid is the most popular unit as it is cheap to store, manipulate and handle and can be of any dimension from the 10 × 10 km unit used for the UK land characteristic data base (Ball *et al.* 1983) to the 79 × 56 m unit of a Landsat MSS pixel.

## 6.4.2 Data management

Once the spatial data have been encoded they must be filed away ready for manipulation. To understand how this filing is managed, it helps to use an analogy from the world of music where inputs to the final recording can be built up from a number of time formated tape tracks.

In recording, track one may be vocals, track two drums and track three guitars, in a geographic information system, element one may be topography, element two geology and element three soils (Fig. 6.36). As is indicated by Farmer (1979) 'Once the recording has been made all sorts of adjustments can be made to the sound on the separate tracks either to improve the overall sound or to alter the volume'. In geographic information systems this adjustment is termed data manipulation (Tilman and Mokma 1980).

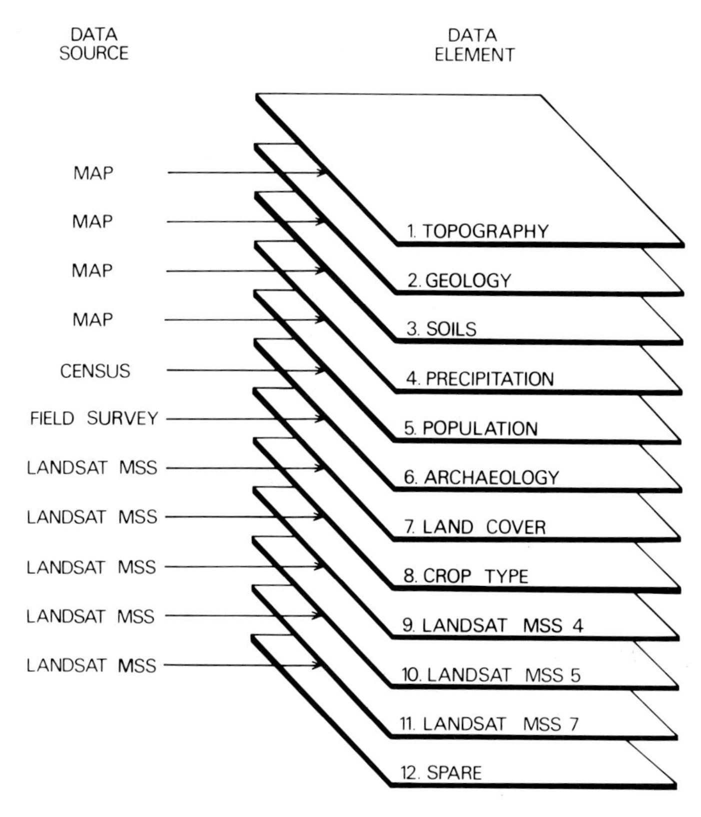


**Fig. 6.36** Integration of Landsat MSS data with other spatial data sources to derive data elements for input to a 12 level geographic information system.

## 6.4.3 Data manipulation

Today there are a number of commercially available software packages designed specifically for geographic information systems (Knapp and Rider 1979; Gregg *et al.* 1980; Whitley *et al.* 1981; Goldberg 1982). Many of these packages are designed around the eight operations of retrieval, transformation, storage, searching, analysis, measurement, compositing and modelling (Table 6.4).

The use of geographic information systems for modelling has great potential. For example, in the mid-1970s a simple grid-based geographic information system was developed for the modelling of water pollution as a result of soil erosion. This contained three elements, an index image of soil erodibility, a distance to water image, with each grid cell having a value that was negatively related to the distance to water and an index of agricultural activity derived from a Landsat MSS image. A composite image of all three elements indicated areas that were likely to contribute to water pollution (Short 1982).

## 6.4.4 Remote sensing and geographic information systems

A geographic information system is a powerful technique for turning large volumes of spatial data into useful information but it lacks ad-

**Table 6.4**
Eight operations performed by a geographic information system.

| *Objective* | *Example* | *Data elements (Fig. 6.36)* |
|---|---|---|
| 1. Retrieval of one or more data elements | Landsat MSS image | 9. Landsat MSS4 |
| 2. Transform, manipulate or combine the values of one or more data elements | Image of landscape attractiveness | 1. Topography<br>7. Land cover |
| 3. Store new data elements | Seasat SAR image | 12. Spare |
| 4. Search and identify particular combinations of data elements | Image of non-urban population distribution | 5. Population<br>7. Land cover |
| 5. Perform statistical analysis on one or more data elements | Image of correlation between crop type and a ratio of Landsat MSS 5 and MSS 7 | 8. Crop type<br>10. Landsat MSS 5<br>11. Landsat MSS 7 |
| 6. Measure area and distance of one or more data elements | Area measurement of limestone grassland | 2. Geology<br>7. Land cover |
| 7. Produce composite images of several data elements | Image of rainfall and topography | 1. Topography<br>4. Precipitation |
| 8. Be capable of modelling using several data elements | If it is suspected that the majority of the archaeological sites are located in unpopulated clay valleys display an image of unpopulated, clay valleys and compare with known archaeological sites | 1. Topography<br>3. Soils<br>5. Population<br>6. Archaeological |

equate spatial information and is static (Tomlinson 1972). By contrast remote sensing is a powerful technique for the collection of multitemporal data sets but there is a gap between data collection and utilisation. Many observers feel that the full potential of both techniques cannot be achieved until they are integrated (Shelton and Estes 1981) and some go even further and see the success of remote sensing firmly linked to its ability to service geographic information systems (Simonett *et al.* 1977).

The integration of remotely sensed data into geographic information systems is no easy task and it has taken a decade to develop operational methodologies (Cicone 1977; Shelton and Estes 1981; Arnberg 1981). Perhaps the best known method is the visual subdivision of both remotely sensed images and maps into polygons (Fig. 6.37a). These polygons can then be composited together to produce terrain units that are suitable for integration into a geographic information system (Townshend and Justice 1981). Currently the most successful method uses a grid square as a base and converts the remotely sensed data into a geometrically corrected thematic map prior to integration with other data elements (Fig. 6.37b). The thematic map is usually of an environmental factor for which there are few traditional data sources, for example land cover, estuarine sediment load, soil moisture, vegetation

**Fig. 6.37**
Methods of integrating remotely sensed data (RS DATA) into a geographic information system. (Modified from Townshend and Justice 1981)

(a)

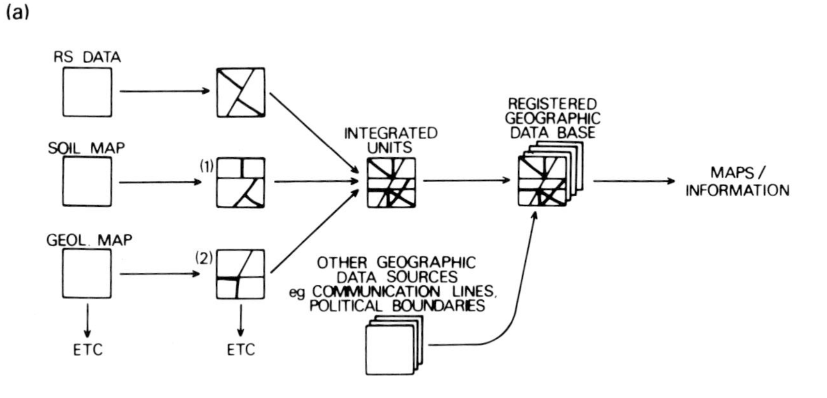


(b)

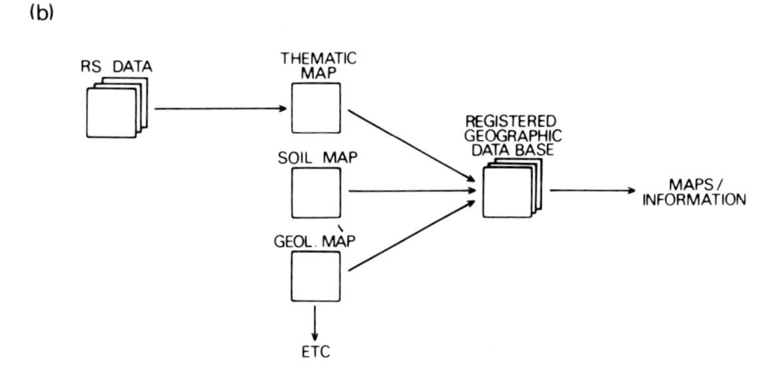


(c)

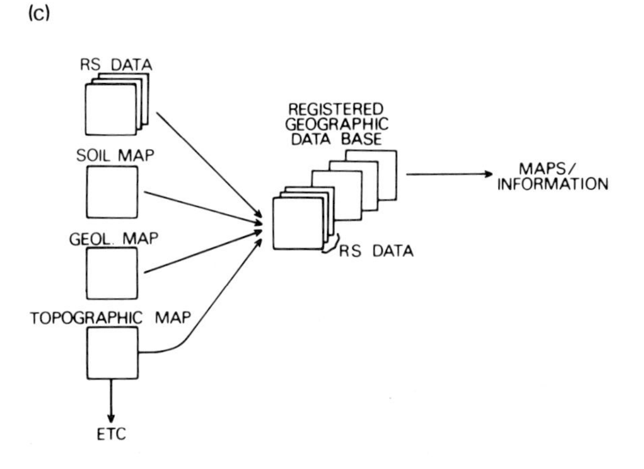


biomass, rainfall or the seasonal change of any of these. A similar method involves the geometric correction of the remotely sensed data and its direct integration with other data elements (Fig. 6.37c). This gives greater flexibility of analysis (Townshend and Justice 1981) and can improve the objectivity with which the remotely sensed data are classified (Shelton and Estes 1981).

The users of many geographical information systems throughout the world do not employ remotely sensed data. This is probably due to the variable accuracy of thematic maps derived from remotely sensed images, the awkward geographic referencing and tape formating systems used for satellite sensor images and the Herculean task of handling the large amount of data associated with multispectral sensors.

In recent years these problems have proved to be surmountable and now powerful geographic information systems in the USA (Richason 1982) and Europe (Rhind 1981) use remotely sensed data (Marble and Peuquet 1983). Some of these systems are static and national in coverage while others, for example those developed for hydrological monitoring (Smith and Blackwell 1980) and desertification modelling (Hellden *et al.* 1982), are non-static and milieu specific (Maher *et al.* 1980; Maggio *et al.* 1983). One thing is certain, they all have an exciting future.

## 6.5 Recommended Reading

**Billingsley, F. C.** (1983) Data processing and reprocessing: *In* Colwell, R. N. (ed.) *Manual of Remote Sensing* (2nd edn). American Society of Photogrammetry, Falls Church, Virginia, pp. 719–92.

**Castleman, K. R.** (1979) *Digital Image Processing.* Prentice Hall Inc, New Jersey.

**Haralick, R.M.** and **Fu, K.S.** (1983) Pattern recognition and classification: *In* Colwell, R. N. (ed.) *Manual of Remote Sensing* (2nd edn). American Society of Photogrammetry, Falls Church, Virginia, pp. 793–805.

**Moik, J.G.** (1980) *Digital Processing of Remotely Sensed Images.* National Aeronautics and Space Administration, Washington, D C.

**Richason, B. F.** (ed.) (1982) *Remote Sensing: An Input to Geographic Information Systems in the 1980s.* Pecora VII Symposium, United States Geological Survey, Sioux Falls, South Dakota.

**Rosenfeld, A.** and **Kak, A.C.** (1976) *Digital Picture Processing.* Academic Press, New York.

**Skaley, J. E.** (1980) Photo-optical techniques of image enhancement: *In* Siegal, B. S. and Gillespie, A. R. (eds) *Remote Sensing in Geology.* Wiley, New York, pp. 119–38.

**Townshend, J. R. G.** (1981) Image analysis and interpretation for land resources survey: *In* Townshend, J. R. G. (ed.) *Terrain Analysis and Remote Sensing.* Allen & Unwin, London; Boston, pp. 59–108.

# Appendices

*'I believe that through the ages man's greatest technologic breakthroughs have resulted from his curiosity about a phenomenon he could not experience directly with his senses.'* (Holz 1973)

## Appendix A Sources of data

### Sources of satellite sensor data

1. Prime international data distribution centre for manned satellite sensor data (other than Space Shuttle) and Landsat MSS, RBV and TM data.
   US Department of Commerce
   National Oceanic and Atmospheric Administration
   National Earth Satellite Service
   EROS Data Center
   Sioux Falls
   South Dakota 57198
   USA

2. Other international Landsat sensor data distribution centres
   (a) Instituto de Pesquisas Espacias (INPE)
   Departmento de Producao de Imagens
   ATUS-Banco de Imagens Terrestres
   Rodoua Presidente Dutra, KM 210
   Cachoeira Paulista – CEP 12.630
   São Paulo
   Brazil
   (b) Canadian Centre for Remote Sensing (CCRS)
   User Assistance and Marketing Unit
   717 Belfast Road
   Ottawa
   Ontario
   K1A OY7
   Canada
   (c) European Space Agency (ESA)
   Earthnet User Services
   Via Galileo Galilei
   00044 Frascati
   Italy
   (d) Remote Sensing Technology Center (RESTEC)
   7–15–7 Rippongi
   Minato-ku
   Tokyo 106
   Japan
   (e) Director
   National Remote Sensing Agency
   No. 4 Sardar Patel Road
   Hyderabad 500–003
   Andhra Pradesh
   India

(f) Australian Landsat Station
14–16 Oatley Court
P.O. Box 28
Belconnen, A.C.T. 2616
Australia

(g) Comision Nacional de Investigaciones Espaciales (CNIE)
Centro de Procesamiento Dorrego 4010
(1425) Buenos Aires
Argentina

(h) Director
National Institute for Telecommunications Research
ATTN: Satellite Remote Sensing Centre
P.O. Box 3718
Johannesburg 2000
Republic of South Africa

(i) Remote Sensing Division
National Research Council
Bangkok 9
Thailand

(j) Academia Sinica
Landsat Ground Station
Beijing
Peoples' Republic of China

3. Examples of two national satellite sensor data distribution centres

(a) National Remote Sensing Centre
Space Department
Royal Aircraft Establishment
Farnborough
Hants
GU14 6TD
UK

(b) Irish National Point of Contact
Snelbourne House
Snelbourne Road
Dublin 4
Eire

These particular centres distribute Landsat, Seasat, and some meteorological satellite sensor data.

4. International Space Shuttle sensor data distribution centre

(a) OSTA data:
National Space Science Data Center
Goddard Space Flight Center
Greenbelt
Maryland 20071
USA

(b) Spacelab data:
European Space Agency
8–10 Rue Mario Nikis
75738
Paris 15
France

5. International SPOT sensor data distribution centre
SPOT Image
18 Avenue Edouard-Belin
F 31055
Toulouse
Cedex
France

6. International HCMM sensor data distribution centre
World Data Center A for Rockets and Satellites
Code 601.4
NASA Goddard Space Flight Center
Greenbelt

Maryland 20071
USA

7. International TIROS/NOAA, nimbus, SMS and Seasat sensor data distribution centre
United States Department of Commerce
National Oceanic and Atmospheric Administration
Environmental Data and Information Service
National Climatic Center
Satellite Services Division
Washington DC. 20233
USA

8. International Meteosat sensor data distribution centre
MDMD Data Service
European Space Operation Centre
Robert-Bosch-Strasse 5
6100 Darmstadt
Germany

9. Example of a national meteorological satellite sensor data distribtuion centre
Department of Electrical Engineering and Electronics
University of Dundee
DU1 4HN
UK

**Sources of aerial photography** There are three sources of aerial photography: government registers, commercial companies and users of aerial photography. The first two will undertake a search to find the aerial photographic cover of your study area and then offer the aerial photographs to you for sale. Most of the aerial photographs recorded in these registers are black and white panchromatic.

Users of aerial photography will often let you use but rarely let you purchase copies of their photographs. Any service they provide would be as a favour to you so it is advisable to visit them in person to help find the aerial photographs you want.

To give the UK as an example:

1. The five main government registers of UK aerial photography are:
   (a) The Air Photographs Officer
   The Air Photographs Unit
   Royal Commission on the
   Historical Monuments of
   England,
   Fortress House
   23 Saville Road
   London
   SW1X 1AB
   (b) Central Register of Air Photographs for Wales
   Welsh Office
   Cathays Park
   Cardiff
   CF1 3NQ
   (c) The Air Photographs Officer
   Central Register of Air Photography
   Scottish Development Department
   New St Andrews House
   St James's Centre
   Edinburgh
   EH1 3SZ
   (d) Department of the Environment (N.I)
   Ordnance Survey of Northern Ireland
   83, Ladas Drive
   Belfast
   BT6 9FT
   (e) Ordnance Survey
   Air Photo Cover Group
   Romsey Road
   Maybush
   Southampton
   SO9 4DH

2. The five main commercial sources of UK aerial photography are:
   (a) Clyde Surveys
   Reform Road
   Maidenhead
   SR6 8BU
   (b) Meridian Air Maps Ltd
   Marlborough Road
   Lancing
   Sussex
   BN15 8TT
   (c) Huntings Surveys Ltd
   Elstree Way
   Boreham Wood
   Hertford
   WD6 1SB
   (d) BKS Surveys Ltd
   Ballycairn Road
   Coleraine
   Co. Londonderry
   BT51 3HZ
   (e) University of Cambridge
   Committee for Aerial Photography
   Mond Building
   Free School Lane
   Cambridge
   CB2 3RF
3. There are also many small aerial photographic companies who hold stocks of small format aerial photography. The companies can be located through a local telephone directory index under 'Aerial Photography and Surveys'.
   The three main users who hold aerial photography are
   (a) Academic institutions: especially the geography, civil engineering, geology and botany departments of Polytechnics and Universities.
   (b) Local government: city councils, county councils, national parks, water boards etc.
   (c) Government funded bodies: Ministry of Agriculture, Air Force Secretariat, Soil Survey, Geological Survey, Director of Historical Monuments, Nature Conservancy Council, National Environment Research Council Institutions etc.
4. In the USA the situation is the same, for example, three government sources of aerial photography in the USA are:
   (a) Aerial Photography Field Office
   ASCS-USDA
   P.O. Box 30010
   Salt Lake City
   Utah 84125
   (b) National Cartographic Information Center
   U.S. Geologic Survey
   507 National Center
   Reston
   Virginia 22092
   (c) United States Geological Survey Aerial Photography
   EROS Data Center
   Sioux Falls
   South Dakota 57198
   For details of commercial companies operating in the USA, peruse the advertisements in *Photogrammetric Engineering and Remote Sensing*.

# Appendix B Remote sensing journals and symposia

## Major remote sensing journals

*Canadian Journal of Remote Sensing*
*IEEE Transactions on Geoscience and Remote Sensing*

*International Journal of Remote Sensing*
*Photogrammetric Engineering and Remote Sensing*
*Remote Sensing of Environment*
*Remote Sensing Quarterly*

### Some journals which frequently carry articles on remote sensing

*Agronomy Journal*
*Applied Optics*
*Bulletin of the American Meteorological Society*
*Geographical Journal*
*Geo-processing*
*Journal of Agronomy*
*Journal of Applied Meteorology*
*Journal of the British Interplanetary Society*
*Journal of Environmental Management*
*Journal of Geophysical Research*
*Journal of the Optical Society of America*
*Pattern Recognition*
*Photogrammetria*
*Photogrammetric Record*
*Proceedings of the Society of Photo-Optical Instrumentation Engineers*
*Progress in Physical Geography*
*Science*
*Scientific American*
*Weather*

### Major remote sensing symposia

*ACSM/ASP Convention.* American Society of Photogrammetry, Falls Church, Virginia.
*EARSeL/ESA Symposium.* European Space Agency ESTEC Netherlands.
*Annual Conference of the Remote Sensing Society.* Remote Sensing Society.
*Proceedings of the International Symposium on Remote Sensing of Environment.* University of Michigan, Ann Arbor.
*Proceedings of the Symposium on Machine Processing of Remotely Sensed Data.* University of Purdue.
*Remote Sensing of Earth Resources.* University of Tennessee.
*Pecora Symposia.* American Society of Photogrammetry.

## Appendix C Abbreviations and acronyms

A selection of the many abbreviations and acronyms to be found in the remote sensing literature.
**ADP**: Automatic Data Processing
**AEI**: Aerial Exposure Index
**AEM**: Applications Explorer Mission satellite
**AFS**: Aerial Film Speed
**AgRISTARS**: Agriculture and Resources Inventory Surveys Through Aerospace Remote Sensing
**ASA**: American Standards Association film speed
**ATS**: Advanced or Applications Technology Satellite
**AVHRR**: Advanced Very High Resolution Radiometer
**BPI**: Bits Per Inch
**CA**: Canonical Analysis
**CCRS**: Canadian Centre for Remote Sensing
**CCT**: Computer Compatible Tape
**CIA**: Central Intelligence Agency
**CNES**: Centre National d'Études Spatiales
**CNIE**: Comision Nacional de Investigaciones Espaciales
**CRT**: Cathode Ray Tube
**CZCS**: Coastal Zone Colour Scanner
**DCP**: Data Collection Platform
**DCS**: Data Collection System or Data Collection Service
**DMSP**: Defence Meteorological Satellite Program
**DN**: Digital Number
**EARSeL**: European Association of Remote Sensing Laboratories

**EDC**: EROS Data Center
**EREP**: Earth Resources Experiment Package
**EDIS**: Environmental Data and Information Service of NOAA
**EROS**: Earth Resources Observing Service
**ERS**: Earth Resources Satellite
**ERTS**: Earth Resources Technology Satellite
**ESA**: European Space Agency
**ESMR**: Electronically Scanning Microwave Radiometer
**ESOC**: European Space Operations Centre
**ESRO**: European Space Research Organisation
**FILE**: Feature Identification and Location Experiment
**FOV**: Field of View
**GARP**: Global Atmospheric Research Programme
**GCP**: Ground Control Point
**GIS**: Geographic Information System
**GMS**: Geostationary Meteorological Satellite
**GOES**: Geostationary Operational Environmental Satellite
**GSFC**: Goddard Space Flight Center
**HCMM**: Heat Capacity Mapping Mission satellite
**HDDT**: High Density Digital Tape
**HRIR**: High Resolution Infrared Radiometer
**HRV**: High Resolution Visible scanner
**IA**: Indium Antimonide detector
**IFOV**: Instantaneous Field of View
**INPE**: Instituto de Pesquisas Espacias
**IR**: InfraRed
**ITOS**: Improved TIROS Observational Satellite
**JPL**: Jet Propulsion Laboratory
**LACIE**: Large Area Crop Inventory Experiment
**LAI**: Leaf Area Index
**LAPR**: Linear Array Pushbroom Radiometer
**LFC**: Large Format Camera
**MbC**: Multiband Camera
**MCT**: Mercury Cadmium Telluride detector
**MDG**: Mercury Doped Germanium detector
**MESSR**: Multispectral, Electronic, Self-Scanning Radiometer
**MMS**: Multimission Modular Spacecraft
**MOS**: Marine Observation Satellite
**MRS**: Multispectral Resource Sampler scanner
**MSS**: MultiSpectral Scanner or Multispectral Scanning System
**NASA**: National Aeronautics and Space Administration
**NERC**: Natural Environment Research Council
**NESS**: National Earth Satellite Service
**NIR**: Near InfraRed
**NOAA**: National Oceanic and Atmospheric Administration
**NPOC**: National Point Of Contact
**OCE**: Ocean Colour Experiment
**OCS**: Ocean Colour Scanner
**OSTA**: Office of Space and Terrestrial Applications
**PCA**: Principal Components Analysis
**RADAR**: RAdio Detection and Ranging
**RAE**: Royal Aircraft Establishment
**RBV**: Return Beam Vidicon camera
**RESTEC**: REmote Sensing TEchnology Centre
**SAR**: Synthetic Aperture Radar
**SERC**: Science and Engineering Research Council
**SI**: Systeme International d'Unites
**SIR**: Shuttle Imaging Radar
**SLAR**: Sideways-Looking Airborne Radar
**SMS**: Synchronous Meteorological Satellite
**S/N**: Signal to Noise ratio
**SOM**: Space Oblique Mercator projection
**SPOT**: Satellite Probatoire de l'Observation de la Terre
**TDRS**: Tracking and Data Relay Satellite
**TIR**: Thermal InfraRed

**TIROS**: Television InfraRed Observation Satellite
**TM**: Thematic Mapper
**UN**: United Nations
**USDA**: United States Department of Agriculture
**USGS**: United States Geological Survey
**UTM**: Universal Transverse Mercator projection
**UV**: UltraViolet
**VDU**: Visual Display Unit
**VHRR**: Very High Resolution Radiometer
**VI**: Vegetation Index
**WWW**: World Weather Watch

### Sources

Barrett and Curtis (1982) and Short (1982).

## Appendix D Glossary

A description of terms used in the remote sensing literature.

Absorptance: A measure of the ability of a substance to absorb incident energy.
Absorption: The process by which radiant energy is absorbed and converted into other forms of energy.
Absorption band: A range of wavelengths in the electromagnetic spectrum within which energy is absorbed by a substance.
Accuracy: Either (i) the success in estimating the true value; or (ii) the closeness of an estimate of a characteristic to the true value of the characteristic in the population.
Active microwave: Usually referred to as radar.
Active system: A remote sensing system that transmits its own electromagnetic radiation, e.g. radar.
Additive colour process: A method for creating colours through the addition of blue, green and red light.
Agfacontour: A specialised type of photographic film suitable for the density slicing of remotely sensed images (trade name).
Albedo: Either (i) the ratio of the amount of electromagnetic radiation reflected by a body, to the amount incident upon it; or (ii) the reflectivity of a body as compared to that of a perfectly diffusing surface at the same distance from the Sun, and normal to the incident radiation.
Algorithm: Either (i) a fixed step-by-step procedure to accomplish a given result; or (ii) a computer-orientated procedure for resolving a problem.
Anaglyph: A stereogram in which the two views are superimposed in complementary colours, usually of red and green.
Analogue: A form of data recording that works on the principle of continuous measurement, rather than discrete counting.
Analogue image processor: Equipment for the manipulation and display of continuous images.
Angle of depression: The angle between the horizontal plane passing through the sensor and the line connecting the sensor and the target.
Angle of incidence: Either (i) the angle between the direction of incoming electromagnetic radiation and the normal to the intercepting surface; or (ii) the angle between the vertical and a line connecting sensor and target.
Angle of reflection: The angle that reflected electromagnetic radiation makes with the surface normal.
Antenna: The device that radiates electromagnetic radiation from a transmitter and receives electromagnetic radiation from other sources.
Apollo: Manned, American satellite.
'Area survey' satellite: American reconnaissance satellite.
Atmospheric windows: Those wavelength ranges in which electromagnetic radiation can pass through the atmosphere with relatively little attenuation.
Attenuation: Any process in which the intensity of a beam of energy decreases with increasing distance from the energy source.
Attitude: The angular orientation of a sensor in relation to a geographical reference system.

Azimuth: The geographical orientation of a line given as an angle measured clockwise from north.

Backscatter: The scattering of radiant energy back towards its source.
Band: A selection of wavelengths.
Band-pass filter: A wave filter that has a transmission band extending between lower and upper cut-off frequencies.
Beam: A focused pulse of energy.
Bhaskara: Indian reconnaissance satellite.
Bilinear interpolation: A resampling technique used for the geometric correction of images.
Big Bird: American reconnaissance satellite.
Bit: A binary digit (0 or 1).
Blackbody: An ideal emitter which radiates energy at the maximum possible rate per unit area at each wavelength for a given temperature. A blackbody also absorbs all the radiant energy incident upon it.
Blackbody radiation: The electromagnetic radiation emitted by an ideal blackbody; it is the theoretical maximum amount of radiant energy at all wavelengths that can be emitted by a body at a given temperature.
Brightness: The attribute of visual perception in accordance with which an area appears to emit more or less light.
Brightness temperature: Either (i) the temperature of a blackbody radiating the same amount of energy per unit area as the observed body; or (ii) the apparent temperature of a non-blackbody determined by measurement with a radiometer.
Browse files: Files of remotely sensed images that are used to locate scenes of interest.
Byte: A group of eight bits of digital data.

Cadastral survey: A survey to define limits of title to land.
Camera: A lightproof chamber in which the image of an exterior object is projected upon a sensitised plate or film.
Canonical analysis: A statistical form of data compression.
Category: A definable characteristic that is of interest to the investigator. See Class.
Cathode ray tube: A vacuum tube capable of producing a black and white or colour image by beaming electrons onto a sensitised screen.
Change detection images: Images prepared by optically or digitally comparing two images acquired at different times.
Characteristic curve: A graph showing the relationship between the logarithm of exposure and the logarithm of photographic density.
Chinasat: Chinese reconnaissance satellite.
Cibachrome: Positive colour printing paper, onto which positive photographic transparencies can be printed (trade name).
Class: A definable characteristic that is of interest to the investigator. See Category.
Classification: The processing of assigning image areas (e.g. pixels) to categories, generally on the basis of reflectance or backscatter characteristics.
'Close look' satellite: American reconnaissance satellite.
Clustering: The analysis of a set of pixels to detect their inherent tendency to form clusters in multi-dimensional measurement space.
Coherent radiation: Electromagnetic radiation that is in phase.
Collateral data: Secondary data pertaining to an area of interest.
Colour: The visual property of an object which is dependent on the wavelength of the light it reflects.
Colour composite: A colour image produced by assigning a colour to each of a number of images of a scene and optically or digitally superimposing the result.
Complex dielectric constant: Electrical property of matter that influences radar returns.
Contrast: The difference in value or tone between highlight and shadow in an image.
Contrast stretching: Increasing the contrast of images by expanding the original range of values or tones to utilise the full contrast range of the recording film or display device.
Convolution: Digital image filtering
Cosmetic: A procedure purporting to improve image beauty.
Cosmos: Unmanned, Soviet satellite.
Crab: Any turning of an aircraft which causes its longitudinal axis to vary from the flight line.

Cubic convolution: A resampling technique used for the geometric correction of images.
Cursor: Aiming device, such as a lens with cross-hairs, on a digitiser or VDU.

Data acquisition system: The devices and media that measure and record physical variables.
Data bank: A well defined collection of data which can usually be accessed by a computer.
Data dimensionality: The number of variables (e.g. wavebands) present in a data set. Intrinsic dimensionality refers to the smallest number of variables that could be used to accurately represent a data set.
Data processing: Application of mechanical, electrical, and or computational procedures to change data from one form into another.
Decimated image: An image display in which only a sample of the total pixels are used.
Decision rule: The criterion used to establish discriminant functions during the training stage of a supervised image classification.
Densitometer: A device used to measure the density of a small area on an image.
Density: A measure of the degree of blackening of an exposed photographic film, plate, or paper after development.
Density slicing: The process of converting the DN or grey tone of an image into a series of intervals, or slices, each corresponding to a specific DN or tonal range.
Detector: A device that produces electrical current in direct proportion to the radiation incident upon it.
Diazo colour printer: Photographic device, that uses Diazo graphic arts film for the production of coloured contact transparencies (trade name).
Dichroic: A beam splitting mirror that efficiently reflects certain wavelengths while efficiently transmitting others.
Diffuse reflection: The type of reflection obtained from a relatively rough surface, in which the reflected rays are scattered in all directions.
Diffuse reflector: Any surface that reflects incident rays in many directions.
Diffuse sky radiation: Solar radiation reaching the Earth's surface after having been scattered from the direct solar beam by molecules or suspended material in the atmosphere. Also called skylight and diffuse skylight.
Digital data: Data displayed, recorded or stored in binary notation.
Digital image filtering: Spatial filtering to smooth, edge enhance or texturally enhance digital image data.
Digital image processor: Equipment for the manipulation of discrete images.
Digital number: Integer value of each point on a discrete remotely sensed image. Both singular and plural.
Digitisation: The process of converting material from a continuous into a discrete format.
Discriminant function: One of a set of mathematical functions which are commonly derived from training samples and a decision rule, and are used to divide the measurement space into decision regions as part of image classification.
Display: An output device that produces a visual representation of a data set.

Edge: The boundary of an object in an image.
Edge enhancement: The use of analytical techniques to emphasise spatial transition of image tone.
Electromagnetic radiation: Energy propagated through space or through material media in the form of an advancing interaction between electric and magnetic fields.
Electromagnetic spectrum: The ordered array of electromagnetic radiation extending from short cosmic waves to long radio waves.
Electronic dodger: Photographic device used to decrease image contrast.
Element: The smallest definable object of interest in a survey.
Emission: The process by which a body emits electromagnetic radiation as a consequence of its temperature.
Emissivity: The ratio of the radiation emitted by a surface to the radiation emitted by a blackbody at the same temperature.
Emittance: The obsolete term for exitance.
Emulsion: A suspension of photosensitive silver halide grains in gelatin that constitutes the image forming layer on photographic materials.
ERTS: Unmanned, polar orbiting, American Earth resources satellite.

Exitance: The radiant flux per unit area emitted by a body.
Explorer: Unmanned, polar orbiting, American, meteorological satellite.

False colour: The use of one colour to represent another.
False colour near infrared film: Photographic colour film sensitive to energy in visible and near infrared wavelengths from 0.4–0.9 μm.
Far range: The portion of a sideways-looking radar image that is furthest from the aircraft flight line.
Fiducial marks: Index marks, usually four, rigidly connected with the camera lens through the camera body and forming images on the negative which can be used to define the principal point of a photograph.
Field of view: The solid angle through which an instrument is sensitive to radiation.
Film: Light sensitive photographic emulsion and its base.
Film speed: A measure of the minimum level of exposure to which a film will respond.
Film writer: Device for writing image data onto photographic film.
Filter: Either (i) any material which, by absorption or reflection, selectively modifies the radiation transmitted through an optical system; or (ii) to remove a certain wavelength from a beam of electromagnetic radiation.
Filtering: The removal of certain spectral or spatial frequencies to enhance features in the remaining image.
Flight line: A line drawn on a map or chart to represent the track over which an aircraft or satellite has been flown or is to fly. See Ground track.
Focal length: The distance along the optical axis of a camera from the optical centre of a lens to the point at which a very distant object is brought into focus.
Focus: The point at which the rays from a point source of light reunite and cross after passing through a camera lens.
Fourier transform: An optical or digital means of transforming an image from the spatial to the frequency domain.
Frequency: The number of wavelengths that pass a point in unit time.

Gain: An increase in signal power during transmission from one point to another.
Gamma: A numerical measure of the extent to which a photographic negative has been developed.
Gemini: Manned, American satellite.
Geocoding: Geographical referencing of data items.
Geometrical transformations: Adjustments made in image data to change its geometrical character.
GOES east: Unmanned, geostationary, American meteorological satellite.
GOES west: Unmanned geostationary, American meteorological satallite.
Grain: One of the discrete silver particles or clumps of silver particles resulting from the development of an exposed light sensitive material.
Grey body: A radiating surface whose radiation has essentially the same spectral energy distribution as that of a blackbody at the same temperature, but whose emissive power is less.
Grey scale: A monochrome strip of shades ranging from white to black with intermediate shades of grey.
Ground control point: A geographical feature of known location that is recognisable on images and can be used during geometrical correction.
Ground data: Supporting data collected on the ground, and information derived therefrom, as an aid to the interpretation of remotely sensed imagery.
Ground range: The distance from the ground track of the sensing vehicle to a given object.
Ground receiving station: A facility that records and transmits data from satellite borne sensors.
Ground sampling element: The area on the terrain that is covered by the instantaneous field of view of a sensor.
Ground track: The vertical projection of the actual flight path of an aerial or space vehicle onto the surface of the Earth. See Flight line.
Ground truth: Jargon term given to the ground data used in a remote sensing investigation to imply that the data are without error. Ground data is to be preferred.

Hadamard transform: A digital means of transforming an image from the spatial to the frequency domain.
Hardware: The physical components of a computer and its peripheral equipment.

HCMM: Unmanned, polar orbiting, American Earth resources satellite.

High pass filter: A spatial filter that enhances high spatial frequencies and as a result sharpens the image.

Himawari: Unmanned, geostationary, Japanese meteorological satellite.

Histogram: The graphical display of a set of data which shows the frequency of occurrence (along the vertical axis) of individual values (along the horizontal axis).

Hot spot: Jargon term for an image region that records the specular reflection of solar radiation.

Hue: That attribute of a colour by virtue of which it differs from grey of the same brilliance, and which allows it to be classed as red, yellow, green, blue, or intermediate shades of these colours.

Illumination: The intensity of light striking a unit surface.

Image: The representation of one object by something else. Usually the recorded representation of a scene by optical, electro-optical, optical mechanical or electronic means.

Image enhancement: Any operations that improve the value of an image for a particular application.

Image pattern: The regularity and characteristic placement of tones or textures.

Image processing: The various operations that can be applied to image data.

Image restoration: A process by which a degraded image is restored to its original condition.

Incident ray: A ray impinging on a surface.

Infrared: That portion of the electromagnetic spectrum lying between the red end of the visible spectrum and microwave radiation. For remote sensing, the infrared wavelengths are often subdivided into near infrared (0.7–1.3 μm), middle infrared (1.3–3.0 μm) and thermal infrared (3.0–14.0 μm).

Insolation: Incident solar energy.

Instantaneous field of view: The field of view of a detector at one instant.

Interactive image processing. The method of data processing in which the operator views preliminary results and can alter the image processing techniques accordingly.

Irradiance: The measure, in power units, of electromagnetic radiation incident upon a surface.

Irradiation: The impinging of electromagnetic radiation upon a surface.

ITOS: Unmanned, polar orbiting, American meteorological satellite.

Joystick: Interactive control lever on an image processor.

Kelvin: The unit of temperature on the SI absolute or thermodynamic or kelvin scale of temperature.

Kernel: A pixel array used for digital image filtering.

KH 11: American reconnaissance satellite.

Kinetic temperature: The internal temperature of an object.

Kirchhoff's law: The radiation law which states that at a given temperature the ratio of the emissivity to the absorptivity for a given wavelength is the same for all bodies and is equal to the emissivity of an ideal blackbody at that temperature and wavelength.

Lambertian surface: An ideal, perfectly diffusing surface, which reflects energy equally in all directions.

Land cover: The surface covering of the land.

Land use: The use to which land is put. Often confused with land cover.

Landsat: Unmanned, polar orbiting, American Earth resources satellite.

Layover: Displacement of the top of an elevated feature with respect to its base on sideways-looking radar images.

Linescanner: A sensor that produces a continuous strip image by means of scanning.

Light: Visible radiation with wavelengths of between 0.4–0.7 $\mu$m.

Low pass filter: A spatial filter that enhances low spatial frequencies and as a result smooths the image.

Mapping camera: Primary camera for aerial photography.

Menu: A list of computer software options.

Mercury: Manned, American satellite.

Meteor: Unmanned, Soviet satellite.

Meteosat: Unmanned, geostationary, European meteorological satellite.

Microwave: That portion of the electromagnetic spectrum lying between the thermal end of the infrared spectrum and radio waves.
Mie scattering: Multiple reflection of light waves by atmospheric particles that have the approximate dimensions of the wavelength of light.
Mixel: Jargon term for a pixel that represents a mix of surface types.
Molynia: Unmanned, Soviet satellite.
MOS: Unmanned, polar orbiting, Japanese oceanographic satellite.
Mosaic: An assemblage of overlapping images whose edges have been matched to form a continuous pictorial representation of a portion of the Earth.
Multi-additive viewer: Optical device for the interactive production of colour composite images.
Multiband camera: A camera that exposes different areas of one film, or more than one film, through two or more lenses equipped with different filters.
Multispectral: Remote sensing in two or more spectral bands.
Multispectral scanner: A scanning sensor that produces images of the Earth's surface in many narrow wavebands from 0.3 μm to 14 μm.

Nadir: That point on the ground vertically beneath the perspective centre of a sensor.
Near range: The portion of a sideways-looking radar image that is closest to the aircraft flight path.
Nearest neighbour: A resampling technique used for the geometric correction of images.
Nimbus: Unmanned, polar orbiting, American meteorological satellite.
NOAA: Either (i) Unmanned, polar orbiting, American meteorological satellite, or (ii) National Oceanic and Atmospheric Administration.
Noise: Random or regular effects in data which degrade its quality.

Orbit: The path of a satellite around a body under the influence of gravity.
Overlap: The area common to two successive images along the same flight line.
Overlay: A transparent sheet giving information to supplement that shown on maps or images.

Pallet: Moveable portion of a platform that can be used to hold a sensor.
Panchromatic: Films that are sensitive to the entire visible region of the electromagnetic spectrum.
Parallax: The apparent displacement of position of a body with respect to a reference point or a system of co-ordinates, caused by moving the point of observation.
Passive microwave scanner: A passive scanning sensor that produces images of the Earth's surface in microwave wavelengths.
Passive system: A sensing system that detects radiation reflected or emitted by the Earth's surface.
Pattern: A regular repetition of values or tones in an image.
Photocentre: The centre of an aerial photograph. See Principal point.
Photogrammetry: The art or science of obtaining reliable measurements by means of photography.
Photograph: A picture formed by the action of light on a base material coated with a sensitised solution that is chemically treated to fix the image points at the desired density.
Photographic infrared: The portion of the electromagnetic spectrum to which photographic film can be sensitised, usually between 0.7 to 1.0 μm.
Photographic interpretation: The act of examining photographs for the purpose of identifying objects and judging their significance.
Pitch: Rotation of an aircraft about the horizontal axis, normal to its longitudinal axis.
Pixel: A picture element having both spatial and spectral aspects.
Pixtrix: Jargon term for a matrix of pixels.
Planck's Law: An expression for the variation of monochromatic exitance as a function of wavelength of blackbody radiation at a given temperature.
Platform: An object from which a sensor operates, e.g. an aircraft.
Polarisation: The direction of vibration of the electrical field vector of electromagnetic radiation.
Polar orbiting: A satellite that circles the Earth passing over or near to the north and south poles as it does so.
Preprocessing: Poorly defined term that often refers to image correction.
Principal point: The centre of an aerial photograph. See Photocentre.
Principal components analysis: A statistical form of data compression.

Pulse: A short burst of electromagnetic radiation.

Quick look Image: Jargon term for a rough first image that is used to assess spatial coverage.

Radar: Acronym for radio detection and ranging. A method, system or technique for using beamed reflected and timed electromagnetic radiation of microwave wavelengths to detect and measure objects, and to acquire an image.
Radarsat: Unmanned, polar orbiting, Canadian Earth resources satellite.
Radar shadow: A dark area on a radar image that extends in the far-range direction from an object on the terrain that intercepts the radar pulse.
Radiance: Total energy radiated by a unit area per solid angle of measurement.
Radiant energy: Energy carried by electromagnetic radiation.
Radiant flux: The time rate of the flow of radiant energy.
Radiometer: An instrument for measuring quantitatively, the intensity of electromagnetic radiation in a band of wavelengths in any part of the electromagnetic spectrum.
Range direction: The direction in which energy is transmitted from an antenna.
Rayleigh scattering: The wavelength-dependent scattering of electromagnetic radiation by particles in the atmosphere that are much smaller than the wavelengths scattered.
Real time: Jargon term for the simultaneous collection and display of remotely sensed images.
Reflectance: The ratio of the radiant energy reflected by a body to that incident upon it. Sometimes incorrectly used in place of radiance.
Reflection: Electromagnetic radiation that is neither absorbed nor transmitted.
Refraction: The bending of rays of electromagnetic radiation when they pass from one medium into another medium, that has a different index of refraction or dielectric coefficient.
Registration: The process of geometrically aligning two or more sets of image data such that resolution cells for a single ground area can be digitally or visually superposed.
Relief: The vertical irregularities of a surface.
Remote sensing: The use of electromagnectic radiation sensors to record images of the environment which can be interpreted to yield useful information.
Resampling: Geometric correction by the reformation of an image onto a new base, usually a map.
Resolution cell: The smallest area in a scene considered as a unit of data.
Return beam vidicon: A modified television camera tube, in which the output signal is derived from the depleted electron beam reflected from the tube target.
Roll: Rotation of an aircraft about its longitudinal axis.

Salut: Manned, Soviet space station.
Sample: A subset of a population selected to obtain information concerning the characteristics of the population.
Sampling rate: The temporal, spatial, or spectral rate at which measurements of physical quantities are taken.
Satellite: An object in orbit around a celestial body.
Scale: The ratio of a distance on an image or map to its corresponding distance on the ground.
Scanner: Any device that scans, to produce an image.
Scanning radiometer: A radiometer, which by the use of a rotating or oscillating plane mirror, can scan a path normal to its movement.
Scattering: Either (i) the process by which small particles suspended in a medium of a different index of refraction, diffuse a portion of the incident radiation in all directions; or (ii) the process by which a rough surface re-radiates electromagnetic radiation incident upon it.
Scatterometer: A device for recording the scattering properties of surfaces.
Scene: Everything occurring spatially or temporally before a sensor.
Sea data: Supporting data collected in the sea and information derived therefrom, as an aid to the interpretation of remotely sensed imagery.
Sea truth: Jargon term given to the sea data used in remote sensing investigations to imply that the data are without error. Sea data is to be preferred.
Seasat: Unmanned, polar orbiting, American oceanographic satellite.
Sensitivity: The degree to which a detector responds to electromagnetic energy incident upon it.
Sensor: Any device that receives electromagnetic radiation converts it into a signal

and presents it in a form suitable for obtaining environmental information.
Sidelap: The extent of lateral overlap between images acquired on adjacent flight lines.
Sideways-looking airborne radar: An all weather, day/night remote sensor that is particularly effective in recording large areas of terrain. It is an active sensor as it generates its own energy which is transmitted and received to produce an image of the Earth's surface.
Sixth line banding: Image fault common to Landsat MSS where every scan line in blocks of six have a different response.
Skin depth: Depth into the Earth's surface at which the amplitude of electromagnetic radiation is reduced to 37% of its value at the surface.
Skylab: Manned, American space station.
Smoothing: The averaging of values or tones in adjacent image areas to produce more gradual transitions.
Software: The computer programs that drive the hardware components of a data processing system.
Soyuz: Manned, Soviet satellite.
Space Shuttle: Manned, American space station.
Spatial filter: An image transformation, used to enhance certain spatial characteristics of an image.
Spatial information: Information conveyed by the spatial variations of spectral response present in the scene.
Spatial resolution: The ability of an entire remote sensor system, including lens, antennae, display, exposure, processing, and other factors, to render a sharply defined image.
Specific heat: The ratio between the thermal capacity of a substance and the thermal capacity of water.
Spectral reflectance: The reflectance of electromagnetic energy at specified wavelength intervals.
Spectral regions: Conveniently designated ranges of wavelengths sub-dividing the electromagnetic spectrum.
Spectral response: The response of a sensor to a particular wavelength of electromagnetic radiation.
Spectral signature: Jargon term referring to the spectral characteristic of an object in a scene, which infers that each object reflects radiation in a unique and identifiable manner. Spectral response is to be preferred.
Spectrometer: A device to measure the spectral distribution of electromagnetic radiation.
Specular reflection: The reflectance of electromagnetic energy without scattering or diffusion from a surface that is smooth in relation to the wavelengths of the incident energy.
SPOT: Unmanned, polar orbiting, French Earth resources satellite.
Stefan-Boltzmann's law: A radiation law which states that the amount of energy radiated per unit time from a unit surface area of an ideal blackbody is proportional to the fourth power of the absolute temperature of the blackbody.
Steradian: The unit solid angle that cuts unit area from the surface of a sphere of unit radius centred at the vertex of the solid angle.
Stereoscope: A binocular optical device for viewing overlapping images to obtain a mental impression of three dimensions.
Substractive colour process: A method of creating many colours through the substraction of proportions of light in the three subtractive colour primaries of cyan, magenta and yellow, from a single white light source.
Sun-synchronous satellite: An Earth satellite orbit in which the orbital plane is near polar with an altitude such that the satellite passes over all places on Earth having the same latitude, twice daily at the same local Sun time.
Supervised classification: A computer-implemented process through which each image pixel is assigned to a class according to a specified decision rule.
Swath: A strip of terrain or ocean, recorded by a sensor.
Synoptic view: The ability to see or otherwise measure widely dispersed areas at the same time and under the same conditions.

Table digitiser: Equipment used to digitise geographic coordinates.
Target: Either (i) an object on the terrain of specific interest in a remote sensing investigation; or (ii) the portion of the Earth's surface that produces by reflection or emission the radiation measured by the remote sensing system.
Telemetry: Transmission of data to a distant station.

Texture: The frequency of change and arrangement of tones in an image.
Thematic map: A map designed to illustrate a particular theme.
Thematic Mapper: Optical sensor carried by the satellites Landsat 4 and D′.
Thermal band: A general term for thermal infrared wavelengths which are transmitted through the atmosphere.
Thermal conductivity: The measure of the rate at which heat passes through a material.
Thermal capacity: The ability of a material to store heat.
Thermal inertia: A measure of the response of a material to temperature changes.
Thermal infrared linescanner: A scanning sensor that produces images of the Earth's surface in thermal infrared wavelengths.
Threshold: The boundary in spectral space beyond which an image pixel, has such a low probability of inclusion in a given class that the pixel is excluded from that class.
TIROS: Unmanned, polar orbiting, American meteorological satellite.
Tone: Each distinguishable shade of grey from white to black.
Trackball: An interactive control on an image processor.
Training: Informing an image processor which sites to analyse for spectral properties as a prerequisite to supervised classification.
Training samples: The data samples of known identity used to determine decision boundaries as part of a supervised classification.
Transmittance: The ratio of the radiant energy transmitted through a body to that incident upon it.
Transparency: Either (i) the light transmitting power of a material; or (ii) a positive image upon glass or film which is viewed using transmitted light.
True colour film: Term to differentiate ordinary colour film from false colour near infrared film. This implies that colours recorded on the film are truthful reproductions of those in the original scene. Colour film is to be preferred.

User friendly: Jargon term for equipment or a technique that is easy to use.

Vanguard: Unmanned, polar orbiting, American meteorological satellite.
Vignetting: A gradual reduction in the density of parts of a photographic image.
Visible wavelengths: The radiation range to which the human eye is sensitive, approximately 0.4–0.7 $\mu$m.
Voshkod: Manned, Soviet satellite.
Vostok: Manned, Soviet satellite.

Waveband: A selection of wavelengths.
Wavelength: The length of a wave of electromagnetic radiation.
Wien's displacement law: Describes the shift of the radiant power peak to shorter wavelengths with increasing temperature.
Window: A band of the electromagnetic spectrum which offers maximum transmission and minimal attenuation through a particular medium when using a specific sensor.

Yaw: Rotation of an aircraft about its vertical axis, causing the longitudinal axis to deviate from the flight line.

Zenith: The point in the celestial sphere that is exactly overhead: opposed to nadir.

### Sources

Curran (1983d); Haralick (1973); Sabins (1978);
Short (1982); Swain and Davis (1978); and Colwell (1983b).

## Appendix E Index of formulae

## Appendix F Remote sensing today and tomorrow

*'Remote sensing has the ability to present many images of one physical-organic-social world, thereby doing much to destroy lingering Newtonian concepts of a fixed (building block) world.'* (G. M. Lewis, 1984, personal communication)

# 1. Introduction

Interest within remote sensing is currently focused upon satellites and the effect of government policy upon remote sensing. Satellites are under scrutiny for four reasons. First, results derived from data collected by the Thematic Mapper and the first Shuttle Imaging Radar (SIR-A) (Ch. 5), are starting to enter the remote sensing literature. Second, Landsat 5 and a Space Shuttle carrying SIR-B have recently been launched. Third, there is rationalisation of the US remote sensing satellite program, for example, there is the demise of the MRS and the well-publicised plan for commercialisation (Ch. 5). Lastly, there is speculation over the future of unmanned Earth resources satellites, as to date there is no successor to Landsat 5 and data collected from the HRV on board the satellite SPOT are as yet, untested.

A realisation of the value of remote sensing by government is evident through all levels of administration. This can be seen in the use of geographic information systems by federal agencies in the USA (Ch. 6), the publication in 1984 of an enthusiastic report on the future of remote sensing and digital mapping in the UK by a House of Lords Select Committee (House of Lords 1984) and increased European collaboration over the first European remote sensing satellite, ERS-1 (Ch. 5).

To look forward beyond 1984 is a hazardous business, but as this of all years reminds us (Orwell 1949) forecasting, regardless of its accuracy, can provide valuable signposts to the way ahead. The two factors that are likely to influence the climate of opinion in remote sensing circles over the next few years are debate on the place of remote sensing within science and the scale at which remote sensing is used.

## 2. The place of remote sensing within science

To date there has been little published discussion on the place of remote sensing within science (Simonett 1983). This can be attributed to the subject's rapid development which has given researchers little time for contemplation.

The current literature on remote sensing gives the impression of a wide range of techniques with little methodological unity. This is because researchers enter the subject from a wide range of backgrounds and describe reality as they have been trained to see it, within the confines of the paradigm to which they subscribe (Eilon 1975). In remote sensing four methodologies happily coexist, these are:

1. **Inductive/qualitative methodology**, where generalisations are derived from individual observations on the basis of professional judgement. For example, the mapping of sites of archaeological interest from aerial photography (Ch. 3).
2. **Inductive/quantitative methodology**, where generalisation are derived from individual measurements using precise, repeatable decision rules, in which the magnitude of error is known. For example, the aerial photographic interpretation of land cover at random points which can then be extrapolated over a region to tested levels of accuracy (Ch. 3).
3. **Deductive/qualitative methodology**, where experiments are designed and data collected to test hypotheses that are based upon professional judgement. For example, the classification of NOAA, AVHRR data using a digital image processor to produce vegetation maps in the absence of ground data (Ch. 5).
4. **Deductive/quantitative methodology**, where experiments are designed and data collected to test hypotheses using precise, repeatable decision rules, in which the magnitude of error is known. For example, the use of calibrated digital airborne thermal infrared linescanner data to predict sea surface temperature to tested levels of accuracy (Ch. 4).

Regardless of methodology, science is usually judged on the basis of originality of ideas, ingenuity of experimentation and clarity of thought, while remote sensing has often been judged on the basis of its utility. The contents of the major remote sensing journals (Appendix B) reveal the extent of this pragmatic approach to science. Fortunately this attitude is changing, as in 1984 alone articles within these journals have posed six scientifically important questions. These are the place of models within remote sensing, the limits to the spatial extrapolation of remotely sensed data, the consistency of results, the errors inherent in remotely sensed data, the spatial frequency of our environment and the limitations of applying data collected at one scale to problems at another.

## 3. The scale of remote sensing studies

We are most comfortable working with information at a human scale. This small slice of reality represents that which is accessible to us through our unaided senses and occurs between less than 1 millimetre to a few kilometres in space, from less than a second to a few decades in

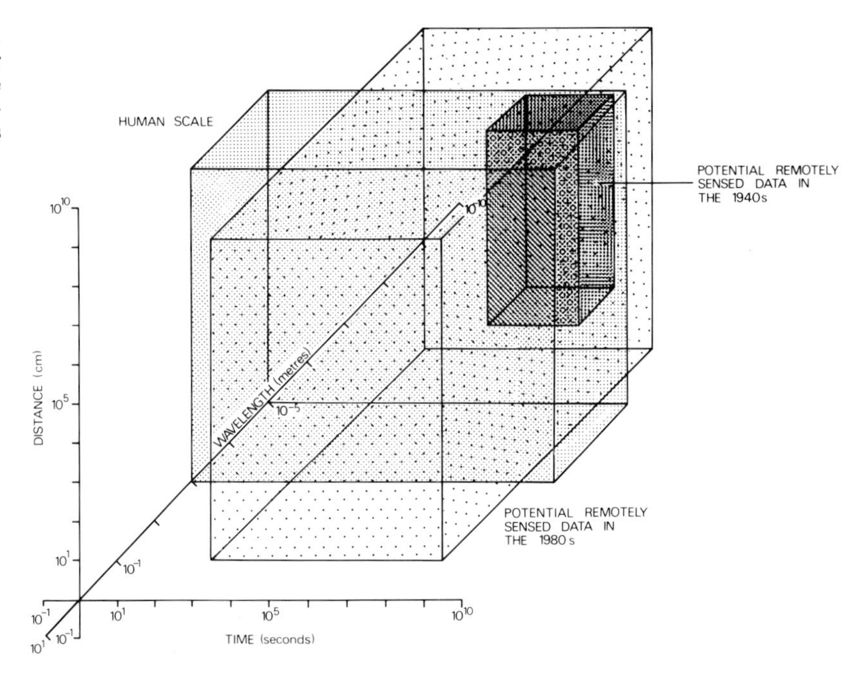


**Figure A**
The relationship between our perception of reality and the potential reality offered by remotely sensed data in the 1940s and 1980s.

time and between visible and thermal wavelengths of electromagnetic radiation.

In the 1940s remotely sensed data were all within the human scale but today they go way beyond it (Fig. A). Driven by experience at the human scale we naturally try to convert remotely sensed data to scales with which we are familiar (Klemes 1983). During the latter half of the 1980s there is likely to be an increased awareness of the value of data collected from beyond the human scale. Slowly but surely remotely sensed data will be applied to rapidly or very slowly changing global phenomena that manifest themselves at various points throughout the electromagnetic spectrum (Salomonson and Rango 1980) and a worldwide public will have the chance to glimpse into macrospace.

February 1984

# 4. Bibliography

**Eilon, S** (1975) Seven faces of research. *Operational Research Quarterly*, **26**: 359–67.

**House of Lords** (1984) *Remote Sensing and Digital Mapping*. House of Lords Select Committe on Science and Technology. 1st Report, HMSO, London.

**Klemes, V.** (1983) Conceptualization and scale in hydrology. *Journal of Hydrology*, **65**: 1–23.

**Orwell, G.** (1949) *Nineteen Eighty-Four*. Martin Secker and Warburg.

**Salomonson, V. V.** and **A. Rango** (1980) Water resources: *In* Siegal, B. S. and A. R. Gillespie (eds) *Remote Sensing in Geology*. John Wiley, New York pp. 607–33.

**Simonett, D. S.** (1983) The development and principles of remote sensing: *In* Colwell, R. N. (ed.) *Manual of Remote Sensing*: (2nd edn). American Society of Photogrammetry, Falls Church, Virginia, pp. 1–35.

# Bibliography


**Abboteen, R. A** (1978) Principal component greenness transformation in multitemporal agricultural Landsat data. *Proceedings of the 12th International Symposium on Remote Sensing of Environment.* University of Michigan, Ann Arbor: 765–74.

**Adams, R. E. W., Brown W. E.** and **Culbert, J. P.** (1981) Radar mapping, archaeology and ancient Maya land use. *Science*, **213**: 1457–63.

**AgRISTARS** (1981) *Agriculture and Resources Inventory Surveys Through Aerospace Remote Sensing.* Annual Report – Fiscal Year 1980. Lyndon B. Johnson Space Center, National Aeronautics and Space Administration. AP-JO-04111.

**AgRISTARS** (1982) *Agriculture and Resources Inventory Surveys Through Aerospace Remote Sensing.* Annual Report – Fiscal Year 1981. Lyndon B. Johnson Space Center, National Aeronautics and Space Administration. AP-J2-04225.

**AgRISTARS** (1983) *Agriculture and Resources Inventory Surveys Through Aerospace Remote Sensing.* Research Report – Fiscal Year 1982. Lyndon B. Johnson Space Center, National Aeronautics and Space Administration. AP-J2-0393.

**Ahmad S. B.** and **Lockwood, J. G.** (1979) Albedo. *Progress in Physical Geography*, **3**: 500–43.

**Aldrich, R. C.** (1979) *Remote Sensing of Wildland Resources: a State-of-the-Art Review.* Rocky Mountain Forest and Range Experiment Station, U.S. Department of Agriculture, Washington DC. General Technical Report RM-71.

**Allum, J. A.E.** (1980) Photogeology – the early days. *Geoscience Canada*, **7**: 155–8.

**Allum, J. A. E.** (1981) Remote sensing in mineral exploration – case histories. *Geoscience Canada*, **8**: 87–92.

**ASP** (1966) *Manual of Photogrammetry.* (3rd edn). American Society of Photogrammetry, Falls Church, Virginia.

**ASP** (1968) *Manual of Colour Aerial Photography.* American Society of Photogrammetry, Falls Church, Virginia.

**ASP** (1978) American Society of Photogrammetry, usage of the international system of units. *Photogrammetric Engineering and Remote Sensing*, **44**: 923–38.

**ASP** (1981) *Manual of Photogrammetry* (4th edn). American Society of Photogrammetry, Falls Church, Virginia.

**Anderson, R., Alsid L.** and **Carter, V.** (1975) Comparative utility of Landsat 1 and Skylab data for coastal wetland mapping and ecological studies. *Proceedings of the National Aeronautics and Space Administration, Earth Resources Symposium.* Houston, Texas. NASA Report TMX 58168, pp. 469–74.

**Anderson, J. R., Hardy, E. E., Roach J. T.** and **Witmer, R. E.** (1976) A land use and Land Cover Classification System for Use with Remote Sensor Data. *U.S. Geological Survey Professional Paper* **964**. US Government Printing Office, Washington DC.

**Andrews, H. C.** (1970) *Computer Techniques in Image Processing.* Academic Press, New York, London.

**Andrews, H. C.** (1974) Digital image restoration: a survey. *IEEE Transactions on Computers*, **7**: 36–45.

**Anon** (1983) ESA satellites will provide remote sensing of oceans. *Industrial Research and Development*, **25**: 78–80.

**Antrop, M.** (1979) The interpretation of soil marks on aerial photographs recorded over the Famenne (Belgium) natural region. *Pedologie*, **29**: 209–40.

**Aranuvachapun, S.** and **LeBlond, P. H.** (1981) Turbidity of coastal water determined from Landsat. *Remote Sensing of Environment*, **84**: 113–32.

**Armstrong, A. C.** (1977) The relative performance of some unsupervised clustering techniques for the per-field classification of Landsat data. *Journal of the British Interplanetary Society*, **30**: 168–71.

**Armstrong, A. C.** and **Brimblecombe, P.** (1979) A conspectus of computer aided and air-photo interpretation techniques for the study of Landsat imagery. *Journal of the British Interplanetary Society*, **32**: 3–8.

**Arnberg, W.** (1981) Integration of map and remote sensing data. *Geografiska Annaler, Series A*, **63**: 319–24.

**Arnoff, S.** (1982) Classification accuracy: a user approach. *Photogrammetric Engineering and Remote Sensing*, **48**: 1299–307.

**Ashley, M., Corcoran, D.** and **Morin, L.** (1978) Vegetation condition estimtes using colour infrared aerial photography: *In* Lund H. G. (ed.) *Integrated Inventories of Renewable Resources: Proceedings of the Workshop.* USDA Forest Service General Technical Report RM-55, pp. 242–7.

**Avery, G.** (1965) Measuring land use changes on USDA photographs. *Photogrammetric Engineering*, **31**: 620–4.

**Axelsson, S. J.** (1980) On the accuracy of thermal inertia mapping by infrared imagery. *Proceedings of the 14th International Symposium on Remote Sensing of Environment.* University of Michigan, Ann Arbor, pp. 359–78.

**Bagchi, A. K.** (1979) Determination of cloud height using Landsat imagery. *Photogrammetric Record*, **9**: 645–58.

**Baker, D.** (1981) *The Shape of Wars to Come – the Hidden Facts Behind the Arms Race in Space.* Patrick Stephens, Cambridge.

**Bakker, P., Church, D. J., Feuchtwanger, T. F., Grootenboer, J., Lee, C. A., Longshaw, T. G.** and **Viljoen, R. P.** (1978) Some practical applications of thermal infrared linescanning. *Mining Magazine*: 398–413.

**Ball, D. F., Hornung, M.** and **Mew, G.** (1971) The use of aerial photography in the study of geomorphology and soils of upland areas: *In* Goodier, R. (ed.) *The Application of Aerial Photography to the Work of the Nature Conservancy.* Natural Environment Research Council, Swindon, pp. 66–77.

**Ball, D. F., Radford, G. L.** and **Williams, W. M.** (1983) *A Land Characteristic Data Bank for Great Britain.* Bangor Occasional Paper **13**, Institute of Terrestrial Ecology, Bangor.

**Balston, D. M.** and **Custance, N. D. E.** (1979) IDP 3000: an image analysis system for Earth resource surveys: *In* Gardner, W. E. (ed.) *Machine Aided Image Analysis.* Institute of Physics, Bristol, London, pp. 287–302.

**Bardinet, C., Monget, J. M.** and **Patoureaux, Y.** (1982) Combined use of daily thermal cycle of Meteosat imagery and multispectral Landsat data: an application to the Bandiagara Plateau, Mali: *In* Longdon, N. and Lévy, G. (eds) *Satellite Remote Sensing for Developing Countries.* Proceedings of an EARSeL-ESA Symposium. Report Number ESA-SP-175. European Space Agency, Paris, pp. 87–101.

**Barnett, M. E.** and **Harnett, P. R.** (1977) Optical processing as an aid in analysing remote sensing imagery: *In* Barrett, E. C. and Curtis, L. F. (eds) *Environmental Remote Sensing 2: Practices and Problems.* Edward Arnold, London, pp. 125–42.

**Barnett, T. L.** and **Thompson, D. R.** (1982) The use of large area spectral data in wheat yield estimation. *Remote Sensing of Environment*, **12**: 509–18.

**Barr, D. J.** and **Miles, R. D.** (1970) SLAR imagery and site selection. *Photogrammetric Engineering and Remote Sensing*, **36**: 1155–70.

**Barrett, E. C.** (1970) Rethinking climatology. *Progress in Geography*, **2**: 155–205.

**Barrett, E. C.** (1974) *Climatology from Satellites.* Methuen and Co. Ltd, Harper & Row Publishers Inc.

**Barrett, E. C.**and **Grant, C. K.** (1978) An appraisal of Landsat 2 cloud imagery and its implications for the design of future meteorological observing systems. *Journal of the British Interplanetary Society*, **31**: 3–10.

**Barrett, E. C.** and **Martin, D. W.** (1981) *The Use of Satellite Data in Rainfall Monitoring.* Academic Press, London.

**Barrett, E. C.** and **Curtis, L. F.** (1982) *Introduction to Environmental Remote Sensing* (2nd edn). Chapman and Hall, London, New York.

**Barrett, E. C.** and **Hamilton, M. G.** (1982) The use of geostationary satellite data in environmental science. *Progress in Physical Geography*, **6**: 159–214.

**Bartholic, J. F., Namken, L. N.**and **Wiegand, C. L.** (1972) Aerial thermal scanner to determine temperatures of soils and crop canopies differing in water stress. *Agronomy Journal*, **64**: 603–8.

**Batson, R. M, Edwards, K.** and **Eliason, E. M.** (1976) Stereo and Landsat pictures. *Photogrammetric Engineering and Remote Sensing*, **42**: 1279–84.

**Battrik, B.** (ed.) (1981) *Introduction to the Meteosat System.* European Space Agency. Toulouse, No. ESA SP-1041.

**Bauer, A., Fontanel, A.** and **Grau, G.** (1967) The application of optical filtering in coherent light to the study of aerial photographs of Greenland glaciers. *Journal of Glaciology*, **6**: 781–93.

**Bauer, M. E., Cipra, J. E., Anuta, P. E.** and **Etheridge, J. B.** (1979) Identification and area estimation of agricultural crops by computer classification of Landsat MSS data. *Remote Sensing of Enviroment*, **8**: 77–92.

**Bauer, M. E., Biehl, L. L.** and **Robinson, B. F.** (eds) (1980) *Field Research on the Spectral Properties of Crops and Soils*. Purdue University, Laboratory for the Applications of Remote Sensing, West Lafayette, Indiana.

**Baylis, P. E.** (1981) University of Dundee satellite image data acquisition and archiving facility: *In* Allan, J. A. (ed.) *Matching Remote Sensing Technologies and their Applications*. Remote Sensing Society, Reading: 517–26.

**Beal, R. C., Deleonibus, P. S.**and **Katz, I.** (eds) (1981) *Spaceborne Synthetic Aperture Radar for Oceanography*. Johns Hopkins Press, Baltimore, Maryland.

**Becking, R. W.** (1959) Forestry applications of aerial colour photography. *Photogrammetric Engineering*, **25**: 559–65.

**Begni, G.** (1982) Selection of the optimum spectral bands for the SPOT satellite. *Photogrammetric Engineering and Remote Sensing*, **48**: 1613–20.

**Bell, T. S.** (1974) Remote sensing for the identification of crops and crop diseases: *In* Barrett, E. C. and Curtis, L. F. (eds) *Environmental Remote Sensing: Applications and Achievements*. Edward Arnold, London pp. 155–66.

**Bell, G. J.** (1981) Fifteen years of satellite meteorology at Hong Kong. *Weather*, **36**: 9–15.

**Benny, A. H.** (1983)Automatic relocation of ground control points in Landsat imagery. *International Journal of Remote Sensing*, **4**: 335–42.

**Benson, A. S., Draeger, W. C. Pettinger, L. R.** (1971) Ground data collection and use. *Photogrammetric Engineering*, **37**: 1159–67.

**Berg, A., Flouzat, G.** and **De Paratesi, S. G.** (eds) (1978) *Agreste Project: Agricultural Resources Investigations in Northern Italy and Southern France*. Commission of the European Communities, Joint Research Centre, Ispra Establishment, Italy.

**Bernard, R., Vauclin, M.** and **Vidal-Madjar, D.** (1981) Possible use of active microwave remote sensing data for prediction of regional evaporation by numerical simulation of soil water movement in the unsaturated zone. *Water Resources Research*, **17**: 1603–10.

**Bernard, R., Martin, P. H., Thony, J. L., Vauclin, M.** and **Vidal-Madjar, D.** (1982) C Band radar for determining surface soil moisture. *Remote Sensing of Environment*, **12**: 189–200.

**Bernstein, R.** (1976) Digital image processing of Earth observation sensor data. *IBM Journal of Research and Development*, **20**: 40–57.

**Bernstein, R.** (1983) Image geometry and rectification: *In* Colwell, R. N. (ed.) *Manual of Remote Sensing* (2nd edn). American Society of Photogrammetry, Falls Church, Virginia, pp. 873–922.

**Best, R. G., Fowler, R., Hause, D.** and **Wehde, M.** (1982) Aerial thermal infrared census of Canada geese in south Dakota. *Photogrammetric Engineering and Remote Sensing*, **48**: 1869–77.

**Billingsley, F. C., Geotz, A. F. H.** and **Lindsley, J. N.** (1970) Colour differentiation by computer image processing. *Photographic Science Engineering*, **17**: 28–35.

**Bjorkland, J., Schmer, F. A.** and **Isakson, R. E.** (1975) A report on the use of thermal scanner data in an operational program for monitoring apparent rooftop temperatures. *Proceedings of the 9th International Symposium on Remote Sensing of Environment*. University of Michigan, Ann Arbor, pp. 1437–46.

**Blagonravov, A. A** and **Dryden, H. L.** (1962) *The Peaceful use of Outer Space* United Nations, Geneva.

**Blanchard, L. E.** and **Weinstein, O.** (1980) Design challenges of the Thematic Mapper. *IEEE Transactions on Geoscience and Remote Sensing*, **18**: 146–60.

**Blankenship J. R.** (1962) An approach to objective nephanalysis from an Earth-orientated satellite. *Journal of Applied Meterology*, **1**: 581–2.

**Blom, R.** and **Daily, M.** (1981) Seasat views California with imaging radar. *California Geology*, **34**: 231–40.

**Blom, R.** and **Elachi, C.** (1981) Spaceborne and airborne imaging radar observations of sand dunes. *Journal of Geophysical Research*, **86**: 3061–73.

**Bodechtel, J.** and **Gierloff-Emden, H. G.** (1974) *The Earth from Space*. David and Charles, Newton Abbot.

**Bonn, F.** (1976) Some problems and solutions related to ground truth measurements for thermal infrared remote sensing. *Proceedings of the American Society of Photogrammetry*, **42**: 1–11.

**Borrowman, G. L.** (1982) Soviet orbital surveillance – the legacy of Cosmos 954. *Journal of the British Interplanetary Society*, **35**: 67–71.

**Boulter, J. F.** (1979) Interactive digital image restoration and enhancement. *Computer Graphics and Image Processing*, **11**: 301–12.

**Bowers S. A.** and **Hanks, R. J.** (1965) Reflection of radiant energy from soil. *Soil Science*, **100**: 130–8.

**Bowman, R. L.** and **Jack, J. R.** (1979) Feasibility of determining flat roof heat losses using aerial thermography. *Proceedings of the 13th International Symposium on Remote Sensing of Environment.* University of Michigan, Ann Arbor, pp. 1199–211.

**Braithwaite, J. G. N.** and **Lowe, D. S.** (1966) A spectrum matching technique for enhancing image contrast. *Applied Optics*, **5**: 893–906.

**Brandli, H. W.** (1978) The night eye in the sky. *Photogrammetric Engineering and Remote Sensing*, **44**: 503–5.

**Brisco, B.** and **Protz, R.** (1980) Corn field identification accuracy using airborne radar imagery. *Canadian Journal of Remote Sensing*, **6**: 15–25.

**Brisco, B.** and **Protz, R.** (1982) Manual and automatic crop identification with airborne radar imagery. *Photogrammetric Engineering and Remote Sensing*, **48**: 101–9.

**Bristor, C. L.** (ed.) (1975) *Central Processing and Analysis of Geostationary Satellite Data.* NOAA Tech. Memo. NESS 64, US Department of Commerce, Washington DC, pp. 1–155.

**Brakke, T. W., Kanemansu, E. T., Steiner, J. L., Ulaby, F. T.** and **Wilson, E.** (1981) Microwave radar response to canopy moisture, leaf area index, and dry weight of wheat, corn and sorgum. *Remote Sensing of Environment*, **11**: 207–20.

**Brooner, W. G., Haralick, R. M.** and **Dinstein, I.** (1971) Spectral parameters affecting automated image interpretation using Bayesian probability techniques. *Proceedings of the 7th International Symposium on Remote Sensing of Environment.* University of Michigan, Ann Arbor, pp. 1929–49.

**Brown, R. J., Chilar, J.** and **Teillet, P. M.** (1981) Quantitative residential heat loss study. *Photogrammetric Engineering and Remote Sensing*, **47**: 1327–33.

**Browning, K. A.** (1982) Extremely low temperatures over England and Wales observed by Meteosat 2. *Weather*, **37**: 79.

**Bryan, M. L.** (1975) Interpretation of an urban scene using multichannel radar imagery. *Remote Sensing of Environment*, **4**: 49–67.

**Bryan, M. L.** (1981) Radar study of a vegetated piedment. *Journal of Soil and Water Conservation*, **36**: 305–8.

**Bryan, M. L.** (1983) Urban land use classification using synthetic aperture radar. *International Journal of Remote Sensing*, **4**: 215–33.

**Buckley, B. A.** (1971) Computerised isodensity mapping. *Photogrammetric Engineering*, **37**: 1039–42.

**Budd, J. T. C.** and **Coulson, M. G.** (1980) *Ecological Mapping of the Intertidal Zone of Chichester Harbour.* Nature Conservancy Council, London.

**Burrough, P. A.** (1980) The development of a landscape information system in the Netherlands, based on a turn-key graphics system. *Geo-Processing*, **1**: 257–74.

**Bush, P. W.** and **Collins, W. G.** (1974) The application of aerial photography to surveys of derelict land in the United Kingdom: *In* Barrett, E. C. and Curtis, L. F. (eds.) *Environmental Remote Sensing: Applications and Achievements.* Edward Arnold, London, pp. 167–81.

**Bush, T. F.** and **Ulaby, F. T.** (1978) An evaluation of radar as a crop classifier. *Remote Sensing of Environment*, **7**: 15–36.

**Byrne, G. F., Crapper, P. F.** and **Mayo, K. K.** (1980) Monitoring land-cover change by principal components analysis of multitemporal Landsat data. *Remote Sensing of Environment*, **10**: 175–84.

**Bryne, G. F., Dabrowska–Zielinska K.** and **Goodrick, G. N.** (1981) Use of visible and thermal satellite data to monitor intermittently flooding marshland. *Remote Sensing of Environment*, **11**: 393–9.

**Calkins, H. W.** and **Tomlinson, R. F.** (1977) *Geographic Information Systems, Methods and Equipment for Land Use Planning.* International Geographical Union (IGU), Commission on Geographical Data Sensing and Processing, Resource and Land Investigations (RALI) Program, Geological Survey, U.S. Department of Interior, Reston, Virginia.

**Campbell, J. B.** (1981) Spatial correlation effects upon accuracy of supervised classification of land cover. *Photogrammetric Engineering and Remote Sensing,* **47**: 355–63.

**Card, D. H.** (1982) Using known map category marginal frequencies to improve estimates of thematic map accuracy. *Photogrammetric Engineering and Remote Sensing*, **48**: 431–9.

**Carleton, A. M.** (1982) Climatological relationships between the cryosphere and synoptic activity in the northern hemisphere from satellite data analysis: *In* Henderson-Sellers, A. and Shine, K. P. (eds) *Remote Sensing and the Atmosphere.* Remote Sensing Society, Reading, pp. 48–54.

**Carlson, T. N.** (1981) The Pennsylvania Barrens: a view from space. *Bulletin of the American Meteorological Society*, **62**: 412–13.

**Carroll, L.** (1865) *Alice's Adventures in Wonderland.* Peal Press, London.

**Carter J.** (1982) *Keeping Faith: Memoirs of a President.* Collins.

**Castleman, K. R.** (1979) *Digital Image Processing.* Prentice Hall, New Jersey.

**Chahine, M. T.** (1983) Interaction mechanisms within the atmosphere: *In* Colwell, R. N. (ed.) *Manual of Remote Sensing* (2nd edn). American Society of Photogrammetry, Falls Church, Virginia, pp. 165–230.

**Chavez, P.** and **Bauer, B.** (1982) An automatic optimum kernel-size selection technique for edge enhancement. *Remote Sensing of Environment*, **12**: 23–38.

**Chen, E., Allen, L. H., Bartholic, J. F., Bill, R. G.** and **Sutherland R. A.** (1979) Satellite-sensed winter nocturnal temperature patterns of the Everglades agricultural area. *Journal of Applied Meteorology*, **18**: 992–1002.

**Chevallier, R., Fontanel, A., Grau, E.** and **Guy, M.** (1970) Application of optical filtering to the study of aerial photographs. *Photogrammetria*, **26**: 17–35.

**Chevrel, M., Courtois, M.** and **Weill, G.** (1981) The SPOT satellite remote sensing mission. *Photogrammetric Engineering and Remote Sensing*, **47**: 1163–71.

**Chizhov, A. N., Glushnev V. G., Slutsker B. D.** and **Borodulin V. V.** (1978) The use of radar to measure ice thickness in rivers, lakes and reservoirs. *Soviet Hydrology: Selected Papers*, **17**: 116–27.

**Choudhury, B. J., Chang, A. T. C., Salomonson, V. V., Schmugge, T. J.** and **Wang, J. R.** (1979) *Preliminary Results of SAR Soil Moisture Experiment, November 1975.* National Aeronautics and Space Administration Technical Paper 1404, Washington DC.

**Cicone, R. C.** (1977) Remote sensing and geographically based information systems *Proceedings of the 11th International Symposium on Remote Sensing of Environment.* University of Michigan, Ann Arbor, pp. 1127–36.

**Cihlar, J.** and **Ulaby, F. T.** (1974) *Dielectric Properties of Soils as a Function of Moisture Content.* CRES Tech. Report 177–47. University of Kansas Centre for Research Inc., Lawrence, Kansas.

**Clark, B. P.** (1981) *Landsat 3 Return Beam Vidicon Response Artifacts.* US Geological Survey, South Dakota.

**Clark, J. R.** and **La Violette, P. E.** (1981) Detecting the movement of oceanic fronts using registered TIROS-N imagery. *Geophysical Research Letters*, **8**: 229–32.

**CNES** (1980) *SPOT Satellite-Based Remote Sensing System.* Centre National d'Études Spatiales, Toulouse.

**CNES/SPOT Image** (1981) *SPOT Newsletter, 1,* Centre National d'Études Spatiales and SPOT Image, Paris.

**Cochrane, G. R.** and **Browne, G. H.** (1981) Geomorphic mapping from Landsat 3 return beam vidicon (RBV) imagery. *Photogrammetric Engineering and Remote Sensing*, **47**: 1305–14.

**Cochrane, G. R.** and **Tianfeng, W.** (1983) Interpretation of structural characteristics of the Taupo Volcanic zone, New Zealand, from Landsat imagery. *International Journal of Remote Sensing*, **4**: 111–28.

**Coiner, J. C.** (1980) Using Landsat to monitor changes in vegetation cover induced by desertification processes. *Proceedings of the 14th International Symposium on Remote Sensing of Environment.* University of Michigan, Ann Arbor, pp. 1341–51.

**Colcord, J. E.** (1981) Thermal imagery energy surveys. *Photogrammetric Engineering and Remote Sensing*, **47**: 237–40.

**Cole, M. M., Owen Jones, E. S.** and **Custance, N. D. E.** (1974) Remote sensing in mineral exploration: *In* Barrett, E. C. and Curtis, L. F. (eds) *Environmental Remote Sensing: Applications and Achievements.* Edward Arnold, London, pp. 49–66.

**Collins, W. G.** and **El-Beck, A. H. A.** (1971) The acquisition of urban land use information from aerial photographs of the city of Leeds. *Photogrammetria*, **27**: 71–92.

**Colvocoresses, A. P.** (1975) Platforms for remote sensors: *In* Reeves, R. G. (ed.) *Manual of Remote Sensing*. American Society of Photogrammetry, Falls Church, Virginia, pp. 539–88.

**Colvocoresses, A. P.** (1979a) Multispectral linear arrays as an alternative to Landsat D. *Photogrammetric Engineering and Remote Sensing*, **45**: 67–8.

**Colvocoresses, A. P.** (1979b) Proposed parameters for Mapsat. *Photogrammetric Engineering and Remote Sensing*, **45**: 501–6.

**Colwell, R. N.** (ed.) (1960) *Manual of Photographic Interpretation*. American Society of Photogrammetry, Falls Church, Virginia.

**Colwell, R. N.** (ed.) (1983a) *Manual of Remote Sensing* (2nd edn). American Society of Photogrammetry, Falls Church, Virginia.

**Colwell, R. N.** (1983b) Glossary: *In* Colwell, R. N. (ed.) *Manual of Remote Sensing* (2nd edn). American Society of Photogrammetry, Falls Church, Virginia, pp. 1183–98.

**Colwell, J. E.** (1974) Grass canopy bidirectional spectral reflectance. *Proceedings of the 9th International Symposium on Remote Sensing of Environment*. University of Michigan, Ann Arbor, pp. 1061–85.

**Condit, C. D.** and **Chavez, P. S.** (1979) Basic concepts of computerised digital image processing for geologists. *Geological Survey Bulletin* **1,462**, Washington DC.

**Corbett, F. J.** (1974) Sensor design performance evaluation of Skylab multispectral photographic facility. *Proceedings of the Society of Photo-Optical Instrumentation Engineers*, **46**: 239–46.

**Corless, K. G.** (1977) Remote sensing by radar: *In* Peel, R. F., Curtis, L. F. and Barrett, E. C. (eds) *Remote Sensing of the Terrestrial Environment*. Butterworths, London, Boston, pp. 38–53.

**Cracknell, A. P., MacFarlane, N., McMillan, K., Charlton J. A., McManus, J.** and **Ulbricht, K. A.** (1982) Remote sensing in Scotland using data received from satellites. A study of the Tay Estuary region using Landsat multispectral scanning imagery. *International Journal of Remote Sensing*, **3**: 113–37.

**Craighead, J. J.** (1976) Studying grizzly bear habitat by satellite. *National Geographic*, **150**: 148–58.

**Crandall, C. J.** (1969) Radar mapping in Panama. *Photogrammetric Engineering*, **35**: 641–8.

**Crane, D.** (1972) *Invisible Colleges: Diffusion of Knowledge in Scientific Communities*. University of Chicago Press, Chicago.

**Croft, T. A.** (1981) Radiometry with nighttime DMSP images in digital form. *Photogrammetric Engineering and Remote Sensing*, **47**: 1319–25.

**Curran, P. J.** (1978a) Oblique false colour photograph of fog in Cheddar Gorge. *Weather*, **33**: 63–4.

**Curran, P. J.** (1978b) A photographic method for the recording of polarised visible light for soil surface moisture indications. *Remote Sensing of Environment*, **7**: 305–22.

**Curran, P. J.** (1979a) Preliminary investigation into the application of oblique multispectral photography for the monitoring of sewage outfalls. *Journal of Environmental Management*, **9**: 41–8.

**Curran, P. J.** (1979b) The use of polarised panchromatic and false colour infrared film in the monitoring of soil surface moisture. *Remote Sensing of Environment*, **8**: 249–66.

**Curran, P. J.** (1980a) Multispectral remote sensing of vegetation amount. *Progress in Physical Geography*, **4**: 315–41.

**Curran, P. J.** (1980b) Multispectral photographic remote sensing of vegetation amount and productivity. *Proceedings of the 14th International Symposium on Remote Sensing of the Environment*. University of Michigan, Ann Arbor, pp. 623–37.

**Curran, P. J.** (1980c) Methodology for quantitative photo-interpretation. *Photo-Interpretation*, **3**: 11–8.

**Curran, P. J.** (1980d) Relative reflectance data from preprocessed multispectral photography. *International Journal of Remote Sensing*, **1**: 77–83.

**Curran, P. J.** (1980e) The Bussex rhyne. *Proceedings of the Somersetshire Archaeological and Natural History Society*, **124**: 167–9.

**Curran, P. J.** (1981a) Remote sensing: the use of polarised visible light (P.V.L.) to estimate surface soil moisture. *Applied Geography*, **1**: 41–53.

**Curran, P. J.** (1981b) The relationship between polarised visible light and vegetation amount. *Remote Sensing of Environment*, **11**: 87–92.

**Curran, P. J.** (1981c) Remote sensing: the role of small format, light aircraft photography. *Geographical Papers*, **75**, Department of Geography, University of Reading.

**Curran, P. J.** (1981d) Infrared: your questions answered. *Amateur Photographer*, **163**: 106–7

**Curran, P. J.** (1981e) Multispectral remote sensing for estimating vegetation biomass and productivity: *In* Smith, H. (ed.) *Plants and the Daylight Spectrum*. Academic Press, pp. 65–99.

**Curran, P. J.** (1981f) The estimation of the surface moisture of a vegetated soil using aerial infrared photography. *International Journal of Remote Sensing*, **2**: 369–78.

**Curran, P. J., Munday, T. J** and **Milton, E. J.** (1981) A comparison between two photographic methods for the determination of relative bidirectional reflectance. *International Journal of Remote Sensing*, **2**: 185–8.

**Curran, P. J.** (1982a) Multispectral photographic remote sensing of green vegetation biomass and productivity. *Photogrammetric Engineering and Remote Sensing*, **48**: 243–50.

**Curran, P. J.** (1982b) Polarised visible light as an aid to vegetation classification. *Remove Sensing of Environment*, **12**: 491–9.

**Curran, P. J.** (1983a) Problems in the remote sensing of vegetation canopies for biomass estimation: *In* Fuller, R. M. (ed.) *Ecological Mapping from Ground, Air and Space*. Institute of Terrestrial Ecology, Symposium No. 10. Cambridge, pp. 84–100.

**Curran, P. J.** (1983b) Crop radiometry. *Analytical Proceedings*, **20**: 517–9.

**Curran, P. J.** (1983c) Multispectral remote sensing for the estimation of green leaf area index. *Philosophical Transactions of the Royal Society, Series A*, **309**: 257–70.

**Curran, P. J.** (1983d) Remote sensing terminology. *International Journal of Remote Sensing*, **4**: 835–6.

**Curran, P. J.** (1983e) Estimating green LAI from multispectral aerial photography. *Photogrammetric Engineering and Remote Sensing*, **49**: 1709–20.

**Curran, P. J.** and **Milton, E. J.** (1983) The relationship between the chlorophyll concentration, LAI and reflectance of a simple vegetation canopy. *International Journal of Remote Sensing*, **4**: 247–56.

**Curran, P. J.** and **Adawy, N.** (1983) Landsat MSS data, its availability and suitability for monitoring the density of date palm in Saudi Arabia: *In* Makki, Y. M. (ed.) *The First Symposium on Date Palm*. King Faisal University, Al-Hassa, Kingdom of Saudi Arabia, pp. 684–90.

**Curran, P. J.** and **Wardley, N. W.** (1983) The contribution of UK geographers to remote sensing. *Area*, **15**: 29–34.

**Curtis, L. F.** and **Mayer, A. .E. S.** (1974) *Remote Sensing Evaluation Flights*, 1971. NERC Publications Series C, No. 12. Natural Environment Research Council, Swindon.

**Cyganiak, N.** (1980) Land use in Mozambique from Landsat imagery. *Remote Sensing Quarterly*, **2**: 33–4.

**Darnell, W. L.** and **Harriss, R. C.** (1983) Satellite sensing capabilities for surface temperature and meteorological parameters over the ocean. *International Journal of Remote Sensing*, **4**: 65–92.

**Davidson, G. J.** (1980) Cover photograph, principal components analysis. *International Journal of Remote Sensing*, **1**: 413–5.

**Davis, C. K.** and **Neal, J. T.** (1963) Descriptions and airphoto characteristics of desert landforms. *Photogrammetric Engineering*, **29**: 621–31.

**Dean, K. G., Forbes, R. B., Turner, D. L., Eaton, F. D.** and **Sullivan, K. D.** (1982) Radar and infrared remote sensing of geothermal features of Pilgrim Springs, Alaska. *Remote Sensing of Environment*, **12**: 391–405.

**Deane, R. A.** (1973) *Side Looking Radar Systems and their Potential Application to Earth-Resources Surveys. Basic Physics and Technology*. Contractor Report 136 from Easams Ltd, UK to European Space Research Organisation, France.

**Dellwig, L. F., MacDonald, H. C.** and **Kirk, J. N.** (1968) The potential of radar in geological exploration. *Proceedings of the 5th International Symposium on Remote Sensing of Environment*. University of Michigan, Ann Arbor, pp. 747–64.

**Deutsch, M.** and **Estes, J.** (1980) Landsat detection of oil from natural seeps. *Photogrammetric Engineering and Remote Sensing*, **46**: 1313–22.

**Deutsch, M., Wiesnet, D. R.** and **Rango, A.** (1981) *Satellite Hydrology*. American Water Resources Association, Minneapolis.

**Dismachek, D. C., Booth, A. L.** and **Leese, J. A.** (1980) *National Environmental Satellite Service, Catalog of Products.* NOAA Technical Memorandum, NESS, 109, Washington DC.

**Dobson, M. C.** and **Ulaby, F.** (1981) Microwave backscatter dependence on surface roughness, soil moisture and soil texture: Part III – soil tension. *IEEE Transactions on Geoscience and Remote Sensing*, **19**: 51–61.

**Domain, F., Citeau, J.** and **Noel, J.** (1980) Sea surface temperatures studied by Meteosat data along the coast of Senegal and Mauritania: *In* Cracknell, A. P. (ed.) *Coastal and Marine Applications of Remote Sensing.* Remote Sensing Society, Reading, pp. 59–67.

**Dosiére, P.** and **Justice, C.** (1983) Spatial and radiometric resolution of the Landsat-3 RBV system. *International Journal of Remote Sensing*, **4**: 447–55.

**Doverspike, G. E., Flynn, F. M.** and **Heller, R. C.** (1965) Microdensitometer applied to land use classification. *Photogrammetric Engineering*, **31**: 294–306.

**Dowman, I. J.** and **Morris, A. H.** (1982) The use of synthetic aperture radar for mapping. *Photogrammetric Record*, **10**: 687–96.

**Doyle, F. J.** (1978) The next decade of satellite remote sensing. *Photogrammetric Engineering and Remote Sensing*, **44**: 155–64.

**Doyle, F. J.** (1979) A large format camera for Shuttle. *Photogrammetric Engineering and Remote Sensing*, **45**: 74–8.

**Doyle, F. J.** (1982) *Status of Satellite Remote Sensing Systems.* United States Department of Interior, Geological Survey, Reston, Virginia. USGS Open-file Report 82–237.

**Dozier, J., Schneider, S. R.** and **McGinnes, D. F.** (1981) Effect of grain size and snowpack water equivalence on visible and near infrared satellite observations of snow. *Water Resources Research*, **17**: 1213–21.

**Dozier, J.** and **Strahler, A. H.** (1983) Ground Investigations in support of remote sensing: *In* Colwell, R. N (ed) *Manual of Remote Sensing* (2nd edn). American Society of Photogrammetry, Falls Church, Virginia, pp. 959–86.

**Duda, R. O.** and **Hart, P. E.** (1973) *Pattern Classification and Scene Analysis.* Wiley, New York.

**Duggin, M. J.** (1977) Likely effects of solar elevation on the quantification of changes in vegetation with maturity using sequential Landsat imagery. *Applied Optics*, **16**: 521–23.

**Easams Ltd.** (1972) *Side Looking Radar Systems and their Potential Application to Earth Resource Surveys.* Easams Reports prepared for ESRO under ESTEC contract 1537/71 EL (in 7 volumes).

**Eastman Kodak** (1965) *Kodak Wratten Filters for Scientific and Technical Use.* (22nd edn.) Eastman Kodak Publication B-3. Rochester, New York.

**Eastman Kodak** (1972) *Color as Seen and Photographed* (2nd edn). Eastman Kodak Company, Rochester, New York.

**Eastman Kodak** (1976) *Kodak Data for Aerial Photography.* No. M-29 (4th edn). Eastman Kodak Company, New York.

**Eckhart, D.** and **Geerders, P. J. F.** (1975) Remote sensing in the Netherlands. *Proceedings of the 10th International Symposium on Remote Sensing of Environment.* University of Michigan, Ann Arbor, pp. 545–8.

**Edelson, B. I.** (1982) *Shuttle Imaging Radar-B. Announcement of Opportunity.* National Aeronautics and Space Administration. A.O. No. OSSA-1-82, Washington, DC.

**Edwards, D. A.** and **Partridge, C.** (1978) *Aerial Archaeology 1977.* Committee for Archaeological Air Photography, Hertfordshire.

**Eigenwillig, N.** and **Fischer, H.** (1982) Determination of midtrophospheric wind vectors by tracking pure water vapour structures in Meteosat water vapour image sequences. *Bulletin American Meteorological Society*, **63**: 45–58.

**Elkington, M. D.** and **Hogg, J.** (1981) The characterisation of soil moisture content and actual evaporation from crop canopies using thermal infrared remote sensing. *University of Leeds, School of Geography, Working Paper*, **292.**

**Ellefsen, R.** and **Davidson, R.** (1980) Monitoring the aerial growth of San Jose, Costa Rica. *Proceedings of the 14th International Symposium on Remote Sensing of Environment.* University of Michigan, Ann Arbor, pp. 1243–9.

**Endlich, R. M.** and **Wolf, D. E.** (1981) Automatic cloud tracking applied to GOES and Meteosat observations. *Journal of Applied Meteorology*, **20**: 309–19.

**Eppler, W. G.** (1974) An improved version of the table look-up algorithm for pattern recognition. *Proceedings of the 9th International Symposium on Remote Sensing of Environment.* University of Michigan, Ann Arbor, pp. 793–812.

**Erb, T. L., Phillipson, W. R., Teng, W. L.** and **Liang, T.** (1981) Analysis of landfills with historic airphotos. *Photogrammetric Engineering and Remote Sensing*, **47**: 1363–9.

**ESA** (1981a) *E R S –1 Announcement of Opportunity.* European Space Agency Report APP (**81**) 1, Paris.

**ESA** (1981b) *E R S –1 Mission Objectives.* European Space Agency Report PB-RS (**80**) 24, Paris.

**Estep, S. D.** (1968) Legal and social policy ramifications of remote sensing techniques. *Proceedings of the 5th International Symposium on Remote Sensing of Environment.* University of Michigan, Ann Arbor, pp. 197–217.

**Estes, J. E.** (1982) Remote sensing and geographic information systems coming of age in the eighties: *In* Richason, B. F. (ed.) *Remote Sensing: As an Input to Geographic Information Systems in the 1980s.* Pecora VII Symposium, United States Geological Survey, Sioux Falls, pp. 23–40.

**Estes, J. E.** and **Senger, C. W.** (1971) An electronic multi-image processor. *Photogrammetric Engineering*, **37**: 577–86.

**Estes, J. E.** and **Simonett, D. S.** (1975) Fundamentals of image interpretation: *In* Reeves, R. G. (ed.) *Manual of Remote Sensing.* American Society of Photogrammetry, Falls Church, Virginia, pp. 869–1076.

**Estes, J. E., Hajic, E. J.** and **Tinney, L. R.** (1983) Fundamentals of image analysis: analysis of visible and thermal infrared data: *In* Colwell, R. N. (ed.) *Manual of Remote Sensing* (2nd edn). American Society of Photogrammetry, Falls Church, Virginia, pp. 987–1124.

**Estes, J. E., Jensen, J. R.** and **Simonett, D. S.** (1980) Impacts of remote sensing on US Geography. *Remote Sensing of Environment*, **10**: 43–80.

**Evans, R.** (1972) Air photographs for soil survey in lowland England: soil patterns. *Photogrammetric Record*, **7**: 302–22.

**Evans, R.** (1974) The time factor in aerial photography for soil surveys in lowland England: *In* Barrett, E. C. and Curtis, L. F. (eds) *Environmental Remote Sensing: Applications and Achievements.* Edward Arnold, London, pp. 69–86.

**Evans, R. M., Hanson, W. T.** and **Brewer, W. L.** (1953) *Principles of Color Photography.* Wiley, New York.

**Eyre, J. A.**. (1981) Meteosat water vapour imagery. *Meteorological Magazine*, **110**: 345–51.

**Eyton, J. R.** (1983) Landsat multitemporal colour composites. *Photogrammetric Engineering and Remote Sensing*, **49**: 231–35.

**Farmer, P.** (1979) *Population.* Longman, London.

**Feder, G. L.** and **Barks, J. H.** (1972) A losing drainage basin in the Missouri Ozarks identified on side-looking radar imagery. *US Geological Survey Professional Paper* **800-C**: C249–C252.

**Feinberg, G.** (1968) Light. *Scientific American*, **219**: 50–9.

**Ferguson, F. L., Jorde, D. G.** and **Sease, J. J.** (1981) The use of 35 mm colour aerial photography to acquire mallard sex ratio data. *Photogrammetric Engineering and Remote Sensing*, **47**: 823–7.

**Finkel, H. J.** (1961) The movement of barchan dunes measured by aerial photogrammetry. *Photogrammetric Engineering*, **27**: 439–44.

**Fischer, W. A.** (1962) Colour aerial photography in geologic investigations. *Photogrammetric Engineering*, **28**: 133–9.

**Fischer, W. A.** (1975) History of remote sensing: *In* Reeves, R. G. (ed.) *Manual of Remote Sensing.* American Society of Photogrammetry, Falls Church, Virginia, pp. 27–50.

**Fischer, W. A, Hemphill, W. R.** and **Kover, A.** (1976) Progress in remote sensing (1972–1976). *Photogrammetria*, **32**: 33–72.

**Fitzgerald, E.** (1972) *Multispectral Scanning Systems and their Potential Application to Earth Resource Surveys.* European Space Research Organisation Report (P)232.

**Foote, R.** and **Draper, L. T.** (1980) TIROS-N Advanced very high resolution radiometer AVHRR: *In* Cracknell, A. P. (ed.) *Coastal and Marine Applications of Remote Sensing.* Remote Sensing Society, Reading, pp. 25–35.

**Ford, K.** (ed.) (1979) *Remote Sensing for Planners.* Centre for Urban Policy Research, Rutgers. The State University of New Jersey.

**Ford, J. P., Blom, R. G., Bryan, M. L., Daily, M. I., Dixon, T. H., Elachi, C.** and **Xenos, E. C.** (1980) *Seasat Views North America, the Caribbean and Western*

*Europe With Imaging Radar*. National Aeronautics and Space Administration, Jet Propulsion Laboratory Publication 80–67. Pasadena, California.

**Ford, G. E., Algazi, V. R.** and **Meyer, D. I.** (1983) A noninteractive procedure for land-use determination. *Remote Sensing of Environment*, **13**: 1–16.

**Forshaw, M. R. B., Haskell, A., Miller, P. F., Stanley, D. J.** and **Townshend, J. R. G.** (1983) Spatial resolution of remotely sensed imagery: review paper. *International Journal of Remote Sensing*, **4**: 497–520.

**Forster, B. C.** (1980) Urban residential ground cover using Landsat digital data. *Photogrammetric Engineering and Remote Sensing*, **46**: 547–58.

**Foster, J. L., Ormsby, J. P.** and **Gurney, R. J.** (1981) Satellite observations of England and north-western Europe. *Weather*, **36**: 252–9.

**Fotheringham, R. R.** (1979) *The Earth's Atmosphere Viewed from Space*. University of Dundee.

**Fraser, R. S.** (1975) Interaction mechanisms within the atmosphere: *In* Reeves, R. G. (ed.) *Manual of Remote Sensing*. American Society of Photogrammetry, Falls Church, Virginia, pp. 181–233.

**Fraser, R. S.** and **Curran, R. J.** (1976) Effects of atmosphere on remote sensing: *In* Lintz, J and Simonett, D. S. (eds) *Remote Sensing of Environment*. Addison-Wesley, Reading, Massachusetts, London, pp. 34–84.

**Frazee, C. J., Myers, V. I.** and **Westin, F. C.** (1972) Density slicing techniques for soil survey. *Soil Science Society of America, Proceedings*, **36**: 693–5.

**Frazee, C. J., Carey, R. L.** and **Westin, F. C.** (1973) Utilising remote sensing data for land use decision for Indian lands in South Dakota. *Proceedings of the 8th International Symposium on Remote Sensing of Environment*. University of Michigan, Ann Arbor, pp. 375–92.

**Freden, S. C.** and **Gordon, F.** (1983) Landsat satellites: *In* Colwell, R. N. (ed.) *Manual of Remote Sensing* (2nd edn). American Society of Photogrammetry, Falls Church, Virginia, pp. 517–70.

**Friedman, J. D., Frank, D., Kieffer, H. H.** and **Sawatzky, D. L.** (1981) The 1980 eruptions of Mount St Helens, Washington. Thermal infrared surveys of the 18 May crater, subsequent lava domes, and associated volcanic deposits. *US Geological Survey Professional Paper*, **1250**: 279–93.

**Frost, V. S., Perry, M. S., Dellwig, L. F.** and **Holtzman, J. C.** (1983) Digital enhancement of SAR imagery as an aid in geologic data extraction. *Photogrammetric Engineering and Remote Sensing*, **49**: 357–63.

**Fu, K. S.** (1976) Pattern recognition in remote sensing of the Earth resources. *IEEE Transactions on Geoscience Electronics*, **14**: 10–11.

**Fuhrer, J., Erismann, K. H., Keller, H. J.** and **Faure, A.** (1981) A system for quantitative determination of species and vitalities of urban trees on colour infrared photographs. *Remote Sensing of Environment*, **11**: 1–8.

**Gagnon, H.** (1975) Remote sensing of landslide hazards on quick clays of eastern Canada. *Proceedings of the 10th International Symposium on Remote Sensing of Environment*. University of Michigan, Ann Arbor, pp. 803–10.

**Gammon, P. T.** and **Carter, V.** (1979) Vegetation mapping with seasonal colour infrared photographs. *Photogrammetric Engineering and Remote Sensing*, **45**: 87–97.

**Garland, G. G.** (1982) Mapping erosion with airphotos: panchromatic or black and white infrared. *ITC Journal*: 309–12.

**Garafalo, D.** and **Wobber, F. J.** (1974) The Nicaragua earthquake: aerial photography for disaster assessment and damage. *Photographic Applications in Science, Technology and Medicine*, pp. 36–8.

**Gausman, H. W.** (1974) Leaf reflectance of near infrared. *Photogrammetric Engineering and Remote Sensing*, **40**: 183–91.

**Gawarecki, S. J., Moxham, R. M., Morgan, J. O.** and **Parker, D. C.** (1980) An infrared survey of Irazu volcano and vicinity, Costa Rica. *Proceedings of the 14th International Symposium on Remote Sensing of Environment*. University of Michigan, Ann Arbor, pp. 1901–72.

**Geiger, R.** (1965).*The Climate Near the Ground*. Harvard University Press.

**Genderen, J. L. van** (1975) Visual interpretation of remote sensing data and electronic image enhancement techniques: *In* Genderen, J. L. van and Collins, W. E. (eds) *Remote Sensing Data Processing*. Remote Sensing Society, Reading, pp. 19–51.

**Genderen, J. L. van, Lock, B. F.** and **Vass, P. A.** (1978) Remote sensing: statistical testing of thematic map accuracy. *Remote Sensing of Environment.* **7**: 3–15.

**Gerbermann, A. H., Gausman, H. W.** and **Wiegand, C. L.** (1971) Colour and colour I.R. films for soil identification. *Photogrammetric Engineering*, **37**: 359–64.

**Gersen, D. J.** and **Rosenfeld, A.** (1975) Automatic sea ice detection in satellite pictures. *Remote Sensing of Environment*, **4**: 187–98.

**Gilbertson, B.** and **Longshaw, T. G.** (1975) Multispectral aerial photography as an exploration tool. I Concepts, techniques and instrumentation. *Remote Sensing of Environment*, **4**: 129–46.

**Ginevan, M. E.** (1979) Testing land use map accuracy: another look. *Photogrammetric Engineering and Remote Sensing*, **45**: 1371–7.

**Goldberg, M.** (1982) *Geographic Information Systems at the Goddard Space Flight Center.* National Aeronautics and Space Administration, Technical Memorandum 83941, Washington DC.

**Gomber, R.** (1980) Soil and land use distribution as discernible on Meteosat 1 imagery over West Africa. *Pedologie*, **30**: 127–36.

**Gonzalez, R. C.** and **Wintz, P.** (1977) *Digital Image Processing.* Addison-Wesley, Reading, Massachusetts.

**Goodier, R.** and **Grimes, B. H.** (1970) The interpretation and mapping of vegetation and other ground surface features from air photographs of mountainous areas in North Wales. *Photogrammetric Record*, **6**: 553–66.

**Goodman, J. W.** (1978) *Introduction to Fourier Optics.* McGraw Hill, San Francisco.

**Gordon, H. R.** and **Clark, D. K.** (1980) Initial coastal zone colour scanner imagery. *Proceedings of the 14th International Symposium on Remote Sensing of Environment.* University of Michigan, Ann Arbor, pp. 517–28.

**Gornitz, V.** (1979) Detection of hydrothermal alteration with 24-channel multispectral scanner data and quantitative analyses of linear features. Monroe geothermal area, Utah. *Proceedings of the 13th International Symposium on Remote Sensing of Environment.* University of Michigan, Ann Arbor, pp. 825–9.

**Gosling, N.** (1976) *Nadar.* Secker and Warburg, London.

**Grainger, J. E.** (1981) A quantitative analysis of photometric data from aerial photographs for vegetation survey. *Vegetation*, **48**: 71–82.

**Green, J. E.** and **Crouch, L. W.** (1979) A remote sensing method for the detection of malfunctioning on-site septic fields: *In* Hanten, E. W. and Otano, J. J. (eds) *The Urban Environment in a Spatial Perspective.* Centre for Urban Studies, University of Akron, Ohio.

**Green, A. A., Whitehouse, G.** and **Outhet, D.** (1983) Causes of flood streamlines observed on Landsat images and their use as indicators of floodways. *International Journal of Remote Sensing.* **4**: 5–16.

**Gregg, T. W. D., Sugarbaker, L. J., Barthmaller, E. W., Scott, R. B.** and **Harding, R. A.** (1980) *Development of a Statewide Geographic Information System in Washington.* Department of Natural Resources, Olympia, Washington.

**Grigg, D.** (1965) The logic of regional systems. *Annals of the Association of American Geographers*, **55**: 465–91.

**Grumstrup, P., Meyer, M., Gustafson, R.** and **Hendrickson, E.** (1982) Aerial photographic assessment of transmission line structure impact on agricultural crop production. *Photogrammetric Engineering and Remote Sensing*, **48**: 1313–7.

**Gurney, C. M.** (1982) The use of contextural information to detect cumulus clouds and cloud shadows in Landsat data. *International Journal of Remote Sensing*, **3**: 51–62.

**Gurney, R. J.** and **Templeman, R. F.** (1981) Using remotely sensed data on a Univac 1108 Computer. *Computer Applications*, **7**: 994–1007.

**Gurney, C. M.** and **Townshend, J. R. G.** (1983) The use of contextural information in the classification of remotely sensed data. *Photogrammetric Engineering and Remote Sensing*, **49**: 55–64.

**Gustafson, G. C.** (1980) Standoff reconnaissance imagery: applications and interpreter training. *Society of Photo-Optical Instrumentation Engineers*, **242**: 108–18.

**Gustafson, G. C.** (1982) Applications of aerial imaging technology to environmental problems: *In* Frazier, J. W. (ed.) *Applied Geography: Selected Perspectives.* Prentice-Hall Inc. New York: 170–96.

**Guyenne, T. O.** and **Levy, G.** (1981) *Coherent and Incoherent Radar Scattering from Rough Surfaces and Vegetated Areas.* European Space Agency Special Report **166**, Paris.

**Hagen, R.** and **Meyer, M.** (1979) *Vegetation Analysis of Red Lake, Minnesota Peatlands by Remote Sensing Methods.* University of Minnesota Remote Sensing

Laboratory, Institute of Agriculture Forestry and Home Economics Research Report, 79–2.

**Hajic, E.J.** and **Simonett, D. S.** (1976) Comparisons of qualitative and quantitative image analysis: *In* Lintz, J. and Simonett, D. S. (eds) *Remote Sensing of Enronment.* Addison-Wesley, Reading, Massachusetts, London, pp. 374–411.

**Hall, D. K., Ormsby, J. P., Johnson, L.** and **Brown, J.** (1980) Landsat digital analysis of the initial recovery of burned tundra at Kokolik River, Alaska. *Remote Sensing of Environment,* **10**: 263–72.

**Hall, R. T.** and **Rothrock, D. A.** (1981) Sea ice displacement from Seasat synthetic aperture radar. *Journal of Geophysical Research,* **86**: 1178–82.

**Hamilton, M. G.** (1981) GOES 10 Image. *Weather,* **36**: 21–2.

**Hamlin, C. L.** (1980) The temporal dimension, monitoring the changing ecology of settlement in Turan. *Expedition,* **22**: 42–7.

**Handley, J. F.** (1980) The application of remote sensing to environmental management. *International Journal of Remote Sensing,* **1**: 181–95.

**Hannover, G. K.** (1981) The European remote sensing program. *Nachrichten aus dem Karten und Vermessungswesen,* **39**: 5–39.

**Haralick, R. M.** (1976) Automatic remote sensor image processing: *In* Rosenfeld A. (ed.) *Digital Picture Analysis.* Springer-Verlag, Berlin, New York, pp. 5–63.

**Haralick, R. M.** (1977) Interactive image processing software: *In* Simon, J. C. and Rosenfeld, A. (eds) *Digital Image Processing and Analysis.* Noordhoff, Leyden.

**Haralick, R. M., Caspall, F.** and **Simonett, D. S.** (1970). Using radar imagery for crop discrimination: A statistical and conditional probability study. *Remote Sensing of Environment,* **1**: 131–42.

**Haralick, R. M., Shanmugam, K. S.** and **Dinstein, I.** (1973) Textural features for image classification. *IEEE Transactions on Systems, Man and Cybernetics,* **3**: 610–21.

**Haralick, R. M.** and **Shanmugam, K. S.** (1974) Combined spectral and spatial processing of ERTS imagery data. *Remote Sensing of Environment,* **3**: 3–13.

**Hardy, J. R.** (1978) Methods and accuracy of Landsat MSS points on maps. *Journal of the British Interplanetary Society,* **31**: 305–11.

**Hardy, N. E.** (1981) A photo interpretation approach to forest regrowth monitoring using side looking airborne radar – Grant County, Oregon. *International Journal of Remote Sensing,* **2**: 135–44.

**Harger, R. O.** (1970) *Synthetic Aperture Radar Systems, Theory and Design.* Academic Press, New York.

**Harnett, P. R., Mountain, G. D.** and **Barnett, M. E.** (1978) Spatial filtering applied to remote sensing images. *Optica Acta,* **25**: 801–9.

**Harris, R.** (1977) Automatic analysis of meteorological satellite imagery: *In* Thomas J. O. and Davey, P. G. (eds) *Texture Analysis.* British Pattern Recognition Association and Remote Sensing Society, Reading, pp. 45–72.

**Harris, R.** (1980) Spectral and spatial image processing for remote sensing. *International Journal of Remote Sensing,* **1**: 361–75.

**Harris, R.** and **Barrett, E. C.** (1978) Towards an objective nephanalysis. *Journal of Applied Meteorology,* **17**: 1258–66.

**Hartl, P.** (1976) Digital picture processing: *In* Schanda, E. (ed.) *Remote Sensing for Environmental Sciences.* Springer Verlag, Berlin/New York; Chapman and Hall, London, pp. 304–50.

**Hatfield, J. L.** (1979) Canopy temperatures: the usefulness and reliability of remote measurements. *Agronomy Journal,* **71**: 889–92.

**Hatfield, J. L., Millard, J. P.** and **Goettelman, R. C.** (1982) Variability of surface temperature in agricultural fields of central California. *Photogrammetric Engineering and Remote Sensing,* **48**: 1319–25.

**Hay, A.** (1979) Sampling design to test land use map accuracy. *Photogrammetric Engineering and Remote Sensing,* **45**: 529–33.

**Heilman, J. L., Kanemasu, E. T., Rosenberg, N. J.** and **Blad, B. L.** (1976) Thermal scanner measurement of canopy temperatures to estimate evaporation. *Remote Sensing of Environment,* **5**: 137–45.

**Heilman, J. L.** and **Moore, D. G.** (1981) HCMM detection of high soil moisture areas. *Remote Sensing of Environment,* **11**: 75–6.

**Heintz, T. W., Lewis, J. K.** and **Walker, S. S.** (1979) Low level aerial photography as a management and research tool for range inventory. *Journal of Range Management,* **32**: 247–9.

**Heiss, K. P., Sand, F. M.** and **Farley, D. E.** (1981) Economic benefits of improved crop information on wheat and all cereals for European countries: *In* Berg, A.

(ed.) *Application of Remote Sensing to Agricultural Production Forecasting*. A. A. Balkema, Rotterdam, pp. 83–102.

**Helgeson, G. A.** (1970) Water depth and distance penetration. *Photogrammetric Engineering*, **26**: 164–72.

**Hellden, U., Olsson, L.** and **Stern, M.** (1982) Approaches to desertification monitoring in the Sudan: *In* Longdon, N. and Levy, G. (eds). *Satellite Remote Sensing in Developing Countries*. Proceedings of an EARSeL-ESA Symposium. Report Number ESA-SP-175. European Space Agency, Paris, pp. 131–44.

**Heller, R. C.** and **Johnson, K. A.** (1979) Estimating irrigated land acreage from Landsat imagery. *Photogrammetric Engineering and Remote Sensing*, **45**: 1379–82.

**Henderson, F. M.** (1975) Radar for small scale land use mapping. *Photogrammetric Engineering and Remote Sensing*, **41**: 307–19.

**Henderson, F. M.** (1982) An evaluation of Seasat SAR imagery for urban analysis. *Remote Sensing of Environment*, **12**: 439–61.

**Henderson, F. M.** and **Merchant, J. W.** (1978) Microwave remote sensing: *In* Richason, B. F. (ed.) *Introduction to Remote Sensing of the Environment*. Kendall Hunt, Iowa.

**Henderson, F. M.** and **Wharton, S. W.** (1980) Seasat SAR identification of dry climate urban land cover. *International Journal of Remote Sensing*, **1**: 293–304.

**Henneberry, T. J., Hart, W. G., Bariola, L. A., Kittock, D. L., Arle, H. F., Davis, M. R.** and **Ingle, S. J.** (1979) Parameters of cotton cultivation from infrared aerial photography. *Photogrammetric Engineering and Remote Sensing*, **45**: 1129–33.

**Hielkema, J. U.** (1977) Desert locust habitat monitoring with satellite remote sensing. *ITC Journal*: 387–417.

**Hirsch, S. N., Kruckeberg, R. F.** and **Madden, F. M.** (1971) The bi-spectral forest fire detection system. *Proceedings of the 7th International Symposium on Remote Sensing of Environment*. University of Michigan, Ann Arbor, pp. 2253–72.

**Hixson, M., Schulz, D., Nancy, F.** and **Akiyama, T.** (1980) Evaluation of several schemes for classification of remotely sensed data. *Photogrammetric Engineering and Remote Sensing*, **46**: 1547–53.

**Hixson, M. M., Davis, B. J.** and **Bauer, M. E.** (1981) Sampling Landsat classification for crop area estimation. *Photogrammetric Engineering and Remote Sensing*, **47**: 1343–8.

**Hoffer, R. M.** (1978) Biological and physical considerations in applying computer-aided analysis techniques to remote sensor data: *In* Swain, P. H and Davis, S. M. (eds) *Remote Sensing the Quantitative Approach*. McGraw Hill, pp. 227–89.

**Hoffer, R. M.** and **Johannsen, C. J.** (1969). Ecological potentials in spectral signature analysis: *In* Johnson, P. L. (ed.) *Remote Sensing in Ecology*. University of Georgia Press, Athens, pp. 1–16.

**Holdermann, F., Bohner, M., Bargel, B.** and **Kazmierczak, H.** (1978) Review of automatic image processing. *Photogrammetria*, **34**: 225–58.

**Holmes, A. L.** (1983) Shuttle imaging radar-A information and data availability. *Photogrammetric Engineering and Remote Sensing*, **49**: 65–6.

**Holt, B.** (1981) Availability of Seasat synthetic aperture radar imagery. *Remote Sensing of Environment*, **11:** 413–7.

**Holz, R. K.** (1973) Introduction: *In* Holz, R. K. (ed.) *The Surveillant Science, Remote Sensing of Environment*. Houghton Mifflin Co, Boston, pp. v–vii.

**Hong, J. K.** and **Ilsaka, J.** (1982) Coastal environment change analysis by Landsat MSS data. *Remote Sensing of Environment*, **12**: 107–16.

**Hord, R. M.** and **Brooner, W.** (1976) Land use map accuracy criteria. *Photogrammetric Engineering and Remote Sensing*, **42**: 301–7.

**Horder, A.** (ed.) (1976) *The Manual of Photography* (6th edn). Focal Press, London.

**Howard, J. A.** (1970) *Aerial Photo Ecology*. Faber and Faber, London.

**Howarth, P. J.** and **Wickware, G. M.** (1981) Procedures for change detection using Landsat digital data. *International Journal of Remote Sensing*, **2**: 277–91.

**Hsu, S. Y.** (1971) Population estimation. *Photogrammetric Engineering*, **37**: 449–54.

**Hsu, S.** (1978) Texture-tone analysis for automated land use mapping. *Photogrammetric Engineering and Remote Sensing*, **44**: 1393–404.

**Hughes, N. A.** and **Henderson-Sellers, A.** (1982) System albedo as sensed by satellites: its definition and variability. *International Journal of Remote Sensing*, **3**: 1–11.

**Huh, O. K.** and **Di Rosa, D.** (1981) Analysis and interpretation of TIROS-N AVHRR infrared imagery, western Gulf of Mexico. *Remote Sensing of Environment*, **11**: 371–83.

**Hung, R. J.** and **Smith, R. E.** (1982) Remote sensing of tornadic storms from geosynchronous satellite infrared digital data. *International Journal of Remote Sensing*, **3**: 69–81.

**Hunt, G. E., Saunders, R. W., Rumball, D. A.** and **Marriage, N.** (1981) Some quantitative measurements from geostationary satellites. *Weather*, **36**: 96–104.

**Hunter, G. T.** and **Bird, S. J. G.** (1970) Critical terrain analysis. *Photogrammetric Engineering*, **36**: 939–52.

**Hutchinson, C. F.** (1982) Techniques for combining Landsat and ancillary data for digital classification improvement. *Photogrammetric Engineering and Remote Sensing*, **48**: 123–30.

**Idso, S. B., Jackson, R. D.** and **Reginato, R. J.** (1976) Compensating for environmental variability in the thermal inertia approach to remote sensing of soil moisture. *Journal of Applied Meteorology*, **15**: 811–7.

**Ilsaka, J.** and **Hegedus, E.** (1982) Population estimation from Landsat imagery. *Remote Sensing of Environment*, **12**: 259–72.

**Imhuff, M. L., Petersen, G. W., Sykes, S. G.** and **Irons, J. R.** (1982) Digital overlay of cartographic information on Landsat MSS data for soil surveys. *Photogrammetric Engineering and Remote Sensing*, **48**: 1337–42.

**Ince, M.** (1981) *Space*. Sphere Books Ltd, London.

**Inkster, D. R., Raney, R. K.** and **Rawson, R. F.** (1979) State-of-the-art in airborne imaging radar. *Proceedings of the 13th International Symposium on Remote Sensing of Environment*. University of Michigan, Ann Arbor, pp. 361–81.

**Inkster, D. R., Lowry, R. T.** and **Thompson, M. D.** (1980) Optimum radar resolution studies for land use and forestry applications. *Proceedings of the 14th International Symposium on Remote Sensing of Environment*. University of Michigan, Ann Arbor, pp. 865–83.

**Iranpanah, A.** and **Esfaniari, B.** (1980) Interpretation of structural lineaments using Landsat 1 images. *Photogrammetric Engineering and Remote Sensing*, **46**: 225–32.

**Irons, J. R.** and **Petersen, G. W.** (1981) Texture transforms of remote sensing data. *Remote Sensing of Environment*, **11**: 359–70.

**Ito, H.** and **Muller, F.** (1982) Ice movement through Smith Sound in northern Baffin Bay, Canada, observed in satellite imagery. *Journal of Glaciology*, **28**: 129–43.

**Jackson, M. J., Carter, P., Smith, T. F.** and **Gardner, W. G.** (1980). Urban land mapping from remotely sensed data. *Photogrammetric Engineering and Remote Sensing*, **46**: 1041–50.

**Jackson, R. D., Reginato, R. S.** and **Idso, S. B.** (1976) Timing of ground truth acquisition during remote sensing assessment of soil water content. *Remote Sensing of Environment*, **4**: 249–56.

**Jackson, R. D., Pinter, P. J., Idso, S. B.** and **Reginato, R. J.** (1979) Wheat spectral reflectance: interactions between crop configuration, Sun elevation and azimuth angle. *Applied Optics*, **18**: 3730–2.

**James, T. H.** (1977) *Theory of the Photographic Process* (6th edn). Macmillan.

**Janza, F. J.** (1975) Interaction mechanisms: *In* Reeves, R. G. (ed.) *Manual of Remote Sensing*. American Society of Photogrammetry, Falls Church, Virginia, pp. 75–179.

**Jasentuliyana, N.** and **Lee, R. S.** (eds) (1979) *Manual on Space Law*. Oceana, New York, pp. 303–46.

**Jensen, H., Graham, L. C., Porcello, L. J.** and **Leith, E. N.** (1977) Side-looking airborne radar. *Scientific American*, **237**: 84–95.

**Jensen, J. R.** (1981) Urban change detection mapping using Landsat digital data. *The American Cartographer*, **8**: 127–47.

**Jensen, J. R.**, (1983) Biophysical remote sensing. *Annals of the Association of American Geographers*, **73**: 111–32

**Jensen, J. R.** and **Hodgson, M. E.** (1983) Remote sensing brightness maps. *Photogrammetric Engineering and Remote Sensing*, **49**: 93–102.

**Jensen, J. R.** and **Dahlberg, R. E.** (1983) Status and content of remote sensing education in the United States. *International Journal of Remote Sensing*, **4**: 235–45.

**Jensen, N.** (1968) *Optical and Photographic Reconnaissance Systems*. Wiley, New York, London.

**Johnston, R. J.** (1980) *Multivariate Statistical Analysis in Geography*. Longman, London, New York.

**Jones, A. D.** (1971) Film emulsions: *In* Goodier, R. (ed.) *The Application of Aerial Photography to the Work of the Nature Conservancy*. Natural Environment Research Council, Swindon, pp. 120–32.

**Jones, R. V.** (1978) *Most Secret War*. Hamish Hamilton, London.

**Judd, D. B.** (1967) Terms, definitions and symbols in reflectrometry. *Journal of the Optical Society of America*, **57**: 445–52.

**Justice, C. O.** and **Townshend, J. R. G.** (1981) Integrating ground data with remote sensing: *In* Townshend, J. R. G. (ed.) *Terrain Analysis and Remote Sensing*. Allen and Unwin London, Boston, pp. 38–58.

**Justice, C.** and **Townshend, J. R.G.** (1982) A comparison of unsupervised classification procedures on Landsat MSS Data for an area of complex surface conditions in Basilicata, southern Italy. *Remote Sensing of Environment*, **12**: 407–20.

**Kahle, A. B.** (1980) Surface thermal properties: *In* Siegal, B. S. and Gillespie, A. R. (eds) *Remote Sensing in Geology*. Wiley, New York, pp. 257–73.

**Kahle, A. B., Gillespie, A. R.** and **Goetz, A. F. H.** (1976) Thermal inertia imaging – a new geologic mapping tool. *Geophysical Research Letters*, **3**: 26–8.

**Kahle, A. B., Schieldge, J. P., Abrams, M. A., Alley, R. E.** and **Levine, C. J.** (1981) *Geologic Application of Thermal Inertia Imaging Using HCMM Data*. Jet Propulsion Laboratory Publication 81–55. National Aeronautics and Space Administration, Washington, DC.

**Kalma, J. D., Byrne, G. F., Johnson, M. E.** and **Laughlin, G. P.** (1983) Frost mapping in Southern Victoria: an assessment of HCMM thermal imagery. *Journal of Climatology*, **3**: 1–19.

**Kaltenecker, H.** and **Lafferranderie, G.** (1977) Thoughts on the legal aspects of remote sensing of the Earth by satellites: *In* Barrett, E. C. and Curtis, L. F. (eds) *Environmental Remote Sensing 2: Practices and Problems*. Edward Arnold, pp. 72–80.

**Kalush, R. J.** (1979) The problem of resolution in the Landsat imagery. *Remote Sensing Quarterly*, **1**: 38–48.

**Kauth, R. G.** and **Thomas, G. S.** (1976) The tasseled cap, a graphic description of the spectral-temporal development of agricultural crops as seen by Landsat. *Proceedings of the Symposium on Machine Processing of Remotely Sensed Data*. University of Purdue, Indiana, pp. 6.23–7.20.

**Kayan, I.** and **Klemas, V.** (1978) Application of Landsat imagery to studies of structural geology and geomorphology of the Mentese region of South West Turkey. *Remote Sensing of Environment*, **7**: 61–72.

**Khorram, S.** (1981) Use of ocean color scanner data in water quality mapping. *Photogrammetric Engineering and Remote Sensing*, **47**: 667–76.

**Kidson, C.** and **Manton, M. M.** (1973) Assessment of coastal change with the aid of photogrammetric and computer-aided techniques. *Estuarine and Coastal Marine Science*, **1**: 271–83.

**Kilford, W. K.** (1975) *Elementary Air Survey* (3rd edn). Pitman, London, New York.

**Killpack, D. P.** and **McCoy, R. M.** (1981) An application of Landsat derived data to a regional hydrologic model. *Remote Sensing Quarterly*, **3**: 27–33.

**Kim, C. E.** and **Strintzis, M. G.** (1980) High speed multidimensional convolution. *IEEE Transactions on Pattern Analysis and Machine Intelligence*, **2**: 269–73.

**Kimes, D. S.** (1983) Remote sensing of row crop structure and component temperatures using directional radiometric temperatures and inversion techniques. *Remote Sensing of Environment*, **13**: 33–55.

**Kimes, D. S., Idso, S. B., Pinter, P. J., Reginato, R. J.** and **Jackson, R. D.** (1980). View angle effects in the radiometric measurement of plant canopy temperatures. *Remote Sensing of Environment*, **10**: 273–84.

**Kimes, D. S., Smith, J. A.** and **Link, L. E.** (1981) Thermal I.R. exitance model of a plant canopy. *Applied Optics*, **20**: 623–32.

**Kindelan, M., Moreno, V.** and **Valverde, A.** (1981) Geometric correction of airborne multispectral scanner images. *Proceedings of the 15th International Symposium on Remote Sensing of Environment*. University of Michigan, Ann Arbor, pp. 1539–49.

**King, R. B.** (1981) An evaluation of Landsat 3 RBV imagery for obtaining environmental information in Tanzania: *In* Allan, J. A. (ed.) *Matching Remote Sensing Technologies and their Applications*. Remote Sensing Society Reading, pp. 85–95.

**King-Hele, D. G., Pilkington, J. A., Hiller, H.** and **Walker, D. M. C.** (1981) *The RAE Table of Earth Satellites*. Macmillan Press Ltd.

**Klaus, S. P., Estes, J. E., Atwater, S. G., Jensen, J. R.** and **Uollmers, R. R.** (1977) Radar detection of surface oil slicks. *Photogrammetric Engineering and Remote Sensing*, **43**: 1523–31.

**Klemas, V.** (1980) Remote sensing of coastal fronts and their effects on oil dispersion. *International Journal of Remote Sensing*, **1**: 11–28.

**Klemas, V.**, **Bartlett, D.** and **Rodgers, R.** (1975) Coastal zone classification from satellite imagery. *Photogrammetric Engineering and Remote Sensing*, **41**: 499–513.

**Klemas, V.** and **Philpot, W. D.** (1981) Drift and dispersion studies of ocean dumped waste using Landsat imagery and current drogues. *Photogrammetric Engineering and Remote Sensing*, **47**: 533–42.

**Knapp, E. M.** and **Rider, D.** (1979) Automated geographic information systems and Landsat data: A survey. *Computer Mapping in Natural Resources and Environment*. Harvard Library of Computer Graphics, Harvard University, pp. 57–68.

**Koopmans, B. N.** (1973) Drainage analysis on radar images. *ITC Journal*: 464–79.

**Koopmans, B. N.** (1982) Some comparative aspects of SLAR and airphoto images for geomorphologic and geological investigations. *ITC Journal*: 330–7.

**Kozak, R. C.**, **Berlin, G. L.** and **Chavez, P. S.** (1981) Seasat radar image of the Phoenix, Arizona region. *International Journal Remote Sensing*, **2**: 295–8.

**Kozma, A.**, **Leith, E. N.** and **Massey, N. G.** (1972) Tilted plane optical processor. *Applied Optics*, **11**: 1766–77.

**Kuilenberg, J. van** (1975) Radar observations of controlled oil spills. *Proceedings of the 10th International Symposium on Remote Sensing of Environment*. University of Michigan, Ann Arbor, pp. 243–50.

**Landgrebe, D. A.** (1976) Machine processing of remotely acquired data. *In* Lintz, J. and Simonett, D. S. (eds) *Remote Sensing of Environment*. Addison-Wesley, Reading, Massachusetts; London, pp. 349–73.

**Landgrebe, D. A.** (1980) The development of a spectral-spatial classifier for Earth observational data. *Pattern Recognition*, **12**: 165–75.

**Lang, R. H.** and **Sidhu, J. S.** (1983) Electromagnetic backscattering from a layer of vegetation: a discrete approach. *IEEE Transactions on Geoscience and Remote Sensing*, **21**: 62–71.

**Lansing, J. C.** and **Cline, R. W.** (1975) The four and five band multispectral scanners for Landsat. *Optical Engineering*, **14**: 312–22.

**La Riccia, M. P.** and **Rauch, H. W.** (1977) Water well productivity related to photo-lineaments in carbonates of Frederik Valley, Maryland: *In* Dilmarker, R. R. and Csallany, S. C. (eds) *Hydrologic Problems in Karst Regions*. Western Kentucky University, pp. 228–34.

**Lasker, R.**, **Peláen, J.** and **Laurs, R. M.** (1981) The use of satellite infrared imagery for describing ocean processes in relation to spawning of the northern Anchovy (*Engraulis mordax*). *Remote Sensing of Environment*, **11**: 439–53.

**de Latil, P.** (1961) Colour photography as an instrument of scientific observation and measurement. *Camera*, **40**: 29–47.

**Leberl, F.** (1978) *Radargrammetry for Image Interpreters* (2nd edn). International Institute for Aerial Survey and Earth Sciences, Enschede, Netherlands.

**Leberl, F.** (1979) Accuracy analysis of stereo side looking radar. *Photogrammetric Engineering and Remote Sensing*, **45**: 1083–96.

**Lee, J.** (1981) Speckle analysis and smoothing of synthetic aperture radar images. *Computer Graphics and Image Processing*, **17**: 24–32.

**Lee, Y. J.**, **Towler, E.**, **Brodatsch, H.** and **Findings, S.** (1977) Computer assisted forest land classification by means of several classification methods on the CCRS Image – 100. *Proceedings of the 4th Canadian Symposium on Remote Sensing*. Canadian Aeronautical and Space Institute, Ottawa, pp. 37–46.

**Lee, K.** and **Weimer, R. J.** (1975) *Geological Interpretation of Skylab Photographs*. Remote sensing report 75–6. NASA Contract NAS 9–13394, Colorado School of Mines.

**Legeckis, R.**, **Legg, E.** and **Limeburner, R.** (1980) Comparison of polar and geostationary satellite infrared observations of sea surface temperatures in the Gulf of Maine. *Remote Sensing of Environment*, **9**: 339–50.

**Lewis, A. J.** (1974) Geomorphic-geologic mapping from remote sensors: *In* Estes, J. E. and Senger, W. (eds) *Remote Sensing Techniques in Environmental Analysis*. Hamilton, Santa Barbara, California, pp. 105–26.

**Lewis, A. J.** and **Waite, W. P.** (1973) Radar shadow frequency. *Photogrammetric Engineering*, **38**: 189–96.

**Lillesand, T. M.** and **Kieffer R. W.** (1979) *Remote Sensing and Image Interpretation*. Wiley, New York.

**Lindgren, D. T.** (1971) Dwelling unit estimation with colour I.R. photos. *Photogrammetric Engineering*, **37**: 373–7.

**Lins, H. F.** (1979) Some legal considerations in remote sensing. *Photogrammetric Engineering and Remote Sensing*, **45**: 741–48.

**Lintz, J., Brennan, P. A.** and **Chapman, P. R.** (1976) Ground truth and mission operations: *In* Lintz, J. and Simonett, D. S. (eds) *Remote Sensing of Environment*. Addison-Wesley, Reading, Massachusetts; London, pp. 412–37.

**Lo, C. P.** (1976) *Geographical Applications of Aerial Photography*. David and Charles, London; Crane, Russak and Company Inc, New York.

**Lo, C. P.** (1981) Land use mapping of Hong Kong from Landsat images. An evaluation. *International Journal of Remote Sensing*, **2**: 231–52.

**Lo, C. P.** and **Chan, H. F.** (1980) Rural population estimation from aerial photographs. *Photogrammetric Engineering and Remote Sensing*, **43**: 337–45.

**Lodge, D. W. S.** (1981) The Seasat-1 synthetic aperture radar: introduction, data reception and processing: *In* Cracknell, A. P. (ed.) *Remote Sensing in Meteorology, Oceanography and Hydrology*. Ellis Horwood Ltd, Chichester; Wiley, New York, pp. 335–56.

**Lodge, D. W. S.** (1983) Surface expressions of bathymetry on Seasat synthetic aperture radar images, *International Journal of Remote Sensing*, **4**: 639–53.

**Logan, T. L., Strahler, A. H.** and **Woodcock, C. E.** (1979) Use of a standard deviation based texture channel for Landsat classification of forest strata. *Machine Processing of Remotely Sensed Data Symposium*. Purdue University, Indiana, pp. 395–404.

**de Loor, G. P.** (1976) Radar methods: *In* Schanda, E. (ed.) *Remote Sensing for Environmental Sciences*. Springer-Verlag, New York; Chapman and Hall, London, pp. 147–86.

**de Loor, G. P., Jurriëns A. A.** and **Gravesteijn, H.** (1974) The radar backscatter from selected agricultural crops. *IEEE Transactions on Geoscience Electronics*, **12**: 70–4.

**Lovell, B.** (1977) The costs and benefits of space observations: *In* Peel, R. F., Curtis, L. F. and Barrett, E. C. (eds) *Remote Sensing of the Terrestrial Environment*. Butterworths, London, Boston, pp. 1–12.

**Lowe, D. S.** (1976) Nonphotographic optical sensors: *In* Lintz, J. and Simonett, D. S. (eds) *Remote Sensing of Environment*. Addison-Wesley, Reading, Massachusetts; London, pp. 155–93.

**Lowe, D.** (1980) Acquisition of remotely sensed data: *In* Siegal, B. S. and Gillespie, A. R. (eds). *Remote Sensing in Geology*. Wiley, New York, pp. 47–90.

**Lowe, D. S., Kelley, B. O., McDevitt, H. I., Orr, G. T.** and **Yates, H. W.** (1975) Imaging and non imaging sensors: *In* Reeves, R. (ed.) *Manual of Remote Sensing*. American Society of Photogrammetry, Falls Church, Virginia, pp. 369–97.

**Lowman, P. D.** (1965) Space photography – A review. *Photogrammetric Engineering*, **31**: 76–86.

**Lowman, P. D.** (1969) Geologic orbital photography, experience from the Gemini program. *Photogrammetria*, **24**: 77–106.

**Lowman, P. D.** (1980) The evolution of geological space photography: *In* Siegal, B. S. and Gillespie, A. R. (eds) *Remote Sensing in Geology*. Wiley, New York, pp. 91–115.

**Lulla, K.** (1983) The Landsat satellites and selected aspects of physical geography. *Progress in Physical Geography*, **7**: 1–45.

**Lyon, J. G.** and **McCarthy, J. F.** (1981) Seasat imagery for detection of coastal wetlands. *Proceedings of the 15th International Symposium on Remote Sensing of Environment*. University of Michigan, Ann Arbor, pp. 1475–85.

**MacDonald, H. C.** (1969) Geologic evaluation of radar imagery from Darien Province, Panama. *Modern Geology*, **1**: 1–63.

**MacDonald, H. C.** (1980) Techniques and applications of imaging radars: *In* Siegal, B. S. and Gillespie, A. R. (eds) *Remote Sensing in Geology*. Wiley, New York, pp. 297–336.

**MacDonald, H. C.** and **Waite, W. P.** (1971) Soil moisture detection with imaging radar. *Water Resources Research Journal*, **7**: 100–9.

**MacDonald, H. C.** and **Waite, W. P.** (1973) Imaging radars provide terrain texture and roughness parameters in semi-arid environments. *Modern Geology*, **4**: 145–58.

**MacDonald, R. B.** (ed.) (1979) *The LACIE Symposium*. Lyndon B. Johnson Space Center, National Aeronautics and Atmospheric Administration, JSC 16015, vols. I and II, Washington DC.

**Maddox, R. A.** and **Reynolds, D. W.** (1980) GOES satellite data maps areas of extreme cold in Colorado. *Monthly Weather Review*, **108**: 116–18.

**Maes, J., Gombeer, R.** and **O'Moor, J.** (1982) Multitemporal soil and vegetation observations by Meteosat over central Africa: *In* Longdon, N. and Levy, G. (eds) *Satellite Remote Sensing in Developing Countries*. Proceedings of an EARSeL-ESA Symposium. Report Number ESA-SP-175. European Space Agency, Paris, pp. 87–94.

**Maggio, R. C., Baker, R. D.** and **Harris, M. K.** (1983) A geographic data base for Texas Pecan. *Photogrammetric Engineering and Remote Sensing*, **49**: 47–52.

**Maher, R. V., Boyd, M.** and **Langford, G.** (1980) Integration of geographic information and Landsat imagery: a case study of vegetation mapping in the white goat wilderness area, Alberta. *Canadian Journal of Remote Sensing*, **6**: 86–92.

**Marble, B. F.** and **Peuquet, D. J.** (1983) Geographic information systems and remote sensing: In Colwell, R. N. (ed.) *Manual of Remote Sensing* (2nd edn). American Society of Photogrammetry, Falls Church, Virginia, pp. 923–58.

**Marmelstein, A D.** (1978) Remote sensing applications to wildlife management in the U.S. Fish and Wildlife Service. *Proceedings of the 12th International Symposium on Remote Sensing of Environment*. University of Michigan, Ann Arbor, pp. 2315–21.

**Marshall, J. R.** and **Meyer, M. P.** (1978) Field evaluation of small scale forest resource aerial photography. *Photogrammetric Engineering and Remote Sensing*, **44**: 37–42.

**Mason, B. D.** (1981) Meteostat – Europe's contribution to the global weather observing system: *In* Cracknell, A. P. (ed.). *Remote Sensing in Meteorology, Oceanography and Hydrology*. Ellis Horwood Ltd., Chichester and Halsted Press, pp. 56–65.

**Mather, P. M.** (1980a) Accessing Landsat MSS data on ICL 1900 series computers I: Fuchino CCTs. *Computer Applications*, **6**: 873–92.

**Mather, P. M.** (1980b) Accessing Landsat MSS data on ICL 1900 series computers II: USGS EDIPs, CCTs. *Computer Applications*, **6**: 903–19.

**Mather, P. M.** (1981) Use of digital Landsat MSS data to determine patterns of suspended sediment concentration in coastal waters. *Computer Applications*, **7**: 1051–64.

**Matte, N. M.** and **DeSaussure, H.** (eds) (1972) *Legal Implications of Remote Sensing from Outer Space*. Sijthoff-Leyden, A. W.

**Mattie, M. G., Lichy, D. E.** and **Beal, R. C.** (1980) Seasonal detection of waves, currents and inlet discharges. *International Journal of Remote Sensing*, **1**: 377–98.

**Mayer, K. E.** and **Fox, L.** (1981) Identification of conifer species groupings from Landsat digital classifications. *Photogrammetric Engineering and Remote Sensing*, **47**: 1607–14.

**Meijerink, A. H. J.** and **Donker, N. H. W.** (1978) The ITC approach to digital processing applied to land use mapping in the Himalayas and Central Java: *In* Collins, W. G. and van Genderen, J. L. (eds). *Remote Sensing Applications in Developing Countries*, Remote Sensing Society, Reading, pp. 75–83.

**Merchant, J. W.** (1982) Employing Landsat MSS data in land use mapping: observations and considerations: *In* Richason, B. F. (ed.) *Remote Sensing: As an Input to Geographic Information Systems in the 1980s*. Pecora VII Symposium, United States Geological Survey, Sioux Falls, pp. 71–91.

**Merson, R. H.** (1983) A composite Landsat image of the United Kingdom. *International Journal of Remote Sensing*, **4**: 521–7.

**McCoy, R. M.** and **Lewis, A. J.** (1976) Use of radar in hydrology and geomorphology. *Remote Sensing of the Electromagnetic Spectrum*, **3**: 105–22.

**McCullagh, M. J.** (1981) The prediction of visibility by optical techniques: an area amenity study. *Computer Applications*, **7**: 1008–130.

**McKenzie, R. L.** and **Nisbet, R. M.** (1982) Applicability of satellite derived sea surface temperatures in the Fiji region. *Remote Sensing of Environment*, **12**: 349–61.

**Millard, J. P., Reginato, R. J., Goettelman, R. C., Idso, S. B., Jackson, R. D.** and **Leroy, M. J.** (1980) Experimental relations between airborne and ground measured wheat canopy temperatures. *Photogrammetric Engineering and Remote Sensing*, **46**: 221–4.

**Millard, J. P., Goettelman, R. C.** and **LeRoy, M. J.** (1981) Infrared temperature variability in a large agricultural field. *International Journal of Remote Sensing*, **2**: 201–11.

**Miller, J. R., Jain, S. C., O'Neill, N. T., McNeil, W. R.** and **Thomson, K. P. B.** (1977) Interpretation of airborne spectral reflectance measurements over Georgian bay. *Remote Sensing of Environment*, **6**: 183–200.

**Milton, E. J.** (1980) A portable multiband radiometer for ground data collection in remote sensing. *International Journal of Remote Sensing*, **1**: 153–65.

**Missallati, A., Prelat, A. E.** and **Lyon, R. J. P.** (1979) Simultaneous use of geological, geophysical and Landsat digital data in uranium exploration. *Remote Sensing of Environment*, **8**: 189–210.

**Mitchell, C. W.** (1981) Soil degradation mapping from Landsat imagery in north Africa and the Middle East: *In* Allan, J. A. and Bradshaw, M. (eds). *Geological and Terrain Analysis Studies by Remote Sensing*. Remote Sensing Society, Reading, pp. 49–69.

**Mitchell, W. B., Guptill, S. C., Anderson, K. E., Fegeas, R. G.** and **Hallam, C. A.** (1977) *GIRAS: A Geographic Information Retrieval and Analysis System for Handling Land Use and Land Cover Data*. United States Geological Survey Professional Paper, **1059**, Washington DC.

**Moik, J. G.** (1980) *Digital Processing of Remotely Sensed Images*. National Aeronautics and Space Administration, Washington DC.

**Moore, E. G.** and **Wellar, B. S.** (1969) Urban data collection by airborne sensors. *Journal American Institute of Planners*, **35**: 35–43.

**Moore, R. K.** (1975) Microwave remote sensors: *In* Reeves, R. G. (ed.) *Manual of Remote Sensing*. American Society of Photogrammetry, Falls Church, Virginia, pp. 399–538.

**Moore, R. K.** (1976) Active microwave systems: *In* Lintz, J. and Simonett, D. S. (eds.). *Remote Sensing of Environment*. Addison-Wesley, Reading, Massachusetts; London, pp. 234–90.

**Moore, R. K.** (1983) Imaging radar systems: *In* Colwell, R. N. (ed.) *Manual of Remote Sensing* (2nd edn). American Society of Photogrammetry, Falls Church, Virginia, pp. 429–74.

**Moore, R. K.** and **Thomann, G. C.** (1971) Imaging radars for geoscience use. *IEEE Transactions on Geoscience Electronics*, **9**: 156–64.

**Morain, S. A.** (1976) Use of radar for vegetation analysis. *Remote Sensing of the Electromagnetic Spectrum*, **3**: 61–78.

**Morain, S. A.** and **Simonett, D. S.** (1967) K band radar in vegetation mapping. *Photogrammetric Engineering*, **33**: 730–40.

**Mott, P. G.** (1966) Some aspects of colour aerial photography in practice and its applications. *Photogrammetric Record*, **5**: 221–37.

**Mulder, N. J.** (1980) A view on digital image processing. *ITC Journal*: 452–77.

**Mumbower, L. E** and **Donoghue, J.** (1967) Urban poverty study. *Photogrammetric Engineering*, **33**: 610–8.

**Murtha, P. A.** (1983) Some air photo scale effects on Douglas-fir damage type interpretation. *Photogrammetric Engineering and Remote Sensing*, **49**: 327–35.

**Myers, V. I.** (1975) Crops and soils: *In* Reeves, R. G. (ed.) *Manual of Remote Sensing*. American Society of Photogrammetry, Falls Church, Virginia, pp. 1715–813.

**Naraghi, M., Stromberg, W.** and **Daily, M.** (1983) Geometric rectification of radar imagery using digital elevation models. *Photogrammetric Engineering and Remote Sensing*, **49**: 195–9.

**NASA** (1967a) *Gemini Summary Conference*. National Aeronautics and Space Administration Report, SP-138, Washington DC.

**NASA** (1967b) *Earth Photographs from Gemini III, IV and V*. National Aeronautics and Space Administration Report SP-129, Washington DC.

**NASA** (1968b) *Earth Photographs from Gemini VI through XII*. National Aeronautics and Space Administration Report SP-171, Washington DC.

**NASA** (1976) *Landsat Data Users Handbook*. NASA document 76SDS4258 Goddard Space Flight Center, Washington DC.

**NASA** (1977) *Skylab Explores the Earth*. National Aeronautics and Space Administration Report, SP-380, Washington DC.

**NASA** (1978) *Skylab EREP Investigations Summary*. National Aeronautics and Space Administration Report SP-399, Washington DC.

**NASA** (1980a) *Landsat Data Users Notes, 12*. US Geological Survey, South Dakota.

**NASA** (1980b) *Landsat Data Users Notes, 14*. US Geological Survey, South Dakota.

**NASA** (1980c) *Landsat Data Users Notes, 15*. US Geological Survey, South Dakota.

**NASA** (1981a) *The Shuttle Era*. National Aeronautics and Space Administration Educational Publication, NF-127/3-81, Washington DC.

**NASA** (1981b) *Landsat Data Users Notes, 17*. US Geological Survey, South Dakota.

**NASA** (1982a) *Landsat Data Users Notes, 21*. US Geological Survey, South Dakota.

**NASA** (1982b) *Landsat Data Users Notes, 22*. US Geological Survey, South Dakota.

**NASA** (1982c) *Landsat Data Users Notes, 23.* US Geological Survey, South Dakota.
**NASA** (1982d) *Landsat Data Users Notes, 24.* US Geological Survey, South Dakota.
**NASA** (1982e) *Heat Capacity Mapping Mission Data Users Bulletin,* 9. National Aeronautics and Space Administration, Goddard Space Flight Center, Greenbelt, Maryland.
**Nelson, R.** and **Grebowsky, G.** (1982) Evaluation of temporal registration of Landsat scenes. *International Journal of Remote Sensing,* **3**: 45–50.
**Nichol, J.** and **Collins, W. G.** (1980) Ecological monitoring of balancing lakes by multispectral remote sensing. *International Archives of Photogrammetry, part B10,* **23**: 580–9.
**Nicodemus, F. E., Richmond, J. C., Hisia, J. J., Ginsberg, I. W.** and **Limperis, T.** (1977) *Geometrical Considerations and Nomenclature for Reflectance.* US Department of Commerce, National Bureau of Standards, Washington DC. 20234.
**Nielson, U.** (1972) Agfacontour film for interpretation. *Photogrammetric Engineering,* **38**: 1099–105.
**Nikolaev, V. A.** (1981) Space photographs as models for regional landscape structure. *Soviet Journal of Remote Sensing,* **1**: 13–20.
**NOAA/NASA** (1980) *Geostationary Operational Environmental Satellite GOES.* NOAA/NASA, Washington DC.
**NOAA** (1982a) *Landsat Data Users Notes, 25.* National Oceanic and Atmospheric Administration, South Dakota.
**NOAA** (1982b) *Summary of NESS Costs.* National Oceanic and Atmospheric Administration, Washington, DC.
**NOAA** (1983) *Landsat Data Users Notes, 26.* National Oceanic and Atmospheric Administration, South Dakota.
**Norwine, J.** and **Greegor, D. H.** (1983) Vegetation classification based on advanced very high resolution radiometer (AVHRR) satellite imagery. *Remote Sensing of Environment,* **13**: 69–87.
**Norwood, V. T.** and **Lansing, J. C.** (1983) Electro-optical imaging sensors: *In* Colwell, R. N. (ed.) *Manual of Remote Sensing* (2nd edn). American Society of Photogrammetry, Falls Church, Virginia, pp. 335–67.
**Nunnally, N. R.** (1969) Integrated landscape analysis with radar imagery. *Remote Sensing Environment,* **1**: 1–6.

**Obukhov, A. E.** and **Orlov, D. S.** (1964) Spectral reflectivity of major soil groups and possibility of using diffuse reflections in soil investigations. *Soviet Soil Science,* **1**: 174–84.
**Offield, T. W.** (1975) Thermal infrared images as a basis for structure mapping, Front Range and adjacent plains in Colorado. *Geological Society America Bulletin,* **86**: 495–502.
**Ohlhorst, C. W.** (1981) The use of Landsat to monitor deep water dumpsite 106. *Environmental Monitoring and Assessment,* **1**: 143–53.
**Olorunfemi, J. F.** (1982) Applications of aerial photography to population estimation in Nigeria. *Geo Journal,* **6**: 225–30.
**Ordway, F. I.** (1975) *Pictorial Guide to Planet Earth.* Thomas Y. Crowell Company, New York.
**Ormsby, J. P.** (1982) The use of Landsat 3 thermal data to help differentiate land covers. *Remote Sensing of Environment,* **12**: 97–105.
**Owens-Jones, E. S.** (1977) Densitometer methods of processing remote sensing data: *In* Barrett, E. C. and Curtis, L. F. (eds) *Environmental Remote Sensing 2, Practices and Problems.* Edward Arnold, London, pp. 101–24.

**Page, N. R.** (1974) Estimation of organic matter in Atlantic coastal plain soils with a colour-difference meter. *Agronomy Journal,* **66**: 652–3.
**Paine, D. P.** (1981) *Aerial Photography and Image Interpretation for Resource Management.* Wiley, New York.
**Parikh, J. O.** and **Ball, J. T.** (1980) Analysis of cloud type and cloud amount using GATE for SMS infrared data. *Remote Sensing of Environment,* **9**: 225–45.
**Parker, D. C.** and **Wolff, M. F.** (1965) Remote sensing. *International Science and Technology,* **43**: 20–31.
**Parker, H. D.** and **Driscoll, R. S.** (1972) An experiment in deer detection by thermal scanning. *Journal of Range Management,* **25**: 480–1.
**Parry, D. E.** and **Trevett, J. W.** (1979) Mapping Nigeria's vegetation from radar. *Geographical Journal,* **145**: 265–81.
**Parry, J. T., Wright, R. K.** and **Thomson, K. P. B.** (1980) Drainage on multiband

radar imagery in the Laurentian area, Quebec, Canada. *Photogrammetria*, **35**: 179–98.

**Peacock, K., Gasparovic, R. F., Tubbs, L. D.** (1981) High precision radiometric temperature measurements of the ocean surface. *Proceedings of the 15th International Symposium on Remote Sensing of Environment*. University of Michigan, Ann Arbor, pp. 793–802.

**Pearson, R. L., Tucker, C. J.** and **Miller, L. D.** (1976) Spectral mapping of shortgrass prairie biomass. *Photogrammetric Engineering and Remote Sensing*, **42**: 317–23.

**Pease, R. W.** (1969) *Plant Tissue and the Colour Infrared Record*. United States Geological Survey report to the National Aeronautics and Space Administration on contract No. R–09–020–024 (A/1).

**Pease, R. W.** and **Bowden, L. W.** (1969) Making colour infrared film a more effective high altitude remote sensor. *Remote Sensing of Environment*, **1**: 23–30.

**Perkins, D. F.** (1971) The Dartmoor survey: *In* Goodier, R. (ed.) *The Application of Aerial Photography to the Work of the Nature Conservancy*. Natural Environment Research Council, Swindon, pp. 21–8.

**Phillips, T. L.** and **Swain, P. H.** (1978) Data processing methods and systems: *In* Swain, P. H. and Davis, S. M. (eds) *Remote Sensing: The Quantitative Approach*. McGraw Hill. London, New York, pp. 188–226.

**Philipson, W. R.** and **Liang, T.** (1982) An airphoto key for major tropical crops. *Photogrammetric Engineering and Remote Sensing*, **48**: 223–33.

**Piech, K. R, Schott, J. R.** and **Stewart, K. M.** (1978) The blue-to-green reflectance ratio and lake water quantity. *Photogrammetric Engineering and Remote Sensing*, **44**: 1303–10.

**Pincus, H. J.** (1969) The analysis of remote sensing displays by optical diffraction. *Proceedings of the 6th International Symposium on Remote Sensing of Environment*. University of Michigan, Ann Arbor, pp. 261–74.

**Press, N., Kampschuur, W.** and **Duncan, W.** (1980) Contribution of photogeology and remote sensing to mineral exploration in Ireland. *Transactions, Institution of Mining and Metallurgy, Section B*, pp. 50–1.

**Price, D. J.** (1963) *Little Science, Big Science*. McGraw Hill, New York.

**Price, J. C.** (1978) Heat capacity mapping mission. *Journal of the British Interplanetary Society*, **31**: 313–6.

**Price, J. C.** (1981) The contribution of thermal data in Landsat multispectral classification. *Photogrammetric Engineering and Remote Sensing*, **47**: 229–36.

**Purdom, J. F. W.** (1976) Some uses of high resolution GOES imagery in the mesoscale forecasting of convection and its behaviour. *Monthly Weather Review*, **104**: 1474–83.

**RAE** (1983) *Users Guide: The United Kingdom National Remote Sensing Centre*. Remote Sensing Unit, Royal Aircraft Establishment, Farnborough.

**Rango, A.** and **Martinec, J.** (1979) Application of a snowmelt runoff model using Landsat data. *Nordic Hydrology*, **10**: 225–38.

**Ranz, E.** and **Schneider, S.** (1971) Progress in the application of Agfacontour equidensity film for geo-scientific photo interpretation. *Proceedings of the 7th International Symposium on Remote Sensing of Environment*. University of Michigan, Ann Arbor, pp. 779–86.

**Ray, R. E.** and **Fischer, W. A.** (1957) Geology from the air. *Science*, **126**: 725–35.

**Remote Sensing Society** (1983) *Remote Sensing and Digital Mapping*. A submission to the House of Lords Select Committee on Science and Technology, Reading.

**Rhind, D. W.** (1981) Geographical information systems in Britain: *In* Wrigley, N. and Bennett, R. J. (eds) *Quantitative Geography*. Routledge and Kegan Paul, pp. 17–35.

**Richards, J. A., Landgrebe, D. A.** and **Swain, P. H.** (1982) A means for utilising ancillary information in multispectral classification. *Remote Sensing of Environment*, **12**: 403–77.

**Richason, B. F.** (ed.) (1982) *Remote Sensing: an Input to Geographic Information Systems in the 1980s*. Pecora VII Symposium, United States Geological Survey, Sioux Falls.

**Robson, A., Morgan, J., Herschy, R. W.** and **Zchav, J.** (1982) Detection of natural disasters via Meteostat. *ESA Bulletin*, **29**: 10–18.

**Robinove, C. J.** (1975) Worldwide disaster warning and assessment with Earth resources technology satellites. *Proceedings of the 10th International Symposium* *on Remote Sensing of Environment.* University of Michigan, Ann Arbor, pp. 811–20.

**Robinove, C. J.** (1981) The logic of multispectral classification and mapping of land. *Remote Sensing of Environment*, **11**: 231–44.

**Robinove, C. J., Chavez, P. S., Gehring, D.** and **Holmgren, R.** (1981) Arid land monitoring using Landsat albedo difference images. *Remote Sensing of Environment*, **11**: 133–56.

**Rodda, J. C.** (1978) *Institute of Hydrology Research Report, 1976–8.* Natural Environment Research Council, Swindon.

**Roessel, J. V. van** and **de Godoy, R. C.** (1974) SLAR mosaics for project RADAM. *Photogrammetric Engineering*, **40**: 583–95.

**Rohde, W. G.** and **Olson, C. E.** (1970) Detecting tree moisture stress. *Photogrammetric Engineering*, **36**: 561–6.

**Rohde, W. G., Lo, J. K.** and **Pohl, R. A.** (1978) EROS data center Landsat digital enhancement techniques and imagery availability, 1977. *Canadian Journal of Remote Sensing*, **4**: 63–76.

**Rose, P.W.** and **Rosendahl, P. C.** (1983) Classification of Landsat data for hydrologic application, Everglades National Park. *Photogrammetric Engineering and Remote Sensing*, **49**: 505–11.

**Rosenfeld, A.** (1969) *Picture Processing by Computer.* Academic Press, New York; London.

**Rosenfeld, A.** and **Kak, A. C.** (1976) *Digital Picture Processing.* Academic Press, New York.

**Rosenfield, G. H.** (1982) Sample design for estimating change in land use and land cover. *Photogrammetric Engineering and Remote Sensing*, **48**: 793–801.

**Rosenfield, G. H., Fitzpatrick-Lins, K.** and **Ling, H. S.** (1982) Sampling for thematic map accuracy testing. *Photogrammetric Engineering and Remote Sensing*, **48**: 131–7.

**Rowan, L. C.** (1975) Application of satellites to geologic exploration. *American Scientist*, **63**: 393–403.

**Rowan, L. C** and **Kahle A. B.** (1982) Evaluation of 0.46–2.36 μm multispectral scanner images of the east tintic mining district, Utah for mapping hydrothermally altered rocks. *Economic Geology*, **77**: 441–52.

**Royal Society** (1975) *Quantities, Units and Symbols*, (2nd edn). The Royal Society, London.

**Royal Society** (1983) *Remote Sensing and Digital Mapping.* A submission to the House of Lords Select Committee on Science and Technology. Royal Society Document C/61 (83) London.

**Ryerson, R. A., Mosher P.** and **Harvie J.** (1980) *Potato Area Estimation Using Remote Sensing Methods.* Canada Center for Remote Sensing. Energy, Mines and Resources, Ottawa.

**Sabins, F. F.** (1969) Thermal infrared imagery and its application to structural mapping in southern California. *Geological Society America Bulletin*, **80**: 397–404.

**Sabins, F. F.** (1976) Geologic aspects of remote sensing: *In* Lintz, J. and Simonett, D. S. (eds) *Remote Sensing of Environment.* Addison-Wesley, Reading Massachusetts; London, pp. 508–71.

**Sabins, F. F.** (1978) *Remote Sensing: Principles and Interpretation.* W. H. Freeman and Co., San Francisco.

**Sabins, F. F., Blom, R.** and **Elachi, C.** (1980) Seasat radar image of San Andreas Fault, California; *American Association of Petroleum Geologists*, **64**: 610–28.

**Sadowski, F. A.** and **Sarno, J.** (1976) *Forest Classification Accuracy as Influenced by Multispectral Scanner Spatial Resolution.* Report 109600-71F, Environmental Research Institute of Michigan, Ann Arbor, Michigan.

**Salomonson, V. V.** (1978) Landsat-D: A systems overview. *Proceedings of the 12th International Symposium on Remote Sensing of Environment.* University of Michigan, Ann Arbor pp. 371–85.

**Salomonson, V. V., Smith, P. L., Park, A. D., Webb, W. C.** and **Lynch, T. J.** (1980) An overview of progress in the design and implementation of Landsat D systems. *IEEE Transactions on Geoscience and Remote Sensing*, **18**: 137–46.

**Sapp, C. P.** (1971) Evaluation of space and high altitude imagery for the Department of Interior. *Proceedings of the 7th International Symposium on Remote Sensing of Environment.* University of Michigan, Ann Arbor, pp. 435–51.

**Sauchyn, D. J.** and **Trench, N. R.** (1978) Landsat applied to landslide mapping. *Photogrammetric Engineering and Remote Sensing*, **44**: 735–41.

**Sauer, E. K.** (1981) Hydrogeology of glacial deposits from aerial photographs. *Photogrammetric Engineering and Remote Sensing*, **47**: 811–22.

**Saunders, R. W., Ward, N. R., England, C. F.** and **Hunt G. E.** (1981) Sea surface temperature measurements around the UK derived from Meteosat and TIROS-N data: *In* Allan, J. A. (ed.). *Matching Remote Sensing Technologies and their Applications.* Remote Sensing Society, Reading, pp. 191–8.

**Saunders, R. W., Ward, N. R., England, C. F.** and **Hunt, G. E.** (1982) Satellite observations of sea surface temperature around the British Isles. *Bulletin of the American Meteorological Society*, **63**: 267–72.

**Scarpace, F., Madding, R.** and **Green, T.** (1975) Scanning thermal plumes. *Photogrammetric Engineering and Remote Sensing*, **41**: 1223–31.

**Schanda, E.** (1976) Passive microwave sensing: *In* Schanda, E. (ed.) *Remote Sensing for Environmental Sciences.* Springer Verlag, New York; Chapman and Hall, London, pp. 187–256.

**Schlosser, M. S.** (1974) Television scanning densitometer. *Photogrammetric Engineering*, **40**: 199–202.

**Schmidt, R. G.** (1975) Evaluation of improved digital image processing techniques of Landsat data for sulphide mineral prospecting: *In* Wall, P. W. and Fischer, W. A. (eds) *Proceedings 1st Annual Pecora Symposium.* United States Geological Survey Professional Paper **1015**: 201–12.

**Schneider, S. E., McGinnis, D. F.** and **Pritchard, J. A.** (1979) Use of satellite infrared data for geomorphology studies. *Remote Sensing of Environment*, **8**: 313–30.

**Schnetzler, C. C.** and **Thompson, L. L.** (1979) Multispectral resource sampler: an experimental satellite for the mid-1980s. *Society of Photo-Optical Instrumentation Engineers*, **183**: 255–62.

**Schott, J. R.** and **Tourin, R. H.** (1975) A completely airborne calibration of aerial infrared water temperature measurements. *Proceedings of the 10th International Symposium on Remote Sensing of Environment.* University of Michigan, Ann Arbor, pp. 477–84.

**Schwalb, A.** (1982) *The TIROS-N/NOAA A-G Satellite Series.* NOAA Technical Memorandum NESS 95, Washington DC.

**Schwarz, D. E.** and **Caspall, F.** (1969) The use of radar in the discrimination and identification of agricultural land use. *Proceedings of the 5th International Symposium on Remote Sensing of Environment.* University of Michigan, Ann Arbor, pp. 233–47.

**Schweitzer, G. E.** (1982) Airborne remote sensing. *Environmental Science and Technology*, **16**: 338–46.

**Seguin, B.** and **Itier, B.** (1983) Using midday surface temperature to estimate daily evaporation from satellite thermal IR data. *International Journal of Remote Sensing*, **4**: 371–83.

**Shaw, H. J. W.** (1981) Development of a Canadian thermal infrared forest fire mapping operational program. *Proceedings of the 15th International Symposium on Remote Sensing of Environment.* University of Michigan, Ann Arbor, pp. 1465–73.

**Sheffield, C.** (1981) *Earthwatch.* Sidgwick and Jackson, London.

**Shelton, R. C.** and **Estes, J. E.** (1981) Remote Sensing and geographical information systems: an unrealised potential. *Geo-Processing* **1**: 395–420.

**Shih, S. F.** (1980) Use of Landsat data to improve the water budget computation in Lake Okeechobee, Florida. *Journal of Hydrology*, **48**: 237–49.

**Shih, E. H. H.** and **Schowengerdt, R. A.** (1983) Classification of arid geomorphic surfaces using spectral and textural features. *Photogrammetric Engineering and Remote Sensing*, **49**: 337–47.

**Short, N. M.** (1982) *The Landsat Tutorial Workbook: Basics of Satellite Remote Sensing.* National Aeronautics and Space Administration Reference Publication 1078, Washington DC.

**Short, N. M., Lowman, P. D., Freden, S. C.** and **Finch, W. A.** (1976) *Mission to Earth: Landsat Views the World.* National Aeronautics and Space Administration Special Publication SP-360, Washington DC.

**Sidorenko, A. V.** (1981) New experiment in Earth studies from space. *Soviet Journal of Remote Sensing*, **1**: v–vii.

**Siegal, B. S.** and **Goetz, A. F. H.** (1977) Effects of vegetation on rock and soil type discrimination. *Photogrammetric Engineering and Remote Sensing*, **43**: 191–6.

**Silva, L. F.** (1978) Radiation and instrumentation in remote sensing: *In* Swain, P. H. and Davis, S. M (eds) *Remote Sensing the Quantitative Approach.* McGraw Hill, pp. 121–35.

**Simon, J. C.** (1978) Clustering and digital image analysis: *In* Gardner, W. E. (ed.) *Machine Aided Image Analysis.* Institute of Physics, Bristol, London, pp. 20–39.

**Simonett, D. S.** (1976) Remote sensing of cultivated and natural vegetation: cropland and forest land: *In* Lintz, J. and Simonett, D. S. (eds) *Remote Sensing of Environment.* Addison-Wesley, Reading Massachusetts, London, pp. 442–81.

**Simonett, D. S., Smith, T. R., Tobler, W., Marks, D. G., Frew, J. E.** and **Dozier, J. C.** (eds) (1977) *Geobase Information System Impacts on Space Image Formats.* Santa Barbara Remote Sensing Unit Technical Report 3. University of California, Santa Barbara.

**Singh, S. M.** (1982) A procedure for atmosphere correction of coastal zone colour scanner (CZCS) data: *In* Henderson-Sellers, A. and Shine, K. P. (eds) *Remote Sensing and the Atmosphere.* Remote Sensing Society, Reading, pp. 169–76.

**SIPRI** (1978) *Outer Space – Battlefield of the Future?* Taylor and Francis, London.

**SIPRI** (1979) *Stockholm International Peace Research Institute (SIPRI) Yearbook.* Taylor and Francis, London.

**SIPRI** (1980) *Stockholm International Peace Research Institute (SIPRI) Yearbook.* Taylor and Francis, London.

**SIPRI** (1981) *Stockholm International Peace Research Institute (SIPRI) Yearbook.* Taylor and Francis, London.

**SIPRI** (1982) *Stockholm International Peace Research Institute (SIPRI) Yearbook.* Taylor and Francis, London.

**Skaley, J. E.** (1980) Photo-optical techniques of image enhancement. *In* Siegal, B. S. and Gillespie, A. R. (eds) *Remote Sensing in Geology.* Wiley, New York, pp. 119–38.

**Skibitzke, H. E.** (1976) Remote sensing for water resources: *In* Lintz, J. and Simonett, D. (eds) *Remote Sensing of Environment.* Addison-Wesley, Reading, Massachusetts; London, pp. 572–92.

**Skolnik, M. I.** (ed.), (1970) *Radar Handbook.* McGraw Hill, New York.

**Slater, P. N.** (1975) Photographic systems for remote sensing: *In* Reeves, R. G. (ed.) *Manual of Remote Sensing.* American Society of Photogrammetry, Falls Church, Virginia, pp. 235–323.

**Slater, P. N.** (1979) A re-examination of the Landsat MSS. *Photogrammetric Engineering and Remote Sensing.* **45**: 1479–85.

**Slater, P. N.** (1980) *Remote Sensing: Options and Optical Systems.* Addison-Wesley Reading, Massachusetts; London.

**Slater, P. N.** (1983) Photographic systems for remote sensing: *In* Colwell, R. N. (ed.) *Manual of Remote Sensing* (2nd edn). American Society of Photogrammetry, Falls Church, Virginia, pp. 231–91.

**Slegrist, A. W.** and **Schnetcler, C. C.** (1980) Optimum spectral bands in rock discrimination. *Photogrammetric Engineering and Remote Sensing,* **46**: 1207–15.

**Smedes, H. W.** (1975) The truth about ground truth. *Proceedings of the 10th International Symposium on Remote Sensing of Environment.* University of Michigan, Ann Arbor, pp. 821–3.

**Smit, G. S.** (1978) Shifting cultivation in tropical rainforests detected from aerial photographs. *ITC Journal*: 603–33.

**Smith, A. Y.** and **Blackwell, R. J.** (1980) Development of an information data base for watershed monitoring. *Photogrammetric Engineering and Remote Sensing,* **46**: 1027–38.

**Smith, J. A., Ranson, K. J., Nguyen, D., Balick, L., Link, L. E., Fritschen, L.** and **Hutchinson, B.** (1981) Thermal vegetation canopy model studies. *Remote Sensing of Environment,* **11**: 311–26.

**Snyder, D. R.** (1982) Integration of Landsat RBV and MSS imagery to produce land-use maps of Soviet cities: *In* Richason B. F. (ed.) *Remote Sensing: As an Input to Geographic Information Systems in the 1980s.* Pecora VII Symposium, United States Geological Survey, Sioux Falls, pp. 94–103.

**Sobotta, W.** (1979) Spacelab – Europe's first manned spacecraft. *Journal of the British Interplanetary Society,* **32**: 334–8.

**Sobur, A. S., Chambers, A. J., Chambers, R., Damopolii, J., Hadi, S.** and **Hanson, A. J.** (1978) Remote sensing applications in the South East Sumatra coastal environment. *Remote Sensing of Environment,* **7**: 281–303.

**Spann, G. W., Hooper, N. J.** and **Cotter, D. J.** (1981) An analysis of user requirements for operational land satellite data. *Proceedings of the 15th International Symposium on Remote Sensing of Environment.* University of Michigan, Ann Arbor, pp. 1297–303.

**Specht, M. R., Needler, D.** and **Fritz, N. L.** (1973) New colour film for water photography penetration. *Photogrammetric Engineering,* **39**: 359–69.

**Spurr, S. H.** (1960) *Photogrammetry and Photo-Interpretation.* The Ronald Press Company, New York.

**Stanley, D. J.** (1977) Texture analysis in real life – an examination of the efficacy of selected digital methods: *In* Thomas, J. O. and Davey, P. G. (eds) *Texture Analysis.* Remote Sensing Society, Reading, pp. 87–108.

**Stanley, R. M.** (1982) *World War II Photo Intelligence.* Sidgwick and Jackson, London; Charles Scribner's Sons.

**Starr, D. W.** and **Mackworth, A. K.** (1978) Exploiting spectral, spatial and semantic constraints in the segmentation of Landsat images. *Canadian Journal of Remote Sensing.* **4**: 101–8.

**Stembridge, J. F.** (1978) Vegetated coastal dunes: growth detection from aerial infrared photography. *Remote Sensing of Environment,* **7**: 73–6.

**Steiner, D.** (1970). Time dimension for crop surveys from space. *Photogrammetric Engineering,* **36**: 187–94.

**Steiner, D.** and **Salerno, A. E.** (1975) Remote sensor data systems, processing, and management: *In* Reeves, R. G. (ed.) *Manual of Remote Sensing.* American Society of Photogrammetry, Falls Church, Virginia, pp. 611–803.

**Stephens, P. R., Hicks, D. L.** and **Trustrum, N. A.** (1981) Aerial photographic techniques for soil conservation research. *Photogrammetric Engineering and Remote Sensing,* **45**: 79–87.

**Stephens, P. R., Daigle, J. L.** and **Cihlar, J.** (1982) Use of sequential aerial photographs to detect and monitor soil management changes affecting cropland erosion. *Journal of Soil and Water Conservation,* **37**: 101–5.

**Steward, W. F., Carter, V.** and **Brooks, R. D.** (1980) Inland (non-tidal) wetland mapping. *Photogrammetric Engineering and Remote Sensing,* **46**: 617–28.

**Stewart, R. B., Mukammal, E. I.** and **Wiebe, J.** (1978) The use of thermal imagery in defining frost prone areas in the Niagara frost belt. *Remote Sensing of Environment,* **7**: 187–202.

**Stoebner, A. W.** (1976) Remote sensing of Earth resources: technique and law: *In* Matte, N. M. and DeSaussure, H. (eds) *Legal Implications of Remote Sensing from Outer Space.* A. W. Sijthoff-Leyden, pp. 33–40.

**Stoner, E. R.** and **Baumgardner, M. E.** (1981) Characteristic variations in the reflectance of surface soils. *Soil Science Society of America Journal,* **45**: 1161–5.

**Stove, G. C.** and **Hulme, P. D.** (1980) Peat resource mapping in Lewis using remote sensing techniques and automated cartography. *International Journal of Remote Sensing,* **1**: 319–44.

**Stowe, R. F.** (1976) Diplomatic and legal aspects of remote sensing. *Photogrammetric Engineering and Remote Sensing,* **42**: 177–80.

**Strahler, A. H.** (1980) The use of prior probabilities in maximum likelihood classification of remotely sensed data. *Remote Sensing of Environment,* **10**: 135–63.

**Strahler, A. H.** (1981) Stratification of natural vegetation for forest and rangeland inventory using Landsat digital imagery and collateral data. *International Journal of Remote Sensing,* **2**: 15–41.

**Strandberg, C. H.** (1967) Photoarchaeology. *Photogrammetric Engineering,* **33**: 1152–7.

**Strome, W. M.** and **Goodenough, D. G.** (1979) The use of array processors in image analysis: *In* Gardner, W. E. (ed.) *Machine-Aided Image Analysis.* Institute of Physics, Bristol, London, pp. 250–9.

**Sturge, J. M.** (ed.) (1977) *Neblette's Handbook of Photography and Reprography Materials, Processes and Systems* (7th edn). Van Nostrand Reinhold Co, New York.

**Sturm, B.** (1981) The atmospheric correction of remotely sensed data and the quantitative determination of suspended matter in marine water surface layers: *In* Cracknell, A. P. (ed.) *Remote Sensing in Meteorology, Oceanography and Hydrology.* Ellis Horwood Ltd., Chichester; Wiley, New York, pp. 163–97.

**Suits, G. H.** (1972) The calculation of the directional reflectance of a vegetative canopy. *Remote Sensing of Environment,* **2**: 117–25.

**Suits, G. H.** (1975) The nature of electromagnetic radiation: *In* Reeves, R. G. (ed.) *Manual of Remote Sensing.* American Society of Photogrammetry, Falls Church, Virginia, pp. 51–73.

**Suits, G. H.** (1983) The nature of electromagnetic radiation: *In* Colwell, R. N. (ed.) *Manual of Remote Sensing* (2nd edn). American Society of Photogrammetry, Falls Church, Virginia, pp. 37–60.

**Sutherland, R. A., Hannah, H. E, Cook, A. F.** and **Martsolf, J. D.** (1981) Remote sensing of thermal radiation from an aircraft – an analysis and evaluation of crop-freeze protection methods. *Journal of Applied Meteorology,* **20**: 813–20.

**Swain, P. H.** and **Davis, S. M.** (eds) (1978) *Remote Sensing: The Quantitative Approach*. McGraw Hill, London; New York.

**Swain, P. H., Vardeman, S. B.** and **Tilton, J. C.** (1981) Contextural classification of multispectral data. *Pattern Recognition*, **13**: 419–41.

**Sweet, H. C., Poppleton, J. E., Shuey, A. G.** and **Peeples, T. O.** (1980) Vegetation of central Florida's East Coast: The distribution of six vegetational complexes of Merritt Island and Cape Canaveral Peninsula. *Remote Sensing of Environment*, **9**: 93–108.

**Talerico, R. L., Walker, J. E.** and **Skratt, T. A.** (1978) Quantifying gypsy moth defoliation. *Photogrammetric Engineering and Remote Sensing*, **44**: 1,385–92.

**Tanaka, S., Murandka, Y., Miyazawa, H.** and **Suga, Y.** (1977) Multiseasonal data analysis and some extensions for environmental monitoring. *Proceedings of the 11th International Symposium on Remote Sensing of Environment*. University of Michigan, Ann Arbor, pp. 545–61.

**Taranik, J. V.** and **Settle, M.** (1981) Space Shuttle: A new era in terrestrial remote sensing. *Science*, **214**: 619.

**Teicholz, E.** (1980) Geographic information systems: the ODYSSEY project. *Journal of the Surveying and Mapping Division*, **106**: 119–35.

**Teillet, P. M., Guindon, B.** and **Goodenough, D. G.** (1981) Forest classification using simulated Landsat-D Thematic Mapper data. *Canadian Journal of Remote Sensing*, **7**: 51–60.

**Thomas, I. L., Howorth R., Eggers A. E.** and **Fowler A. D. W.** (1981) Textural enhancement of a circular geological feature. *Photogrammetric Engineering and Remote Sensing*, **47**: 89–91.

**Thomas, J. O.** (1977) Texture analysis in imagery processing: *In* Thomas, J. O. and Davey, P. G. (eds) *Texture Analysis*. British Pattern Recognition Association and Remote Sensing Society, Reading, pp. 1–43.

**Thomas, J. O.** (1980) Digital imagery processing – with special reference to data compression in remote sensing: *In* Haralick, R. M. and Simon, J. C. (eds). *Issues in Digital Image Processing*. Sijthoff and Noordhoff, pp. 247–90.

**Thompson, L. L.** (1979) Remote sensing using solid-state linear array technology. *Photogrammetric Engineering and Remote Sensing*, **45**: 47–55.

**Tilman, S. E.** and **Mokma, D. L.** (1980) Description of a user orientated geographic information system: the resource analysis program. *Machine Processing of Remotely Sensed Data Symposium*. Purdue University, pp. 248–58.

**Timon, I.** (1980) An improved method for geographical rectification of satellite imagery. *Idojaras*, **4**: 67–77.

**Todd, W. J.** and **Gehring, D. G.** (1980) Landsat wildland mapping accuracy. *Photogrammetric Engineering and Remote Sensing*, **46**: 509–20.

**Tomlins, G. F.** (1981) Canadian experience in wetland monitoring by satellite: *In* Smith, H. (ed.) *Plants and the Daylight Spectrum*. Academic Press, London, pp. 102–13.

**Tomlinson, R. F.** (1968) A geographic information system for regional planning: *In* Stewart, G. A. (ed.) *Land Evaluation*. Macmillan of Australia.

**Tomlinson, R. F.** (1972) Introduction: *In* Tomlinson, R. F. (ed.) *Geographical Data Handling*. UNESCO/IGU Second Symposium on Geographical Information Systems, Ottawa, Canada, pp. 3b–3m.

**Townshend, J. R. G.** (1981a) The spatial resolving power of Earth resources satellites. *Progress in Physical Geography*, **5**: 32–55.

**Townshend, J. R. G.** (1981b) Image analysis and interpretation for land resources survey: *In* Townshend, J. R. G. (ed.) *Terrain Analysis and Remote Sensing*. Allen and Unwin, London, Boston, pp. 59–108.

**Townshend, J. R. G.** and **Justice C.** (1981) Information extraction from remotely sensed data. A user view. *International Journal of Remote Sensing*, **2**: 213–29.

**Townshend, J. R. G.** and **Tucker, C. J.** (1981) Utility of AVHRR of NOAA 6 and 7 for vegetation mapping: *In* Allan, J. A. (ed.) *Matching Remote Sensing Technologies and their Applications*. Remote Sensing Society, Reading, pp. 97–109.

**Tucker, C. J.** (1980) Radiometric resolution for monitoring vegetation. How many bits are needed? *International Journal of Remote Sensing*, **1**: 241–54.

**Tucker, C. J., Elgin J. H.** and **McMurtrey, J. E.** (1979) Temporal, spectral measurements of corn and soybean crops. *Photogrammetric Engineering and Remote Sensing*, **45**: 643–53.

**Ulaby, F. T.** (1975) Radar response to vegetation. *IEEE Transactions on Antennas and Propagation*, **23** : 36–45.

**Ulaby, F. T., Bush, T. F** and **Batlivala, P. P** (1975) Radar response to vegetation II: 8–18 GHZ 2Band. *IEEE Transactions on Antennas and Propagation*, **23**: 608–22.

**Ulaby, F. T.** and **McNaughton, J.** (1975) Classification of physiography from ERTS imagery. *Photogrammetric Engineering* 4: 1019–27.

**Ulaby, F. T., Batlivala, P. P.** and **Dobson, M. C.** (1978) Microwave backscatter dependence on surface roughness, soil moisture and soil texture. Part 1 – bare soil. *IEEE Transactions on Geoscience Electronics*, **16**: 286–95.

**Ulaby, F. T., Bradley, G. A.** and **Dobson, M. C.** (1979). Microwave backscatter dependence on surface roughness, soil moisture and soil texture. Part II, vegetation covered soil. *IEEE Transactions on Geoscience Electronics*, **17**: 33–40.

**Ulaby, F. T.** and **Bare, J. E.** (1979). Look direction modulation function of the radar backscattering coefficient of agricultural fields. *Photogrammetric Engineering and Remote Sensing*, **45**: 1507–12.

**Ulaby, F. T., Batlivala, P. P.** and **Bare, J. E.** (1980) Crop identification with L band radar. *Photogrammetric Engineering and Remote Sensing*, **46**: 101–5.

**Ulaby, F. T., Moore, R. K.** and **Fung, A. K.** (1981) *Microwave Remote Sensing, Active and Passive, Volume 1, Fundamentals and Radiometry* Addison-Wesley Reading, Massachusetts; London.

**Ulaby, F. T., Moore, R. K.** and **Fung, A. K.** (1982) *Microwave Remote Sensing, Active and Passive, Volume II, Radar Remote Sensing and Surface Scattering and Emission Theory*. Addison-Wesley Reading, Massachusetts; London.

**Ulaby, F. T., Razani, M.** and **Dobson, M. C.** (1983) Effects of vegetation cover on the microwave sensitivity to soil moisture. *IEEE Transactions on Geoscience and Remote Sensing*, **21**: 51–61.

**Valerio, C.** and **Llebaria, A.** (1982) A quantitative multispectral analysis system for aerial photographs applied to coastal planning. *International Journal of Remote Sensing*, **3**: 181–97.

**Vette, J. I.** and **Vostreys, R. W.** (1977) *Report on Active and Planned Spacecraft and Experiments.* NSSUC/WWC-A-R and S77-03. National Aeronautics and Space Administration, Greenbelt, Maryland.

**Vick, C. M.** and **Handley, J. F.** (1977) Survey of damage to trees surrounding a chemical factory, emitting phosphoric and hydrochloric acid pollution. *Environmental Health*: 115–7.

**Vincent, R. K.** (1973) An ERTS multispectral scanner experiment for mapping iron compounds. *Proceedings of the 8th International Symposium on Remote Sensing of Environment.* University of Michigan, Ann Arbor, pp. 1239–47.

**Vlasov, A. A., Yegorov, S. T.** and **Plyushchev, V. A.** (1981) A comparative analysis of satellite microwave and I.R. images. *Soviet Journal of Remote Sensing*, **1**: 47–52.

**Vlcek, J.** (1982) A field method for determination of emissivity with imaging radiometers. *Photogrammetric Engineering and Remote Sensing*, **48**: 609–14.

**Waite, W. P.** (1976) Historical development of imaging radar. *Remote Sensing of the Electromagnetic Spectrum*, **3**: 1–23.

**Walker, F.** (1964) *Geography from the Air.* Methuen and Co. Ltd.

**Wallace, G. A.** (1973) Remote sensing for detecting feedlot runoff. *Photogrammetric Engineering*, **39**: 949–57.

**Wallace, D. E.** (1981) A report on the use of remote sensing techniques for the supervision of New England coastal marshes. *Remote Sensing Quarterly*, **3**: 45–53.

**Walling, D. E.** (1983) Physical hydrology. *Progress in Physical Geography*, **7**: 97–112.

**Wang, J. R., O'Neill, P. E., Jackson, T. J.** and **Engman E. T.** (1983) Multifrequency measurements of the effects of soil moisture, soil texture and surface roughness. *IEEE Transactions on Geoscience and Remote Sensing*, **21**: 44–51.

**Wannamaker, B., Seynaeve, R., Boni, P.** and **Forber, A.** (1980) The use of digitised APT data in oceanographic research: *In* Cracknell, A. P. (ed.) *Coastal and Marine Applications of Remote Sensing.* Remote Sensing Society, Reading, pp. 37–46.

**Wardley, N. W.** and **Curran, P. J.** (1983) Green leaf area index estimation from remotely sensed airborne MSS data: *In* Keech, M. (ed.) *The Application of Remote Sensing Techniques to Aid Range Management.* Remote Sensing Society, Reading, pp. 1–12.

**Waters, M. P.** (1976) Application of geosynchronous meteorological satellite data in fire assessment. Fort Collins, Colorado. *United States Forest Service*, Rocky Mountain Forest and Range Experimental Station, General Technical Report, RM-32: 54–8.

**Watkins, T.** (1978) The economics of remote sensing. *Photogrammetric Engineering and Remote Sensing*, **44**: 1167–72.

**Watson, K.** (1971) Geophysical aspects of remote sensing. *Proceedings of the International Workshop on Earth Resources Survey Systems.* National Aeronautics and Space Administration Special Report, **2**, pp. 409–28.

**Weber, F. P., Aldrich, R. C., Sadowski, F. G.** and **Thompson, F. J.** (1972) Land use classification in the southeastern forest region by multispectral scanning and computerised mapping. *Proceedings of the 8th International Symposium on Remote Sensing of Environment.* University of Michigan, Ann Arbor, pp. 351–73.

**Weber, F. P.** and **Polcyn, F. C.** (1972) Remote sensing to detect stress in forests. *Photogrammetric Engineering*, **38**: 163–75.

**Wechler, H.** and **Citron, T.** (1980) Feature extraction for texture classification. *Pattern Recognition*, **12**: 301–11.

**Weeks, W. F.** (1981) Sea ice: the potential of remote sensing. *Oceanus*, **24**: 39–48.

**Weisblatt, E. A., Zaitzeff, J. B.** and **Reeves, C. A.** (1973) Classification of turbidity levels in the Texas marine coastal zone. *Proceedings of the Symposium on Machine Processing of Remotely Sensed Data*, pp. 3A.42–3A.59.

**Welby, C. W.** (1976) Landsat – 1 imagery for geologic evaluation. *Photogrammetric Engineering and Remote Sensing*, **42**: 1411–9.

**Welch, R.** (1968) Film transparencies vs. paper prints. *Photogrammetric Engineering*, **36**: 490–501.

**Welch, R.** (1980) Monitoring urban population and energy utilisation patterns from satellite data. *Remote Sensing of Environment*, **9**: 1–9.

**Welch, R.** and **Howarth, P. J.** (1968) Photogrammetric measurement of glacial landforms. *Photogrammetric Record*, **6**: 75–96.

**Welch, R., Lo, H. C.** and **Pannell, C. W.** (1979) Mapping China's new agricultural lands. *Photogrammetric Engineering and Remote Sensing*, **45**: 1211–28.

**Welch, R.** and **Pannell, C. W.** (1982) Comparative resolution of Landsat 3 MSS and RBV image data of China. *Photogrammetric Record*, **10**: 575–86.

**Welsted, J.** (1979) Air photo interpretation in coastal studies. Examples from the bay of Fundy, Canada. *Photogrammetria*, **35**: 1–28.

**Wenderoth, S.** and **Yost E.** (1975) *Multispectral Photography for Earth Resources.* Remote Sensing Information Center, New York.

**Wert, S. L., Miller, P. R.** and **Larsh, R. N.** (1970) Colour photos detect smog injury to forest trees. *Journal of Forestry*, **68**: 536–9.

**Westin, F. C.** and **Frazee, C. J.** (1976) Landsat data, its use in a soil survey program. *Proceedings of Soil Science Society of America*, **49**: 81–9.

**Weszka, J. S., Dyer, C. R.** and **Rosenfeld, A.** (1976) A comparative study of texture measures for terrain classification. *IEEE Transactions on Systems, Man and Cybernetics*, **6**: 269–85.

**Wharton, S. W., Irons, J. R.** and **Hueger, F.** (1981) LAPR: An experimental push-broom scanner. *Photogrammetric Engineering and Remote Sensing*, **47**: 631–9.

**Wharton, S. W.** and **Turner, B. J.** (1981) ICAP: An interactive cluster analysis procedure for analyzing remotely sensed data. *Remote Sensing of Environment*, **11**: 279–93.

**White, L. P.** (1977) *Aerial Photography and Remote Sensing for Soil Survey.* Clarendon Press, Oxford.

**Whitley, S. L., Pearson, R. W., Seyfarth, B. R.** and **Graham, M. H.** (1981) ELAS: A geobased information system that is transferable to several computers. *Proceedings of the 15th International Symposium on Remote Sensing of Environment.* University of Michigan. Ann Arbor; pp. 1091–9.

**Whittingham, C. P.** (1974). *The Mechanisms of Photosynthesis.* Edward Arnold, London.

**Wiesnet, D. R.** and **Matson, M.** (1983) Remote sensing of weather and climate: *In* Colwell, R. N. (ed.) *Manual of Remote Sensing* (2nd edn). American Society of Photogrammetry, Falls Church, Virginia, pp. 1305–69.

**Wignall, B. L.** (1977) A critical review of the Quantimet 720 image analyser in remote sensing: *In* Peel, R. F., Curtis, L. F. and Barrett, E. C. (eds) *Remote Sensing of the Terrestrial Environment.* Butterworths, London; Boston, pp. 71–9.

**Wilkes, Q. L.** (1975) Weather and climate: measurement and analysis. *In* Reeves, R. G. (ed.) *Manual of Remote Sensing.* American Society of Photogrammetry, Falls Church, Virginia, pp. 1623–713.

**Williams, D., Curran, P. J.** and **Wardley, N. W.** (1983) The use of airborne

Thematic Mapper simultation data for the estimation and mapping of green leaf area index (cAI): *In* Guyene T. O. and G. Levy (eds) *EARSL-ESA* Symposium, Brussels ESA-SP-188, pp. 157–60.

**Williams, R. S** and **Carter, W. D.** (1976) *ERTS, 1, A New Window on Our Planet.* US Geological Survey Professional Paper, **929**, Washington DC.

**Williams, Lee, T. H.** and **Goodman, J. M.** (1980) Terrain analysis from Landsat using colour TV enhancement system. *Remote Sensing of Environment*, **10**: 213–37.

**Williams, Lee, T. H., Siebert, J.** and **Gunn, C.** (1981) Instructional image processing on a university mainframe – the Kansas system. *Conference on Remote Sensing Education*, Purdue University/NASA/NOAA, pp. 1–5.

**Wing, R. S** (1971) structural analysis from radar imagery of eastern Panama isthmus. *Modern Geology*, **2**: 1–21, 75–127.

**Wolf, P. F.** (1974) *Elements of Photogrammetry.* McGraw Hill, New York.

**Wolfe, E. W.** (1971) Thermal IR for geology. *Photogrammetric Engineering*, **37**: 43–52.

**Wolfe, W. L.** and **Zissis, G. J.** (1978) *The Infrared Handbook.* Office of Naval Research, Department of the Navy, Washington, DC.

**Yanchinski, S.** (1980) Thorny questions over remote sensing. *New Scientist*, **86**: 150–2.

**Yost, E. F.** and **Wenderoth, S.** (1967) Multispectral colour aerial photography. *Photogrammetric Engineering*, **33**: 1020–32.

**You-Ching, F.** (1980) Aerial photo and Landsat image use in forest inventory in China. *Photogrammetric Engineering and Remote Sensing*, **46**: 1421–4.

**Zickler, A.** (1977) Soyuz 22 spacecraft and the MKF-6 Multispectral camera of VEB Carl Zeiss Jena. *Jena Review*, **6**: 263–6.

**Zwick, H. J., Jain, S. C.** and **Bukata, R. P.** (1979) A satellite/airborne and *in situ* water quality experiment in Lake Ontario: *In* Sørensen, B. M. (ed.) *Workshop on the Eurasep ocean colour scanner experiments 1977.* Commission of the European Communities, Ispra, Italy, pp. 181–98.

# Index